Advanced Remote Sensing Technology for Tsunami Modelling and Forecasting

Maged Marghany

School of Humanities
Geography Division
Universiti Sains Malaysia
11800 USM
Pulau Pinang
Malaysia

CRC Press is an imprint of the
Taylor & Francis Group, an **informa** business
A SCIENCE PUBLISHERS BOOK

Cover illustrations provided by the author of the book, Dr. Maged Marghany.

CRC Press
Taylor & Francis Group
6000 Broken Sound Parkway NW, Suite 300
Boca Raton, FL 33487-2742

First issued in paperback 2021

Version Date: 20180515

ISBN 13: 978-0-367-78111-8 (pbk)
ISBN 13: 978-0-8153-8639-1 (hbk)

Visit the Taylor & Francis Web site at
http://www.taylorandfrancis.com

and the CRC Press Web site at
http://www.crcpress.com

Dedicated to

My mother Faridah
my wife
my daughter who assisted me to
model 4-D image
and
Nikola Tesla who is teaching me to be a real scientist

Preface

Remote sensing scientists attempt to promote advanced technology of satellite developments to forecast the future tsunamis. There were various satellite images that showed the horror of the tragic Asian tsunami of the 26 December 2004. Regardless of advanced remote sensing technology, remote sensing scientists could not prevent the occurrence of the savage tsunami of 11 March 2011 in Japan. In this regard, the tsunami of 11 March 2011, was felt world wide, from Norway's fjords to Antarctica's ice sheet. Tsunami debris continued to wash up on North American beaches, even two years later. Conversely, the unexpected disaster record goes to the boxing day of 2004 tsunami where the wave transmissions finished within three days into the Atlantic Ocean and changed the Earth's rotation speed. Besides, double than 300,000 people were killed or went missing behind those remote sensing advanced technologies.

Nowadays, the innovation in space technologies has generated a new trend for the tsunami's observation and monitoring from space. Consequently, the rapid innovation of sensor developments allows high resolution of less than 1 m for optical satellite such as GeoEye-1 satellite, which collects images at nadir up to 60° with 0.41-meter panchromatic (black and white) and 1.65 m multispectral resolution. Synthetic aperture radar (SAR), as a result, also delivered 1 m high resolution image which is assembled by TerraSAR-X spotlight mode. In these contexts, advanced tsunami disaster observation from space has commenced novel perceptions of environmental research. The restricted ranges of satellite sensors had detected the 2004 tsunami. In this regard, optical satellite data in low resolution, with 250 to 500 m for instance, TERRA and AQUA MODIS were operated. Furthermore, medium resolution satellite data, within higher than 10 m for instance, LANDSAT, IRS, and SPOT-5 monitored the damages of the 2004 tsunami. The high resolution satellite data with less than 1 m such as IKONOS and QuickBird of pre and post tsunami period were implemented to detect the damage. Besides, the altimeter satellite data had given precise performance to monitor the tsunami wave propagation from the epicentre towards the Indian Ocean. Nevertheless, the microwave imaging satellite data of Synthetic Aperture Radar (SAR) were totally absent to monitor the first tsunami wave propagation. The shortage of SAR data during the 2004 tsunami event could be due to non-visiting of the satellite of epicentre zone and other affected areas.

The majority of the tsunami remote sensing studies were focused on using fundamental image processing such as classical image processing tools or conventional edge detection procedures. None of those studies integrated modern physics, applied mathematics, signal and communication with remote sensing data. Without this logical integration, alarm warning system cannot build perfect security matrix. Indeed, remote sensing technology is not an art to produce colourful maps but it is advanced technology based on modern and applied physics. The main question that can be raised up is what are the uses of multi satellite sensors advanced remote sensing technology to prevent the occurrence of the tsunami disasters.

Satellite remote sensing has numerous promising applications in a wide range of tsunami disciplines. Exploiting satellite data, the status and temporal growth of the tsunami disasters over large areas at short time intervals can be monitored accurately. Integrating this with *in situ* data and mathematical models allows us to monitor and emphasize the vital processes at work in huge areas, such as snow cover evolution, vegetation development or landslide movements.

The scope of the book is restricted to remote sensing data available for the 2004 tsunami. The first chapters of the book discuss the tsunami theories, and the fundamentals of optical remote sensing theories, as the major remote sensing data covering the 2004 tsunami events were optical data with different resolutions. The specific issue of alimentary satellite data which tracked the tsunami wave propagation has been addressed in this book. This book has also attempted to introduce a new image processing tool for monitoring tsunami disaster. These tools are advanced image processing tools which involve fuzzy B-spline for 3-D reconstruction of tsunami wave propagation in high resolution satellite data of QuickBird. This assists in understanding the spelling of Arabic words of Allah along Sri Lanka coastal waters. In addition, the new study involved in this book talks about the mechanisms of internal wave generated by tsunami, specially in Andaman Sea using SAR satellite data.

More advanced studies and a new work of 4-D image reconstructions of non-coherence optical satellite data of QuickBird are presented. This work is done by implementing a new concept of modern hologram interferometry by computer hologram generation. It is impossible to retrieve the hologram fringes from QuickBird data or other optical remote sensing due to an absence of phase information. Consequently, it may be possible to implement the hologram interferometry from optical satellite data by considering the procedures of incoherent hologram.

The impact of the tsunami on the physical properties of the coastal waters, especially Aceh is also addressed. In addition, the utilization of microwave ENVISAT SAR data to investigate the generation of internal wave in the Adaman Sea post tsunami is also discussed. In this regard, we implemented a new approach of image processing of Particle swarm optimization (PSO).

Finally, the last chapter involves quantum mechanics theory based on Schrödinger equation to forecast the future tsunami in the Malacca Straits due to plate tectonic activity north of the Sunda Trench. Beside the impact of Grand Ethiopian Renaissance Dam (GERD) for causing massive destructive tsunami along the Indian Ocean, Red Sea, Arabian Gulf, Sudan and Egypt, the split of the plate of African Horn is also discussed.

The main question is: can the Malaysian and Singaporean coastal waters be striked by huge tsunami wave heights? What is the travelling period for the tsunami to reach Johor and Singaporean coastal waters and what are the breaking wave heights that can strike the Malacca Straits with future tsunami? The answers of these questions are addressed scientifically in the last chapter of this book.

Maged Marghany,Ph.D
Associate Professor
School of Humanities
Universiti Sains Malaysia
11800 USM

Contents

Chapter 1

Principles of Tsunami

A tsunami is understood to be a comprehensive scientific thought owing to its huge destructive potential. We all have recently become aware of the penalties of living on a dynamic earth. An earthquake that was initially estimated at 8.5 on the Richter scale and was subsequently revised to 9.0 occurred on 26th December 2004. The common tsunami principles are crucial to comprehend the nature of the disaster. Consequently, the copiously comprehensive tsunami principles are delivered in this chapter. In this context, the chapter proceeds to provide a scientific definition of tsunami and its terminology. This chapter also explains the physical characteristics of the tsunami. The foremost aspects that need to be taken into account when classifying tsunami and how do tsunamis differ from other water waves are also introduced.

1.1 Definition of Tsunami

Tsunami or tidal wave is well-known because the natural phenomena comprising of a series of waves is created once the waves have rapidly preceded on an enormous scale. Tsunami (pronounced soo-NAH-mee) is a Japanese word which means, of the harbor ("tsu") and wave ("nami"). Tsunamis are fairly common in Japan and in recent centuries, thousands of Japanese have been killed by them. The term, consequently, was made-up by fishermen who returned to port to find the area close to the harbor devastated, although they had not been awake to any wave on high seas [1].

In other words, a tsunami is an incredibly massive ocean wave that is caused by an underwater earthquake or volcanic eruption and infrequently causes extreme destruction once it strikes land. Tsunamis can have heights of up to 30 m (98 ft) and reach speed of 950 kilometers (589 mi) per hour. They are described by long wavelengths of up to 200 kilometers (124 mi) and long periods, typically between 10 minutes and an hour.

1.1.1 Comments on Tsunami Definition

In earlier times, seismic ocean waves were called "tidal" waves, incorrectly implying that they had some direct connection to the tides. In fact, when the tsunami approached coastal zone, they began to be characterized by a violent onrushing tidal waves rather than the sort of cresting waves that are generated by wind stress upon the sea surface. Conversely, to eliminate this confusion, the Japanese word "tsunami" is used to describe the giant wave (Figure 1.1) which refers to a seismic wave and meaning harbor wave to replace the misleading term tidal wave. This tsunami is a synonym for seismic sea wave. In this regard, a tsunami is a seismic sea wave containing tremendous amounts of energy as a result of its mode of formation, i.e., the factor that causes a seismic wave. Consequently, tsunamis are temporary oscillations of sea level with periods longer than wind, waves and shorter than tides for the tsunami, and shorter than a few days of storm surge [1, 5].

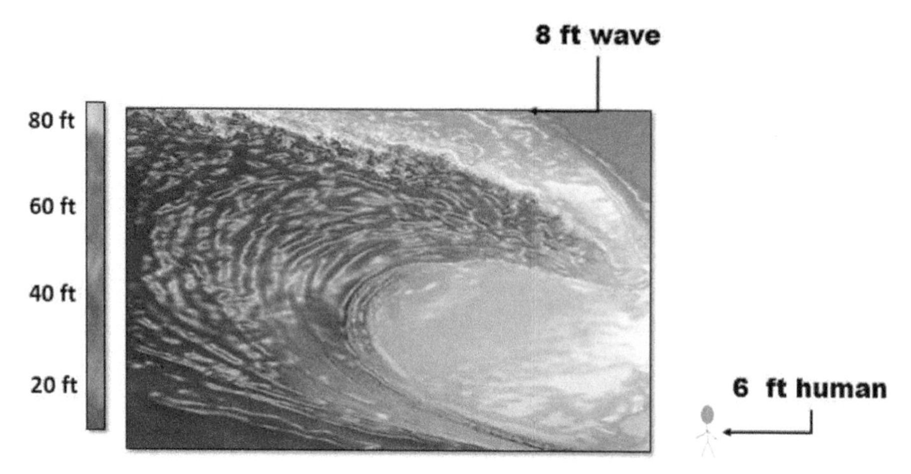

Figure 1.1 Giant tsunami wave.

1.2 Tsunami Terminology

Tsunami terminology is a keystone to comprehend the physics of the tsunami and how the tsunami is propagating across the water body from tiny heights of extremely massive waves. The tsunami terminology can be explained briefly as follows [7].

Amplitude: The swelling exceeds or falls below the ambient sea water level on record as tide gage (Figure 1.2) [6].

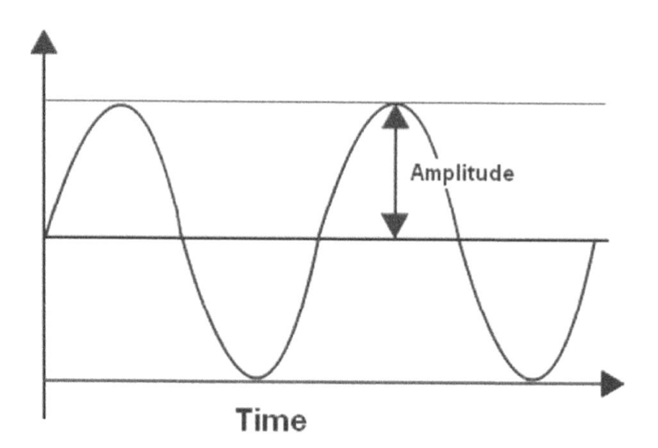

Figure 1.2 Tsunami amplitude.

Arrival time: Period of arrival, normally the first wave of the tsunami at a special geographical location. In this regard, Estimated Time of Arrival **(ETA)** is the estimated arrival time of the first tsunami wave at coastal zones where a particular earthquake has occurred. Therefore, initial motion of the first wave is known as first motion, which is a rise in the water level and is denoted by R, and a fall by F (Figure 1.3) [6, 7].

Period: The length of time (*T*) between two successive peaks or troughs. It may vary due to complex interference of waves (Figure 1.4). Tsunami periods generally range from 5 to 60 minutes [7].

Run-up: Maximum height of the water onshore observed above a reference sea level. Usually measured at the horizontal inundation limit (Figure 1.5).

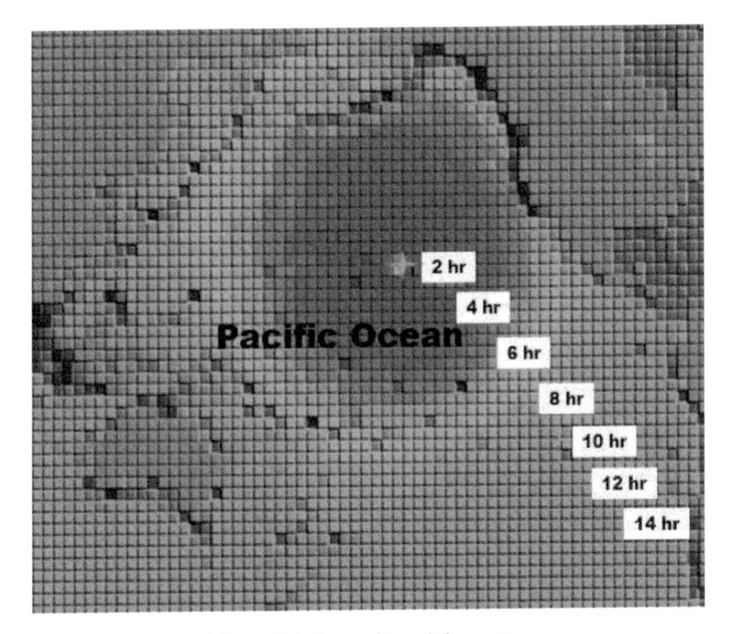

Figure 1.3 Tsunami travel time map.

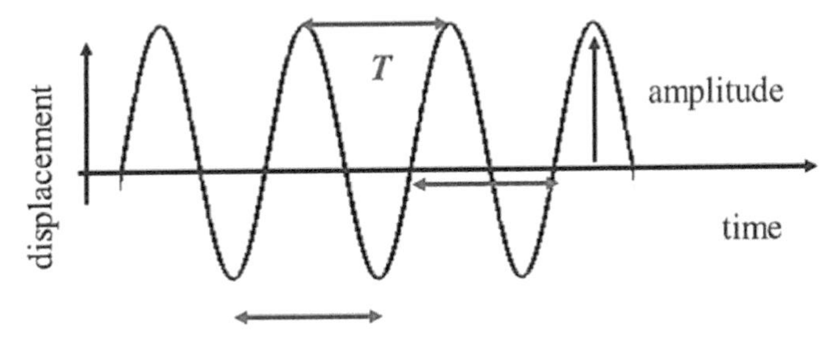

Figure 1.4 Tsunami period concept.

Figure 1.5 Tsunami run-up.

Inundation zone: A zone which is flooded with water (Figure 1.6).

Inundation: The depth, relative to a stated reference level, to which a particular location is covered by water (Figure 1.7) [7].

Inundation line (limit): The inland limit of wetting measured horizontally from the edge of the coast defined by the mean sea level (MSL) (Figure 1.7).

Horizontal inundation distance: The distance that a tsunami wave penetrates onto the shore, measured horizontally from the mean sea level position of the water's edge. Usually measured as the maximum distance for a particular segment of the coast (Figure 1.7).

Figure 1.6 Inundation zone.

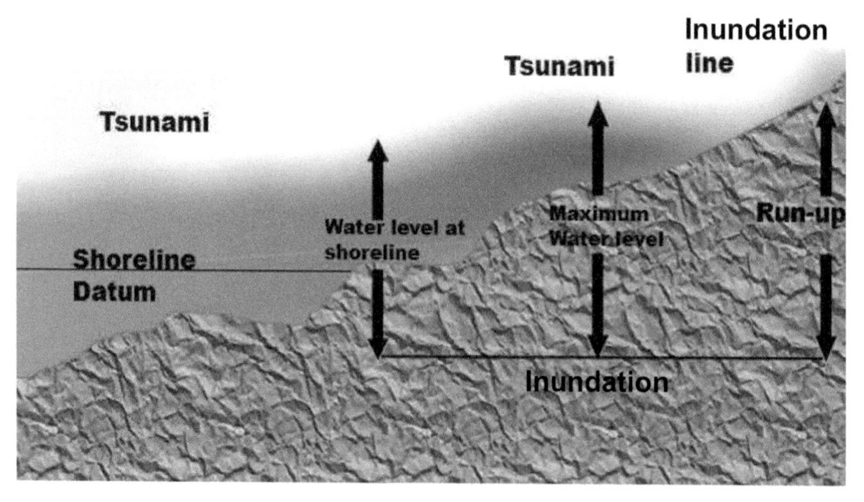

Figure 1.7 Tsunami inundation concepts.

Bore: Propagation wave with an abrupt wall of water or vertical front. Under certain terms, the running edge of a tsunami wave might develop a bore as it approximates and extends onshore. A bore may also be formed when a tsunami wave enters a river channel, and may travel upstream to a superior space inland than the wide-ranging inundation (Figure 1.8).

Teletsunami: It refers to the tsunami which is over 1000 km away from the area of interest. Variously referred to as a distant-source or far-field tsunami (Figure 1.9) [8]. The 2004 Indian Ocean tsunami was a teletsunami.

Free field offshore profiles: A profile of the wave measured far enough offshore so that it is unaffected by interference from the harbor and shoreline effects [8].

Figure 1.8 Tsunami bore.

Figure 1.9 Example of teletsunami.

Harbor resonance: The continued reflection and interference of waves from the edge of a harbor or narrow bay which can cause amplification of the wave heights, and extend the duration of wave activity from a tsunami (Figure 1.10) [8].

Leading-depression wave: Initial tsunami is a trough, causing a drawdown of water levels (Figure 1.11).

Leading-positive wave: Initial tsunami is a crest, causing a rise in water level. Also called a leading-elevation wave (Figure 1.12) [8].

Local/regional tsunami: Source of the tsunami is within 1000 km of the area of interest. Local or near-field tsunami has a very short travel time (30 minutes or less) Midfield or regional tsunamis have travel times in the order of 30 minutes to 2 hours. Note: "local" tsunami is sometimes used to refer to a tsunami of landslide origin [8].

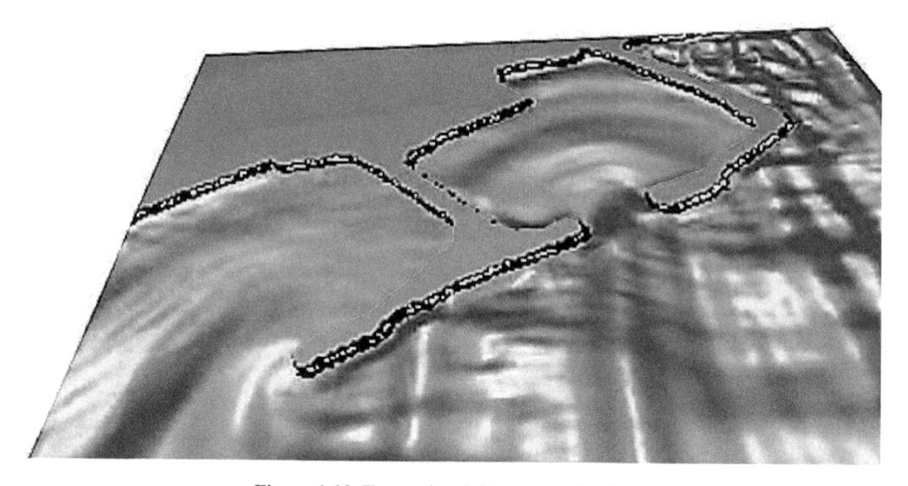

Figure 1.10 Tsunamis might resonate in a bay.

Figure 1.11 Receding very quickly and jetting westward away from the beach.

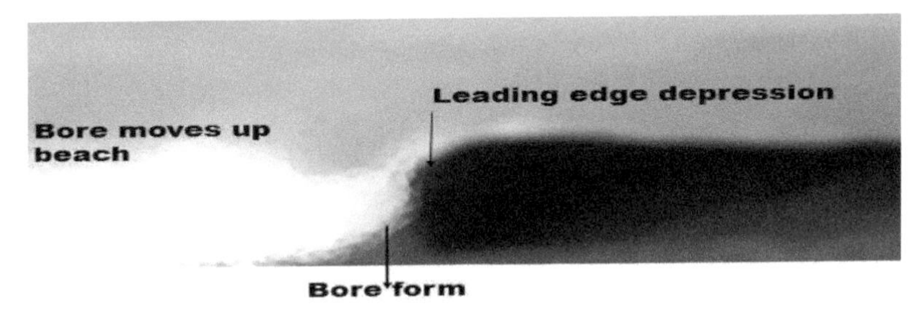

Figure 1.12 Leading-positive tsunami.

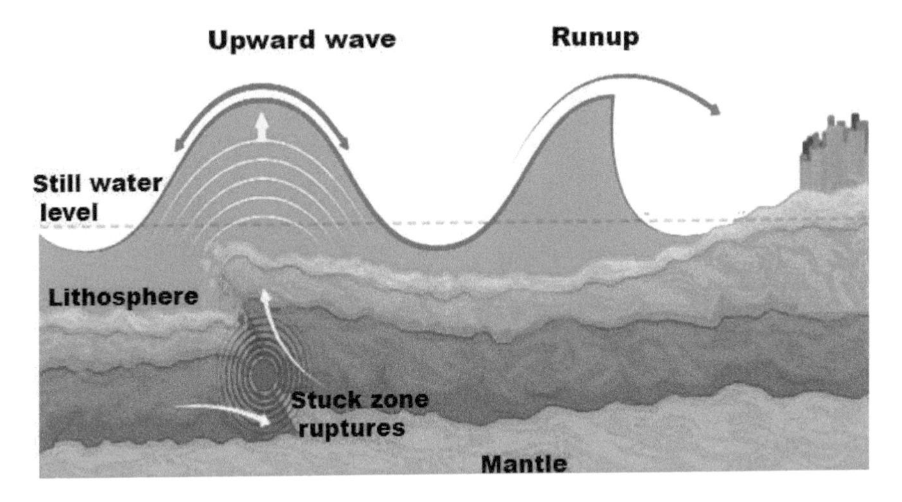

Figure 1.13 Earthquake tsunami.

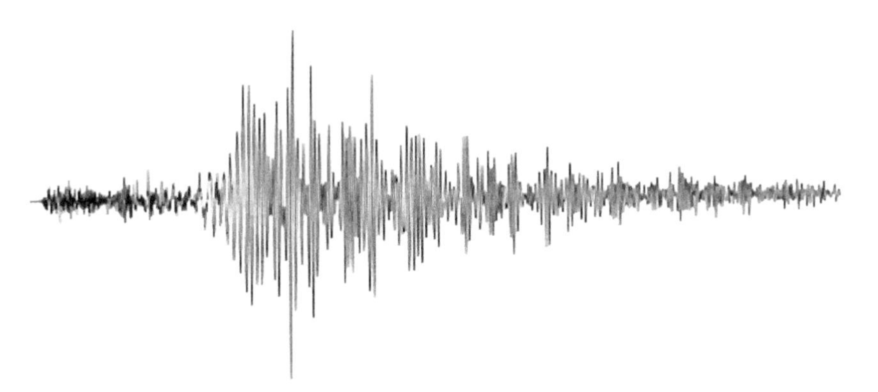

Figure 1.14 Tsunami 2004 magnitude.

Tsunami earthquake: A tsunamigenic earthquake which produces a much larger tsunami than expected from its magnitude (Figure 1.13).

Tsunamigenic earthquake: Any earthquake which causes a measurable tsunami (Figure 1.13).

Tsunami magnitude: A number which characterizes the strength of a tsunami based on the tsunami amplitudes. Several different tsunami magnitude determination methods have been proposed (Figure 1.14).

Maremoto: Spanish term for tsunami.

Marigram: Tide gage documenting wave height as a function of time (Figure 1.15).

Marigraph: The instrument which records wave height.

Mean Lower Low Water (MLLW): The average low tide water elevation often used as a reference to measure run-up (Figure 1.16).

Surface Wave Magnitude: The magnitude of an earthquake as gaged from the amplitude of seismic surface waves. Regularly, it is referred to as "Richter" magnitude (Figure 1.17) [6].

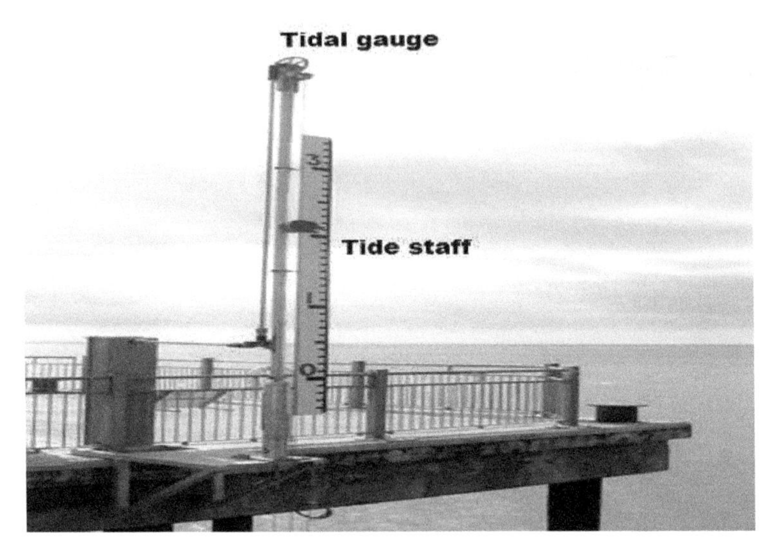

Figure 1.15 Tide gauge.

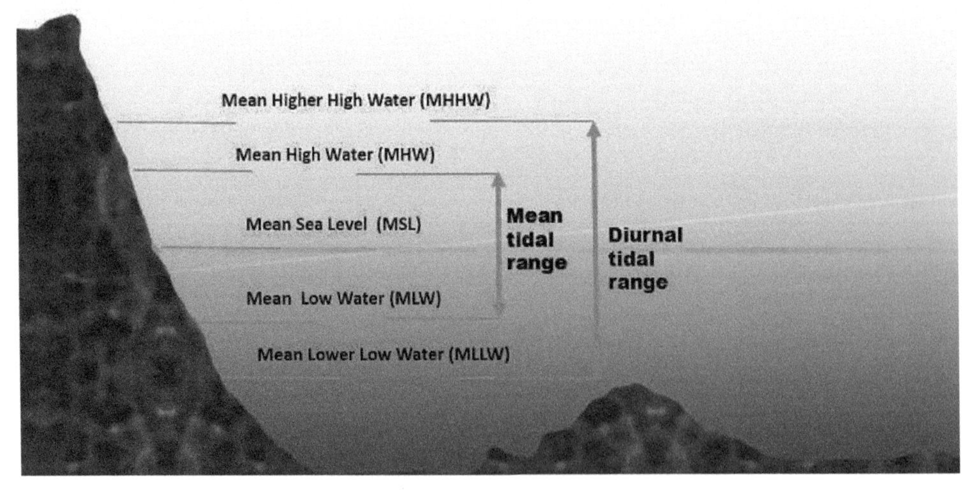

Figure 1.16 Water levels referenced to a tidal datum such as Mean Lower Low Water (MLLW) or Mean Higher High Water (MHHW).

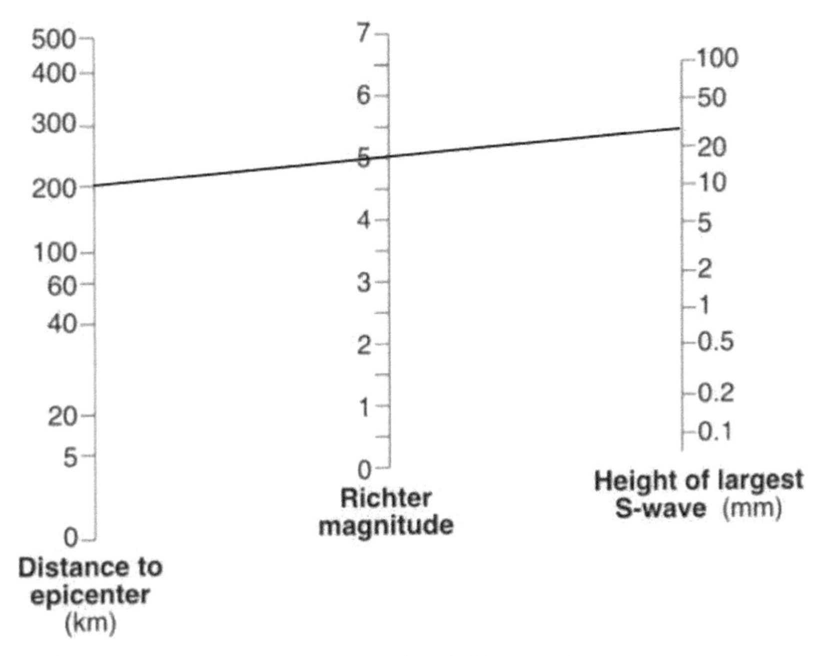

Figure 1.17 Magnitude in Richter scale.

1.3 Physical Characteristics of Tsunami

This tsunami is a synonym for seismic sea wave. In this context, a tsunami is a seismic sea wave containing tremendous amounts of energy as a result of its mode of formation, i.e., the factor that causes a seismic wave. It could be a submarine landslide, a shifting of rocks triggered by an earthquake or volcanic explosion. Further, tsunamis are flat waves with long periods and long wavelengths; they grow in height in shallow water and flood the shoreline, sometimes causing catastrophic destruction. Consequently, tsunamis are temporary oscillations of sea level with periods longer than the wind waves and shorter than tides of the tsunami, and shorter than a few days of storm surge [5].

In a somewhat similar fashion, dropping a stone into a puddle of water creates a series of waves which radiate away from the impact point. In this context, the impact point of a puddle of water represents a sudden shift of rocks or sediments on the ocean floor caused by cataclysmic events, such as a volcanic eruption, an earthquake, or a submarine landslide, which can force the water level to drop ≥ 1 m, generating a tsunami—a series of low waves, with long periods and long wavelengths. This indicates that the tsunami wavelength is bigger than its amplitude in the open ocean. Thus, the tsunami height in the open ocean is approximately less than 1 m which is not noticeable in the open ocean. The tsunami, however, crosses the oceans at a rate of ~ 750 km/hr (Figure 1.18) which is equivalent to the speed of a modern jet aircraft! Despite this phenomenal speed, tsunami poses no danger to vessels in the open ocean. Indeed, the regular ocean swell would probably mask the presence of these low sea waves. The tsunami, however, grows to a height of ≥ 10 m as it impinges on a shoreline and floods the coast, sometimes with catastrophic results, including widespread property damage and loss of life [5].

1.4 Tsunami Classifications

A tsunami is often classified by the gap from their source point to the realm of impact, i.e., local and remote tsunami. Regionally generated tsunamis have short warning times and comparatively short wave periods.

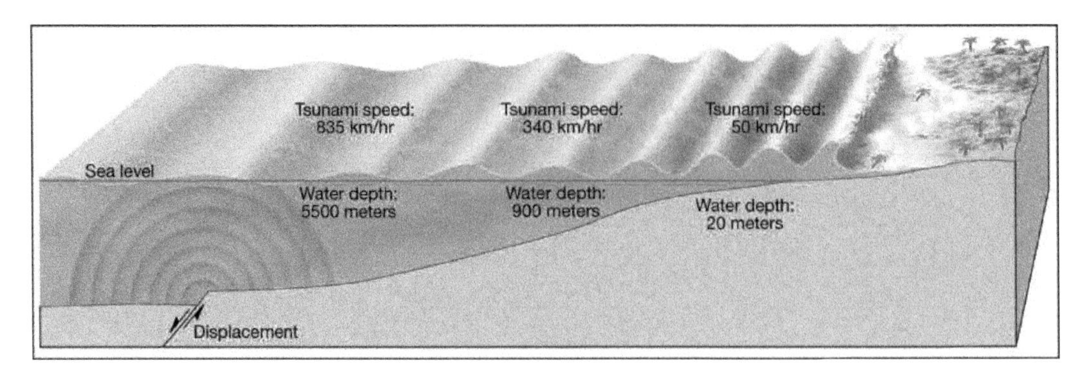

Figure 1.18 Different tsunami velocities.

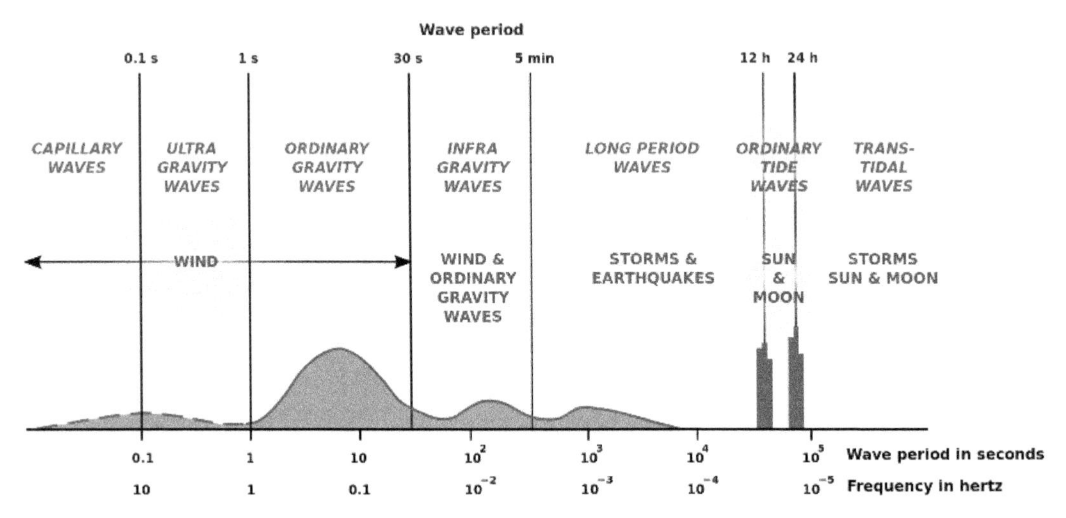

Figure 1.19 Tsunami and other wave classifications.

Remote tsunamis, nevertheless, have longer warning times and relatively long periods. Typical periods for tsunami fluctuates from the quarter of an hour for regionally generated tsunami to countless hours for remote tsunami. Typical run-up height of tsunami varies up to 15 m at the coast, though most are quite smaller. Storm surges on the contrary, are caused by variations in barometric pressure and wind stress across the ocean. Decreasing barometric pressure causes an inverse barometer impact wherever water level rises. This can be usually a slow and large-scale impact and therefore does not usually generate waves within the frequency very typical of the tsunami. Conversely, there may be short-period meteoric events (such as meteoric moving ridge–rissaga) with time-scales of a couple of hours which can be necessary. Wind stress, on the opposite hand, includes a wide range of time-scales and causes coastal sea-level setup as well as wind waves. In fact, the sea-level setup depends on the wind direction, strength, and wave height. Storm surge periods vary from many hours to many days. Typical heights for storm surge alone varies up to 1.0 m as an example, on the coast of New Zealand though most are typically less than 0.5 m. Wind waves, on the opposite hand, are often quite massive, producing wave setup and wave run-up of many meters in height (Figure 1.19) [5].

Generally, a tsunami can be categorized into: (i) microtsunami which has a small amplitude and can not even be noticed visually and (ii) local tsunami which has a destructive impact due to its widespread run-up on coastal zone within hundreds of kilometers and is usually caused by plate tectonic movements.

It might be the internal wave named by internal tsunami owing to its traveling along thermocline layer [5] below the sea surface.

1.5 How do Tsunamis Differ from other Water Waves?

Unlike wind waves, a tsunami has long wavelength and small height less than 1 m (Figure 1.20).

In this view, a tsunami is categorized as shallow-water waves, as well as long periods and wave lengths. The swell is generated by the wind and a storm out in the sea (Figure 1.20) and steadily rolls in, one wave after another. They might have a period of about 10 seconds and a wavelength of 150 m. A tsunami, in other words, can have a wavelength of more than 100 km and a period on the order of one hour (Figure 1.21).

As a result of their long wavelengths, tsunami behaves as shallow-water waves. A wave becomes a shallow-water wave when the ratio between the water depth and its wavelength gets very small. Shallow-water waves move at a speed that is equal to the square root of the product of the acceleration of gravity (9.8 m/s/s) and the water depth. Let us see what this implies: in the Pacific Ocean, where the typical water depth is about 4000 m, a tsunami travels at about 200 m/s, or over 700 km/hr. Because the rate at which a wave loses its energy is inversely related to its wavelength, tsunami not only propagates at high speeds, they can also travel inordinate, transoceanic distances with limited energy losses [3, 5].

Figure 1.20 Wind wave.

Figure 1.21 Ocean surface motion due to tsunami.

In general, the primary differences in tsunami are because of size, speed and source. Let us look at what creates a normal wave. Waves in the ocean are created by a number of things (gravitational pull, underwater activity, atmospheric pressure), but the most common source of waves is the wind (Table 1.1) (Figure 1.22).

Additionally, the mechanism of generating the wind wave or sea is completely different as compared to the tsunami. The ocean wave is principally generated by the impact of wind stress on the ocean surface. Once the wind blows across a swish water surface, the air molecules grab water molecules as they are carried across the water by the wind. The friction between the air and water stretches the water's surface, making ripples within the water called capillary waves. The capillary waves move in a circle. This circular motion of water continues vertically underwater, although the facility of this motion decreases in deeper water (Figure 1.23). Because the wave travels, additional and more water molecules are collected, increasing the scale and momentum of the wave. The foremost vital factor to grasp regarding waves is that they do not represent the movement of water; they instead show the movement of energy through water. In traditional waves, the wind is the source of that energy. The scale and speed of wind waves rely on the strength of the wind [7].

Table 1.1 Characteristics of tsunami wave *vs*. characteristics of wind-generated wave.

Wave Features	Wind-generated Wave	Tsunami Wave
Wave Speed	5–60 mph (8–100 kph)	500–600 mph (800–1,000 kph)
Wave Period (time required for two waves to pass a single point in space)	5 to 20 seconds apart	10 minutes to 2 hours apart
Wave Length (horizontal distance between two waves)	300–600 feet apart (100–200 meters apart)	60–300 miles apart (100–500 km apart)

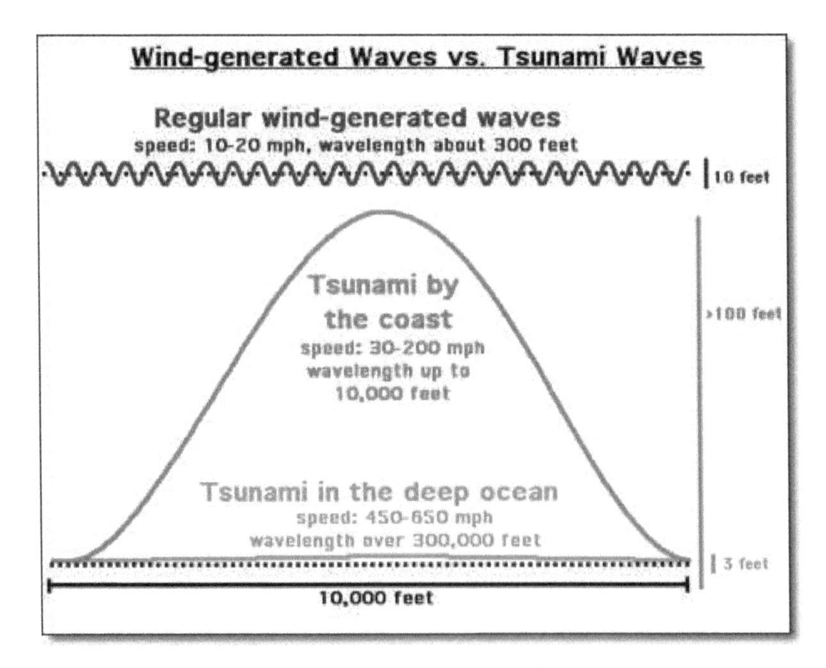

Figure 1.22 Wind waves *vs*. tsunami waves.

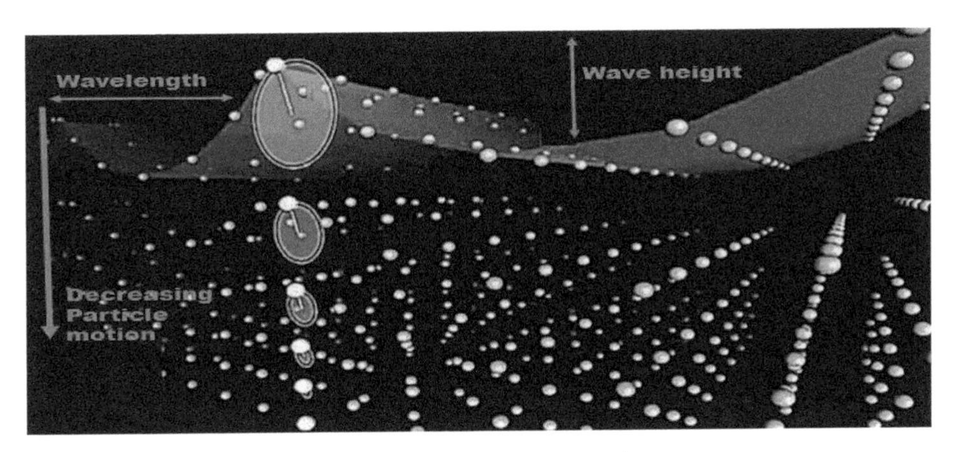

Figure 1.23 Wave Water particle motion.

1.6 The Wave Train

A single tsunami event can create a series of the wave trains which are variants of heights. After the initial wave, numerous waves can arise. For instance, four or five smaller waves may immediately tail the first arrival wave into the shallow water. Tsunami waves spread into the shore as a series of successive "crests" (high water levels) and "troughs" (low water levels). These successive crests and troughs can occur anywhere and the period (the time between the crests of the waves) (Figure 1.24) generally varies from 5 to about 90 minutes. They usually occur 10 to 45 minutes apart. After a powerful tsunami event, multiple waves may arrive over a prolonged time period. Long after the initial event, smaller tsunamis may continue over several days [5, 7].

1.7 The Shoaling Effect

Tsunami becomes hazardous the minute they extend to the shallow waters approaching the coast, in an exceedingly wave shoaling process (Figure 1.25). In coastal zones, wherever water levels steadily become shallower, the tsunamis can curtail dramatically, become compressed and propagate into steeper slope

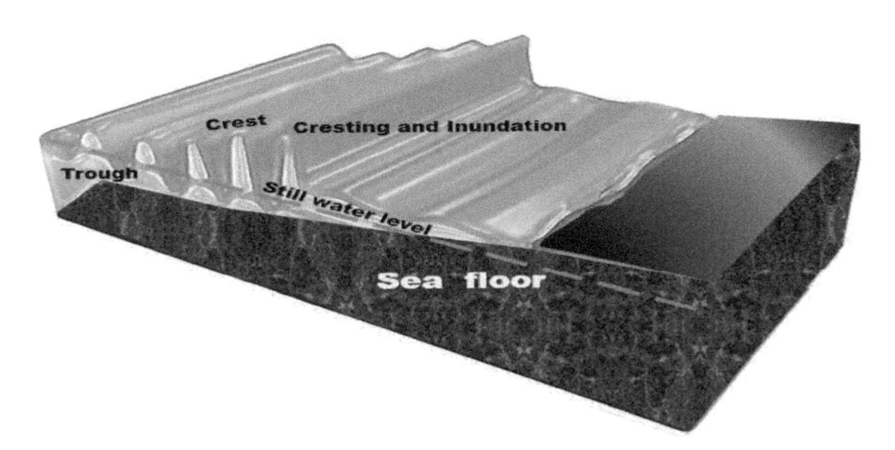

Figure 1.24 Successive tsunami crests and troughs.

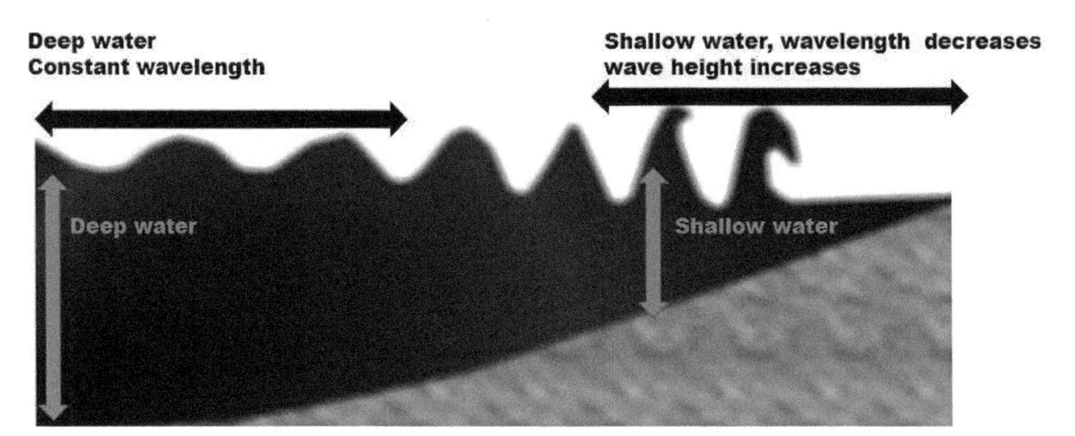

Figure 1.25 Shoaling process.

owing to the reducing water depth. The process of steepening of the tsunami surface is sort of a whip action. Because the wave descends the whip from handle to tip, the energy is discharged into a smaller and smaller mass. This energy is transferred just like the cracking of a whip in an exceedingly violent swelling.

Once the tsunami moves in shallow water, it decelerates down and its height (amplitude) grows. Shortly after the tsunami arrives at the shallow water, its length reduces and its height upsurges. In this regard, tsunamis have been known to surge vertically as high as 100 feet (30 meters). Most tsunamis cause the sea to rise no more than 10 feet (3 meters) (Figure 1.25) [10]. This is often attributable to the mass and energy of a tsunami. In the wind-generated wave, the wind shear stress is stirring the upper layer of the ocean surface with perturbation of a tsunami wave through the water column. This forces the ocean surface into massive wave propagations. While these massive waves breakdown, they regularly abolish docks, constructions, and coasts and vanish the human and natural life [9].

Chapter 2

Tsunami Generation Mechanisms

Tsunami generation mechanisms emerge as questionable due to its massive destructive potential. Indeed, scientists and engineers are attempting to minimize the excessive hazardous impact of tsunami. In this context, the major causes of the tsunami are required to plan specific defence coastal structures. In this sense, the deep details and comprehensive perception of tsunami generation mechanisms are required. This chapter is devoted to explicate the different factors that can generate a tsunami. Further, this chapter will answer the question as to why are tsunamis not seen at sea or from the air?

2.1 Causes of Tsunami

It is established that massive earthquakes that cause vertical displacement of the massive extent of seawater are genuinely generated by the tsunami. Scientists have agreed that underwater nuclear explosion, large meteorites falling into the sea, volcano eruption and rock slides are the likely explanations of the advent and transmission of the tsunami. Consequently, scientists are convinced that the massive earthquakes are the most accepted reason of tsunamis, for instance, the Sumatra-Andaman earthquake of 2004. Under these circumstances, the tsunami dynamics are characteristic of: (i) generation of a tsunami; (ii) propagation of tsunami; and (iii) tsunami run-up and inundation.

2.1.1 Tsunami Generation

The main question that arises is how do earthquakes generate the tsunami? The reply of the query requires a comprehensive grasp of the principle of plate tectonics. Truthfully, plate tectonic theory is the beginning point to grab the mechanisms of tsunami generation. Consequently, the plate tectonic principles involve the generation of earthquakes and volcano which are the main causes of the tsunami. Consequently, there are three versions of the plate tectonic theory. The first version that has been revealed is that the lithosphere is broken into strong, rigid moving plates that carry the continents on their original ocean basins. This mechanism, therefore, created mid-ocean ridges and subduction zone,respectively (Figure 2.1) [60].

Further, the second version has accepted that two actual plate boundaries fluctuate substantially alongside strike combined plate boundary. In this view, large continental features like mountain belts and long faults are indicators of the deformations of rocky mountains, basins, and the disturbances of mid-plate volcanic of the Hawaiian Islands (Figure 2.2) [61].

In the third version of plate tectonics, the lithosphere is considered as a far-from-equilibrium open thermodynamic system of self-organizing semi-rigid plates appearing to dissipate crust and mantle heat (Table 2.1). The plates are susceptible to soften from the underlying, barely solid asthenosphere. Currently, top-down cooling and gravity are counted as the distinctive driving forces behind plate motions. The mostly

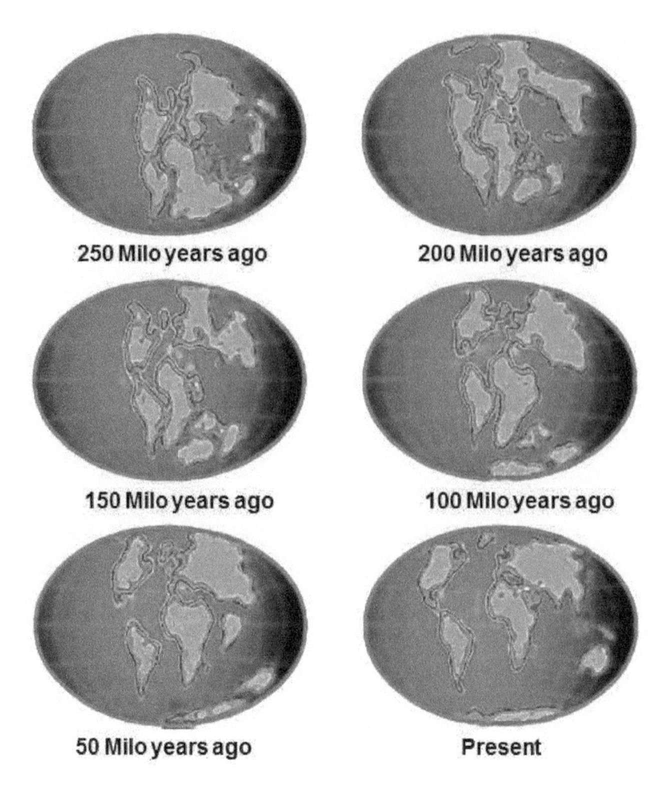

Figure 2.1 First version of plate tectonic theory.

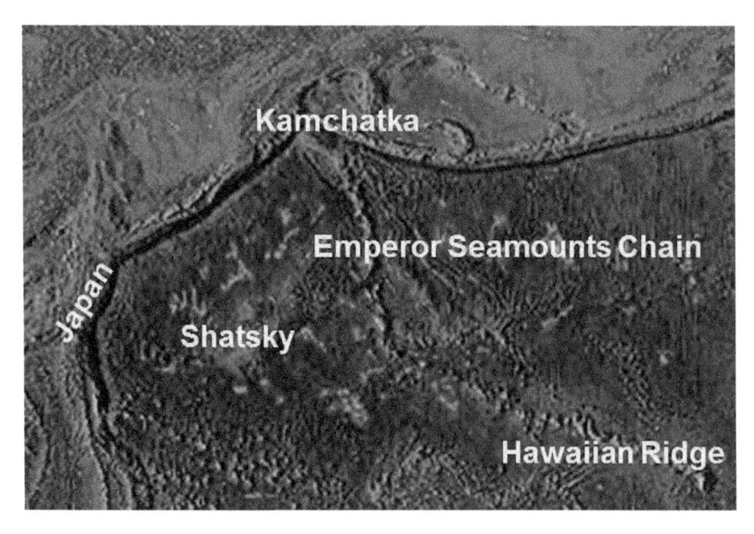

Figure 2.2 Northern Pacific floor with Hawaii-Emperor and many other volcanic chains.

passive upper mantle receives warmth from the lower mantle solely through conduction and cools notably via seafloor spreading (60%) and subduction. In this understanding, subduction-related hinge rollback and overriding plate extension have emerged as the imperative shapes of the face of the earth. Hot spots are now considered as excessive volcanism targeted through extensional plate disasters or, in uncommon cases,

Table 2.1 Mantle heat sources.

Sources	Timing	Origin
Radiogenic	On-going	Decay of radioactive elements (i.e., thorium, uranium, potassium) concentrated for the most part in the upper mantle and crust after ~ 4.4 Ga.
Residual formation of heat	One-shot deal largely complete by ~ 4.4 Ga, stalking off ~ 3.8 Ga	Gravitational, kinetic and thermal energy accumulated by the initial solar system wreckages dropping into the earth via influence warming.
Chemical	One-shot deal ending ~ 4.4 Ga	Discrimination and separation of earth materials into lower energy minerals and layers well harmonized to the temperatures and depths to which they deposited as the initially molten planet sorted itself out.

through local and particularly shallow thermal disturbances associated with upper mantle temperature versions precipitated through close by plate motions [62].

2.1.2 Plate Tectonics: The Main Features

Plate tectonics are related to essential features which are as follows:

- The Earth's surface is made up of a sequence of massive plates (like portions of a large jigsaw puzzle).
- These plates are in steady movement travelling at a few centimetres per year.
- The ocean floors are consistently moving, spreading from the center and sinking at the edges.
- Convection currents of the Earth below plate which derive the plates in different directions.

Consistent with the above perspective, the source of warmth driving the convection currents is radioactive decay, which occurs deep in the Earth. Convection takes place due to the fact that the density of a fluid is associated with its temperature. Hot rocks decrease in the mantle which are much less dense than their cooler counterparts above. The warm rock rises and the cooler rock sinks due to gravity [63] (Figure 2.3).

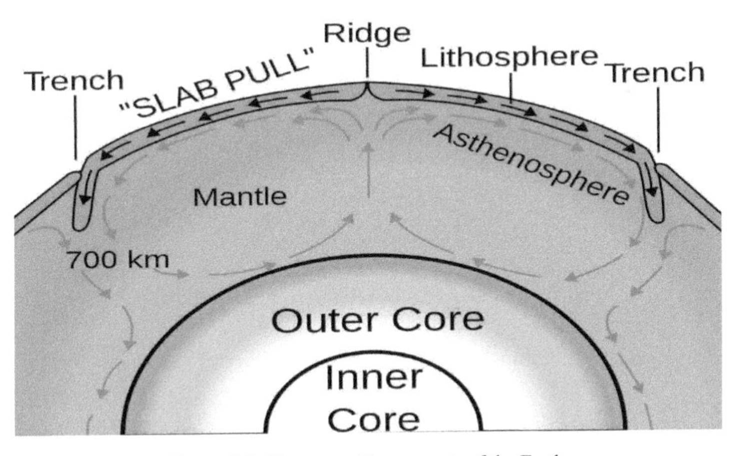

Figure 2.3 The convection currents of the Earth.

2.1.3 Type of Plate Tectonic Boundaries

The comprehensive perception of tectonic plate interactions is the keystone of earthquakes, volcano and tsunami generation. In this view, the concepts of tectonic plate interactions are (i) divergent boundaries; (ii) convergence boundaries and; (iii) radically changed boundaries (Figure 2.4).

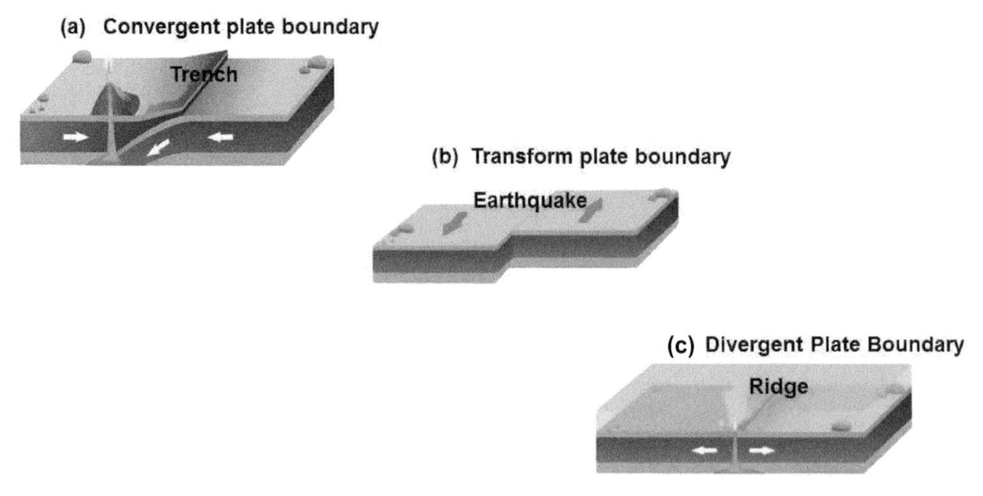

Figure 2.4 Three types of plate boundaries (a) Convergent; (b) Transform; and (c) Divergent boundaries.

2.1.3.1 Divergent Boundaries

In plate tectonics, a divergent obstacles represent zones where plates deviate from each other that is formed as an instance (Figure 2.5), both mid-oceanic ridges or rift valleys. In this regard, divergent boundaries are additionally recognized as positive barriers [63]. The Atlantic Ocean was created via this procedure. The Mid-Atlantic Ridge is a place where the new sea ground is being created.

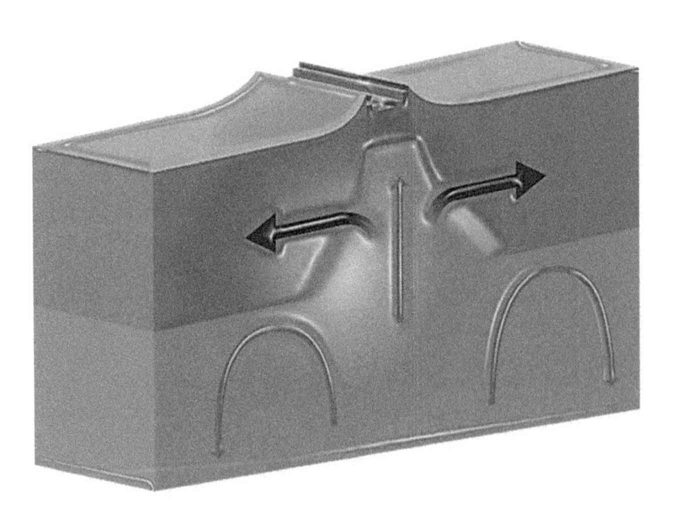

Figure 2.5 Divergent boundaries.

The divergence boundaries are scattered world wide and the most recognized of them are listed as follows [65]:

- The East African Rift (Great Rift Valley) is located in eastern Africa.
- The Mid-Atlantic Ridge widget splits the North American Plate and South American Plate in the west from the Eurasian Plate and African Plate in the east.
- The Gakkel Ridge is a sluggish spreading ridge which is positioned in the Arctic Ocean.

- The East Pacific Rise, extends from the South Pacific to the Gulf of California.
- The Baikal Rift Zone is located in eastern Russia.
- The Red Sea Rift.
- The Aden Ridge is located alongside the southern shore of the Arabian Peninsula.
- The Carlsberg Ridge is found in the eastern Indian Ocean.
- The Gorda Ridge off the northwest coast of North America.
- The Explorer Ridge off the northwest coast of North America.
- The Juan de Fuca Ridge off the northwest coast of North America.
- The Chile Rise of the southeast Pacific [65].

2.1.3.2 Convergence Boundaries

Convergent boundaries are properly identified as devastating margins. Along these boundaries, the plates approach each other and crash (Figure 2.6). These boundaries involve (i) Subduction zones; (ii) Obduction; and (iii) Orogenic belts. Subduction zones occur where an oceanic plate convenes a continental plate and is broken beneath it. Subduction zones are indicated through oceanic trenches (Figure 2.4). The subsequent completion of the oceanic plate softens down and constructs compression in the mantle, triggering volcanoes to create earthquake.

Obduction arises when the continental plate is plugged beneath the oceanic plate. Nonetheless, this is unfamiliar; for instance, the relative densities of the tectonic plates allows subduction of the oceanic plate. This causes the oceanic plate to fasten and commonly penalties in a new-fangled mid ocean ridge growing and revolving the obduction into subduction. The subduction zones are listed as follows [65]:

- The oceanic Nazca Plate subducts underneath the continental South American Plate at the Peru–Chile Trench.
- Just north of the Nazca Plate, the oceanic Cocos Plate subducts beneath the Caribbean Plate and forms the Middle America Trench.
- The Cascadia subduction quarter is where the oceanic Juan de Fuca, Gorda and Explorer Plates subduct underneath the continental North American plate.
- The oceanic Pacific Plate subducts below the North American Plate (composed of both continental and oceanic sections) forming the Aleutian Trench.
- The oceanic Pacific plate subducts underneath the continental Okhotsk Plate at the Japan Trench.
- The oceanic Philippine Sea Plate subducts underneath the Eurasian Plate at the Ryukyu Trench.
- The oceanic Pacific Plate subducts beneath the oceanic Philippine Sea Plate forming the Mariana Trench.
- The oceanic Philippine Sea Plate is subducting underneath the Philippine Mobile Belt forming the Philippine Trench and the East Luzon Trench.
- The Eurasian Plate is subducting beneath the Philippine Mobile Belt at the Manila Trench.
- The Sunda Plate is subducting underneath the Philippine Mobile Belt at the Negros Trench and the Cotobato Trench.
- The oceanic Indo-Australian Plate is subducted underneath the continental Sunda Plate alongside the Sunda Trench.
- The oceanic Pacific Plate is subducting beneath the Indo-Australian Plate north and east of New Zealand; however, the path of subduction reverses south of the Alpine Fault where the Indo-Australian Plate which causes subducting zone under the Pacific Plate.
- The South American Plate is subducting under the South Sandwich Plate, forming the South Sandwich Trench [65].

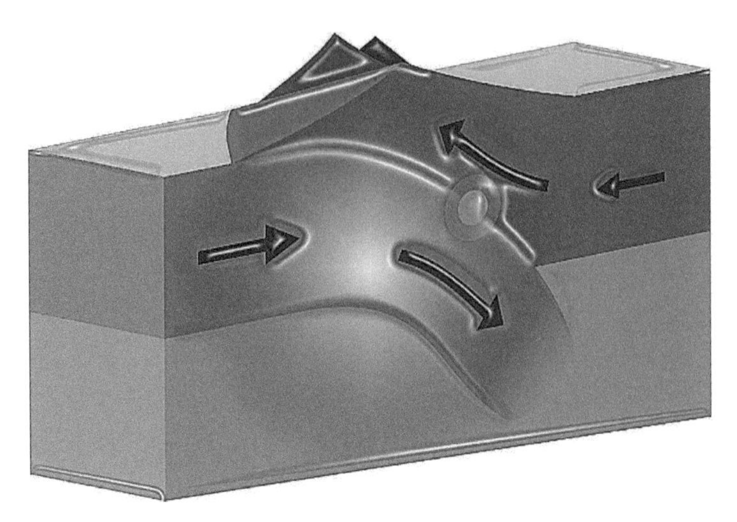

Figure 2.6 Convergent boundaries.

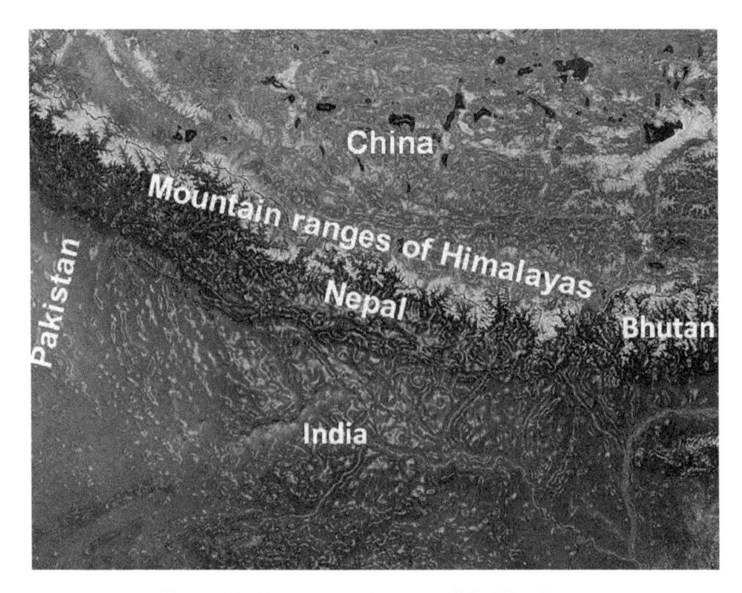

Figure 2.7 Huge mountain range of the Himalayas.

Consequently, the collision boundaries are recognized as orogenic belts [64]. These collision boundaries are produced because of a collision of two plates which thrust upwards to create cumbersome mountain stages, for instance, the Himalayas, the highest mountain range on Earth (Figure 2.7).

The most wonderful orogenic belt of the Earth is the one between the Indo-Australian Plate and African Plate on one side (to the South) and the Eurasian Plate on the other side (to the North). This belt extends from New Zealand in the east-south-east, through Indonesia, alongside the Himalayas, through the Middle East up to the Mediterranean in the west-northwest. It is also known as the "Tethyan" Zone, as it constitutes the quarter alongside which the ancient Tethys Ocean used to be bent and disappeared. The following mountain belts can be exclusive [65]:

- The European Alps
- The Carpathians

- The Pyrenees
- The Apennines
- The Dinarides
- The North African mountain belts such as the Atlas Mountains
- The Karst Plateau of the Balkan Peninsula
- The Caucasus
- The Zagros
- The Himalayas
- The Indonesian Archipelago
- The Southern Alps of New Zealand
- The Andes orogenic belt is the contemporary of a sequence of pre-Andean orogenies alongside the western margin of the South American Plate.

2.1.3.3 Transform Boundaries

Figure 2.8 indicates the occurrence of radically changed boundaries when two plates suppress each other with solely partial convergent or divergent movement.

The radically changed boundaries are documented as follows [65]:

- The San Andreas Fault in California is an active transform boundary. The Pacific Plate (carrying the city of Los Angeles) is shifting northwards with respect to the North American Plate.
- The Queen Charlotte Fault on the Pacific Northwest coast of North America.
- The Motagua Fault, which crosses through Guatemala, is a radical change boundary between the southern facet of the North American Plate and the northern part of the Caribbean Plate.
- New Zealand's Alpine Fault is some other energetic transform boundary.
- The Dead Sea Transform (DST) fault which runs via the Jordan River Valley in the Middle East.
- The Owen Fracture Zone alongside the southeastern boundary of the Arabian Plate [65].

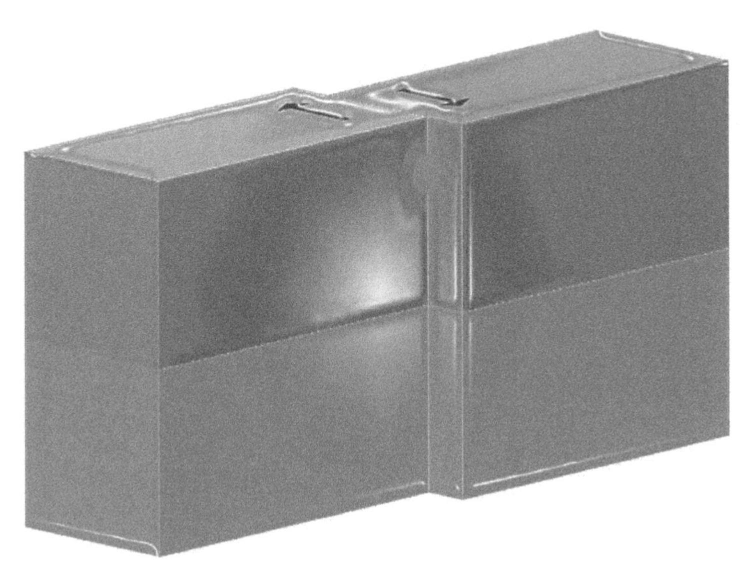

Figure 2.8 Transform boundaries.

2.1.4 Where is the Evidence for Plate Tectonics?

The continents appear to be in shape together like a massive jigsaw puzzle: If you seem to be on a map, Africa appears to snuggle properly into the east coast of South America and the Caribbean Sea. In 1912, a German Scientist known as Alfred Wegener proposed that these two continents were once joined together, and then in some way drifted apart. He proposed that all the continents had been once stuck together as one massive land mass known as Pangea. He believed that Pangea used to be intact until about 200 million years ago. The map of plate tectonics that has mounted through NASA are given in Figure 2.9 with notable evidence of the existence of plate tectonics.

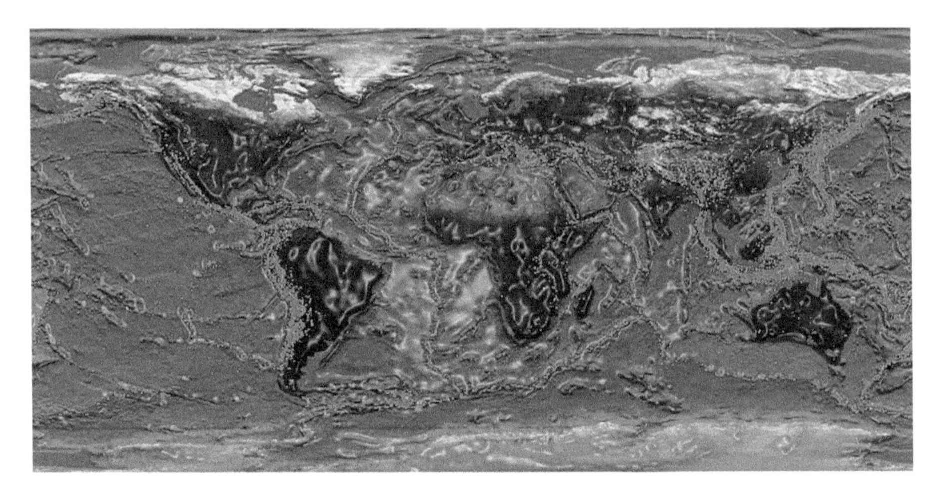

Figure 2.9 Current map of plate tectonics.

2.1.5 Continental Drift

In 1915, the belief of continental drift was once proposed through the German geologist and meteorologist Alfred Wegener, which publicizes that components of the Earth's crust slowly go with the flow atop a liquid core. The fossil document verifies and affords credibility to the theories of continental float and plate tectonics. The precept of continental flow has been counted via manner of the idea of plate tectonics, which clarifies how the continents shift.

The impression that continents can drift about is termed, unsurprisingly, CONTINENTAL DRIFT. The ancient (AND VERY WRONG!!) theory was called the "Contraction theory" which endorsed that the Earth was once a molten ball and in the process of cooling, the surface cracked and folded upon itself. The imperative difficulty with this clue was that all mountain levels have to be roughly the identical age, and this cannot be true. It used to be thought that as the continents relocated, the principal part of the continent would confront resistance and consequently constrict and bend upwards creating mountains close the foremost edges of the drifting continents.

According to Frankel [66], the key argument about the continental drift is absent. In this context, the plate tectonics are identified as a barring mechanism to cause the continental drift.

East Asian Sea Plate was once an unknown tectonic plate which had been swallowed by the Earth. It was discovered in the Philippine Sea. In fact, the Philippine Sea is located at the juncture of numerous predominant tectonic plates: (i) the Pacific; (ii) Indo-Australian; and (iii) Eurasian plates. These formed several minor plates, containing the Philippine Sea Plate. Since 55 million years ago, the Plate of the Philippine sea was drifted northwest due to the continental drift [74].

In the progression of plate tectonic theory, the Philippine Sea Plate crashed in the northern area of the East Asian Sea Plate and pushed it into the Earth's mantle. In this regard, the southern region of the East Asian Sea Plate was once eventually subducted by, or enforced below, different adjoining plates. Under

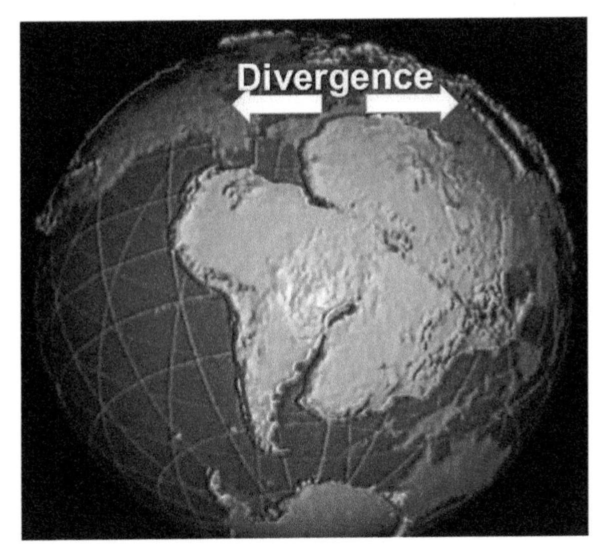

Figure 2.10 Continental drift.

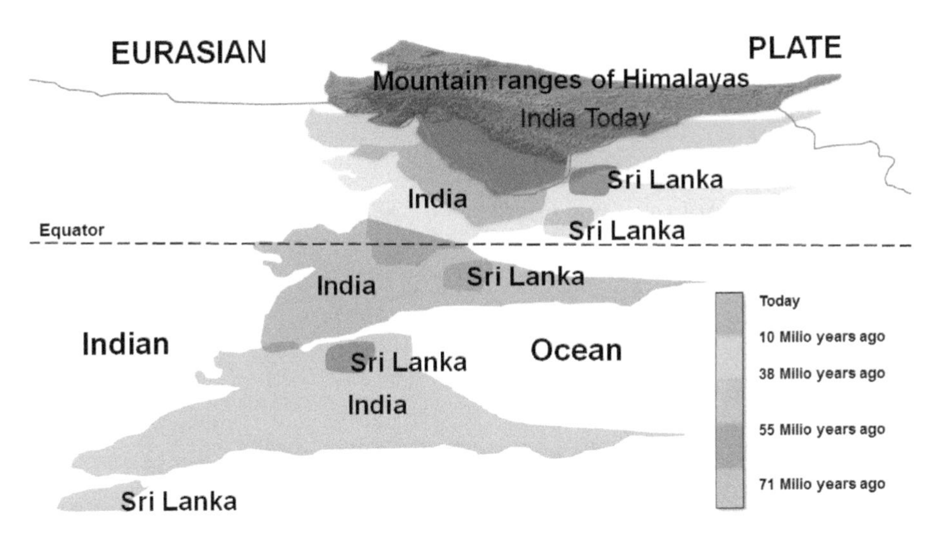

Figure 2.11 Himalayas theory of drifting tectonic.

this circumstance, there is an evidence of the numerous ocean substances locked up in a mineral referred to as ringwoodite lurking deep inside the Earth's rocky mantle. The mantle sits between Earth's crust (at the surface) and its core (Figure 2.13) [67]. Further, Philippine Sea Plate is viewed as micro plates which incorporate numerous comparatively smaller tectonic plates [67].

The mid-oceanic ridges rise 3000 meters from the ocean floor and are greater than 2000 km, extensively surpassing the Himalayas in size. The mapping of the seafloor, additionally, revealed that these massive underwater mountain ranges (Figure 2.14) have a deep trench which bisects the size of the ridges and, in some locations, is deeper than 2000 meters. These contain the biggest warmth float that was once situated at the crests of these mid-oceanic ridges. Seismic investigations show that the mid-oceanic ridges experience an increased quantity of earthquakes. All these observations point out excessive geological endeavour at the mid-oceanic ridges. Further, the deep waters are discovered in oceanic trenches, which plunge as deep as 35,000 ft beneath the ocean's surface [67].

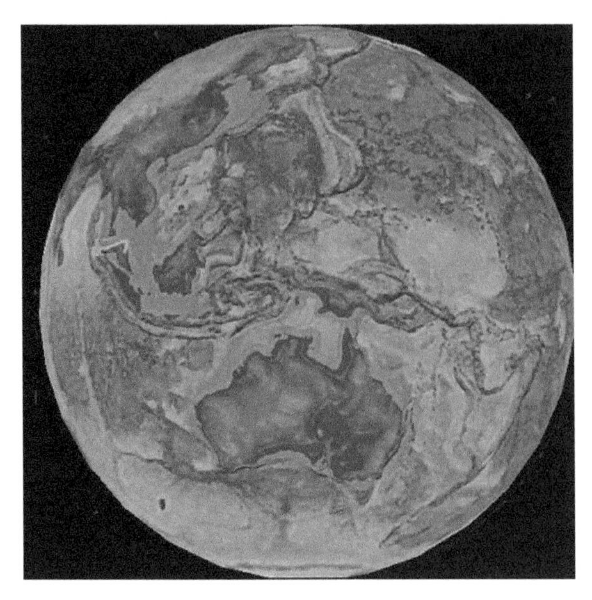

Figure 2.12 Philippine sea plate.

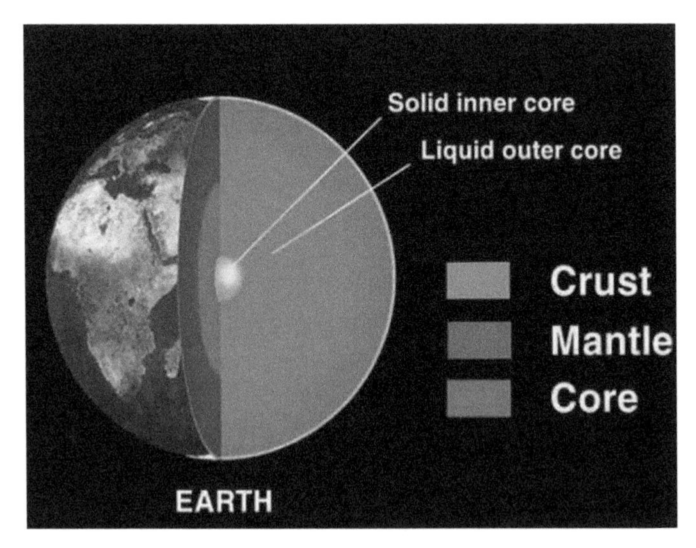

Figure 2.13 Ringwoodite lurking deep within the Earth's rocky mantle.

These trenches are frequently long and slender and run parallel to and close to the ocean margins. They might also be regularly associated with and parallel to massive continental mountain levels. There is, additionally, a discovered parallel association of trenches and island arcs. Just like the mid-oceanic ridges, the trenches are seismically live; however, no longer like the ridges, they have low levels of heat flow. Scientists moreover have started comprehending that the youngest areas of the sea floor have been alongside the mid-oceanic ridges and that the age of the sea floor extended as the space from the ridges increased. Further, it has been decided that the oldest seafloor frequently ends in the deep-sea trenches. Moreover, Figure 2.15 shows the motion of plates in which the fastest motion is taking place along Mid-Pacific Ocean Ridge and east of the Indian Ocean ridge where the Sundae trench locates [66].

Figure 2.14 Underwater mountain.

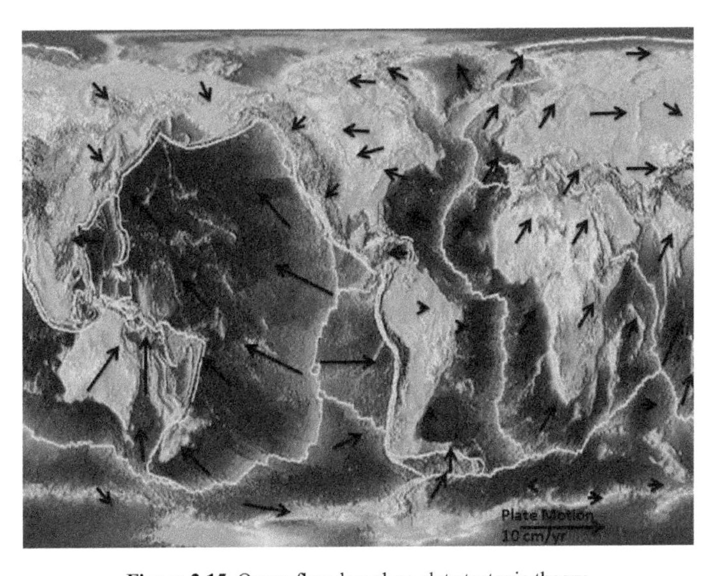

Figure 2.15 Ocean floor based on plate tectonic theory.

Evidently, the volcano locations suit carefully the regions of subduction zones, i.e., trenches and oceanic ridges. In this context, the trenches have crustal friction close to the floor and deep into the crust as subduction forces one plate beneath every other. At oceanic ridges, most of the crustal movement is close to the floor whereas the spreading of the sea floor causes plates to move aside and new crust is fashioned by magma pushing up from the mantle. These data from volcanoes, as well as earthquake locations, help to substantiate the model of plate tectonics. Earthquakes which are deep inside the crust are located inside the identical regions because of the deep-sea trenches (Figure 2.16). Shallow earthquakes (close to the surface of the crust) are discovered in the deep-sea trenches (outlining the subduction zone), in addition to oceanic ridge/upward push areas. Earthquakes and volcanoes define plate boundaries. The movement of Earth's twelve predominant plates may also appear gradually; however, it has a fantastic effect on the placement of the continents over millions of years. Some plate barriers are very energetic at the same time as others are very gradual with millimetres per year. A few plates are totally oceanic crust, for instance the Pacific Plate, and a few are oceanic crust with continental crust, such as the North American Plate [66].

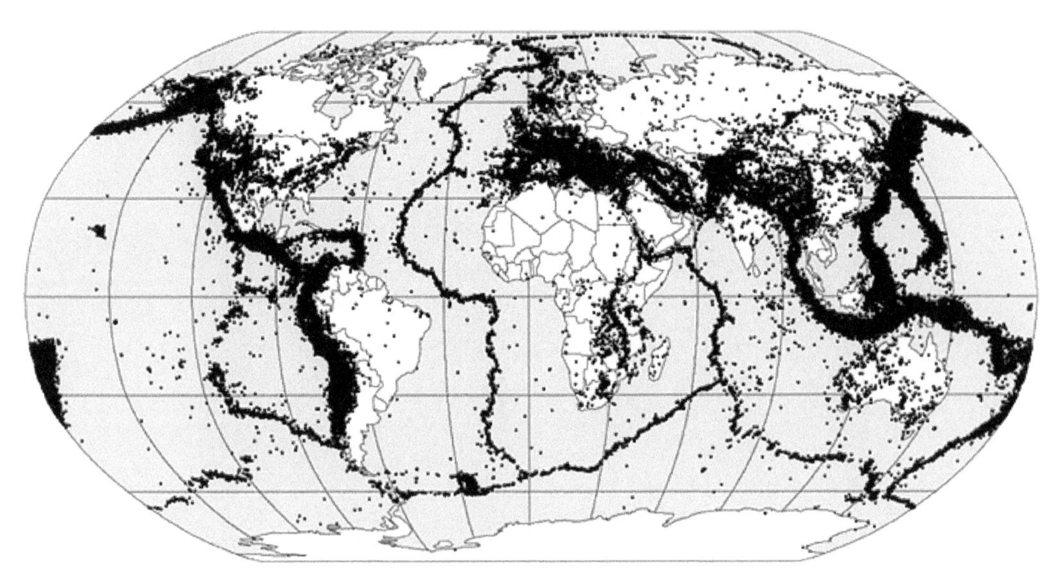

Figure 2.16 Earthquake zone along mid oceanic ridges.

2.2 How do Earthquakes Generate Tsunami?

The sudden collapse of the sea floor that vertically displaces the overlying water can generate tsunami. In fact, tectonic earthquakes are a precise variety of earthquake that is related to the earth's crustal deformation; when these earthquakes take place underneath the sea, the water above the deformed location is displaced from its equilibrium position. Waves are fashioned as the displaced water mass, which acts beneath the impact of gravity, attempts to regain its equilibrium. When large areas of the seafloor bring up or subside, a tsunami can be created. Large vertical movements of the earth's crust can take place at plate boundaries. Plates interact along these boundaries referred to as faults. Around the margins of the Pacific Ocean, for example, denser oceanic plates slip beneath continental plates in a manner recognized as subduction. In this understanding, the tsunami usually can be generated by the subduction earthquakes [69].

2.3 How do Landslides, Volcanic Eruptions, and Cosmic Collisions Generate Tsunamis?

A tsunami can be generated through any disturbance that displaces a massive water mass from its equilibrium position. In the case of earthquake-generated tsunamis, the water column is disturbed through the uplift or subsidence of the sea floor. Submarine landslides, which frequently accompany large earthquakes, as properly as collapses of volcanic edifices, can additionally disturb the overlying water column as sediment and rock slump downslope and are redistributed throughout the sea floor (Figure 2.17). Similarly, a violent submarine volcanic eruption can create an impulsive pressure that uplifts the water column and generates a tsunami. Conversely, super-marine landslides and cosmic-body influences disturb the water from above, as momentum from falling debris is transferred to the water into which the particle falls. Generally speaking, tsunamis generated from these mechanisms, in contrast to the Pacific-wide tsunamis triggered by some earthquakes, dissipate rapidly and not often shake coastlines far-off from the epicenter's location [71].

Though in a given vicinity of the Earth's surface, the hazard of a "direct" hit from an asteroid is slight, two researchers realized that an ocean impact had the potential to be greatly more destructive due to the results of a tsunami [72]. An airburst explosion is a three-dimensional tournament and strength decreases in accordance with the square of the distance; however, a radiating ocean wave is a two-dimensional phenomenon and, in theory, the strength decreases in proportion to distance [71, 73] (Figure 2.18).

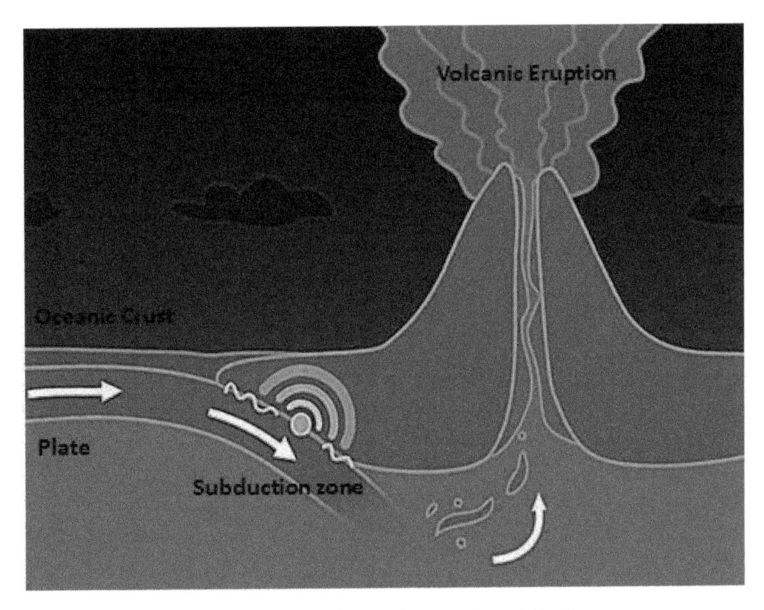

Figure 2.17 Volcanic eruption causing subduction.

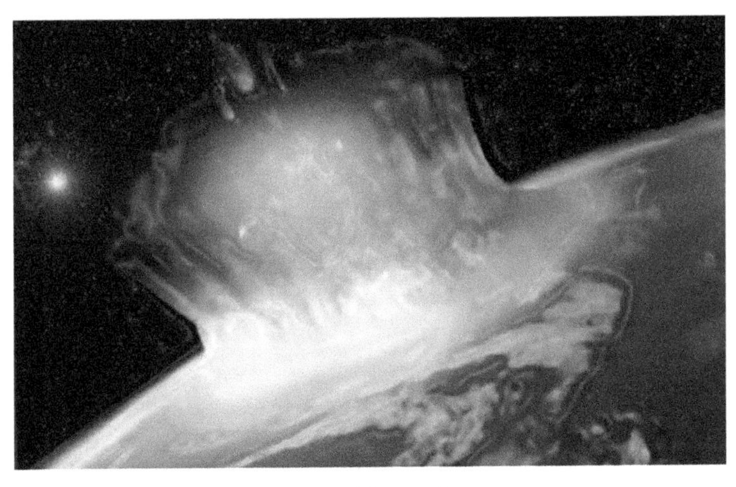

Figure 2.18 Asteroid generating tsunami.

Since the early 1990s, some advanced model simulations have been performed to estimate the consequences of two asteroid influences above deep oceans. Since an asteroid hits the ocean at 70,000 km/h, there is a tremendous explosion. The asteroid and water vaporize and leave a massive crater—commonly 20 times the diameter of the asteroid (that is, a 100 m asteroid will create a 2 km diameter crater). The water rushes lower back in, overshoots create a mountain of water in the center and this spreads out as a huge tsunami. The center of the "crater" oscillates up and down numerous times and a series of waves radiate out. The concept of the mechanism can be demonstrated through bursting a balloon in a bathtub [70, 72, 73].

2.4 What Happens When a Tsunami Encounters Land?

Similar to different water waves, tsunami start to lose energy as they rush onshore–phase of the wave strength is mirrored offshore, at the same time as the shore ward-propagating wave power is dissipated

Figure 2.19 Tsunami's impact on coastal infrastructures.

through backside friction and turbulence. However, tsunami still reach the coastal zone with the massive destructive energy. Tsunami have an incredible erosion ability, stripping beaches of sand which takes years to accumulate and harmfully impacting timber and different coastal flora. Capable of inundating, or flooding, hundreds of meters of inland past the typical high-water level, the short-transferring water associated with the inundating tsunami can overwhelm residences and other coastal systems (Figure 2.19). A tsunami may additionally expand a most vertical peak onshore above sea level, frequently known as a run-up peak, of 10 m, 20 m, and even 30 m.

2.5 Tsunami Generation Mechanisms

The panels are related to tsunami generation mechansims that are fault slip, cut up, and amplification. It has been observed that as a fault occurs along the trench, the vertical water displacement commenced and took a place at the surface. This is similar as pulling the water upward in the container by both hands. The ending of the tsunami propagation is related to run-up mechanisms which are stated below.

2.5.1 Fault Slip

Earthquakes are usually associated with floor shaking that is a result of elastic waves travelling through the solid earth. However, close to the source of submarine earthquakes, the seafloor is "permanently" uplifted and down-dropped, pushing the entire water column up and down (Figure 2.20). The conceivable energy that is a consequence of pushing water above suggests that sea level is then transferred to the horizontal propagation of the tsunami (kinetic energy). For the case proven above, the earthquake rupture took place at the base of the continental slope in relatively deep water. Situations can also occur where the earthquake rupture happens underneath the continental shelf in shallower water [68, 71].

In the Figure 2.20, the wave is greatly exaggerated compared to water depth. In the open ocean, the waves are at most several meters high, spread over tens to hundreds of kilometers in length.

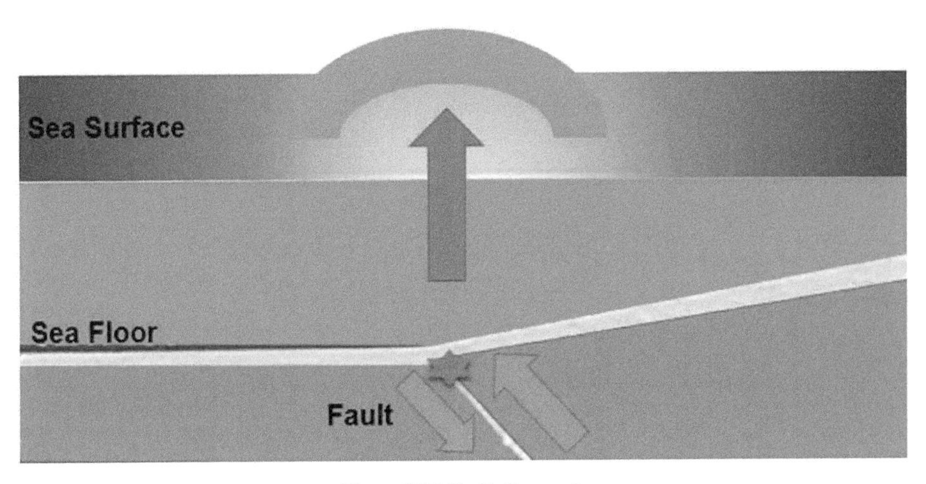

Figure 2.20 Fault slip panel.

2.5.2 Split

Within minutes of the earthquake, the preliminary tsunami is split into a tsunami that travels out to the deep ocean (distant tsunami) and another tsunami that travels towards the nearby coast (local tsunami). The peak above mean sea level [MSL] of the two opposite travelling tsunamis is about half of the original tsunami. This is extremely modified in three dimensions; however, the identical concept holds. The velocity at which each tsunami travels varies as the square root of the acceleration due to gravity, which is a constant value of 9.81 m/s multiplies by the water depth at which the disturbance occurred on the tsunami wave propagation (Figure 2.21). Consequently, the deep-ocean tsunami travels faster than the local tsunami close to shore [68].

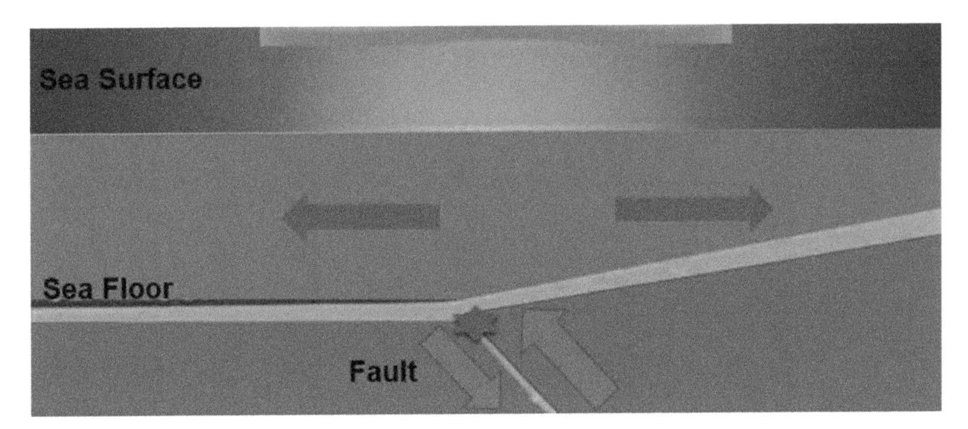

Figure 2.21 Mechanism of split.

2.5.3 Amplification

Several things happen as the local tsunami travels over the continental slope. The most obvious is that the amplitude increases. In addition, the wavelength decreases. This results in steepening of the leading wave—an important control of wave run-up at the coast (next panel). Note that the first part of the wave reaching the local shore is a trough, which will appear as the sea receding far from shore. This is a common natural warning sign for tsunamis. Note also that the deep ocean tsunami has travelled much further

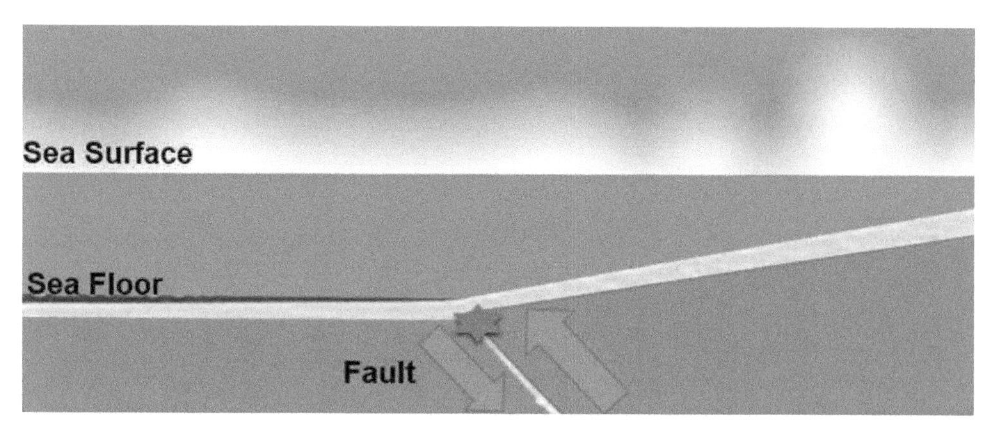

Figure 2.22 Mechanism of tsunami propagation.

than the local tsunami because of the higher propagation speed. As the deep ocean tsunami approaches a distant shore, amplification and shortening of the wave must occur, just as with the local tsunami shown in Figures 2.17 and 2.22.

2.5.4 Tsunami Run-up

Tsunami run-up occurs when a peak in the tsunami wave travels from the near-shore region onto shore. Run-up is a measurement of the height of the water onshore, observed above a reference sea level. Except for the largest tsunamis, such as the 2004 Indian Ocean event, most tsunamis do not result in giant breaking waves (like normal surf waves at the beach that curl over as they approach shore) (Figure 2.23). Rather, they come in much like the very strong and fast-moving tides (i.e., strong surges and rapid changes in sea level). Much of the damage inflicted by tsunamis is caused by strong currents and floating debris. The small number of tsunamis that do break often form vertical walls of turbulent water called bores. Tsunamis will often travel much farther inland than normal waves [71].

Figure 2.23 Tsunami breaking and run-up.

2.5.5 Do Tsunamis Stop Once on Land?

No! After run-up, part of the tsunami energy is reflected back to the open ocean and scattered by sharp variations in the coastline. In addition, a tsunami can generate a particular type of coastal trapped wave called edge waves that travel back-and forth, parallel to shore. These effects result in many arrivals of the tsunami at a particular point on the coast rather than a single wave as suggested by Panel 3. Because of the complicated behavior of a tsunami near the coast, the first run-up of a tsunami is often not the largest, emphasizing the importance of not returning to a beach many hours after a tsunami first hits [68, 71].

2.6 Historical Tsunami Records

Historically, tsunamis have occurred in many parts of the world, especially in the place where there is active seismic activity. Table 2.2 below shows the main earthquakes triggered Tsunami disaster all over the world.

Table 2.2 The main earthquakes, which resulted in Tsunami all over the world.

Location of Earthquake and Tsunami Area	Date of Event	Magnitude (Richter)
Atlantic Coast (Earthquake in Burin Peninsula, Newfoundland)	18 Nov 1929	7.3
Aleutian Tsunami (Muir Beach, Arena Cove, Santa Cruz, California)	1 Apr 1946 1957	7.8 9.1
Chilean Tsunami (Crescent City, California)	22 May 1960	9.5
Alaskan Tsunami (Crescent City) (Point Arguello, California)	27 Mar 1964 1927	9.2 7.3
Alaskan Tsunami Good Friday Earthquake Hawaii (Southeastern Coast) (Kamkatka)	1788 (earliest) 1946 1957 1964 1965 1869, 1975 1952	– 7.3 9.1 9.2 8.7 7.2 9.0
Indian Ocean Earthquake South of Java Island Kuril Islands Solomon Islands Samoa Chile Sumatra New Zealand Pacific coast of Japan Solomon Islands Chile	26 Dec 2004 17 July 2006 15 Nov 2006 2 April 2007 29 Sept 2009 27 February 2010 25 October 2010 22 Feb 2011 11 March 2011 6 Feb 2013 16 September 2015	9.0 7.7 8.3 8.1 8.1 8.8 7.7 6.3 9.0 8.0 8.3

The greatest recorded giant sea waves caused by earthquakes are listed as follows [75]:

The oldest known giant marine earthquake wave, called "tsunami" by the Japanese and "hungtao" by the Chinese, is that which took place in the eastern Mediterranean on **21 July, 365 AD** and killed thousands of people in the Egyptian city of Alexandria.

The Portuguese capital was destroyed in the Great Lisbon Earthquake of **1 November, 1775**. The Atlantic ocean wave, 6 m high, devastated the Portuguese, Spanish and Moroccan coasts.

27 August 1883: The Indonesian volcano Krakatoa erupted, and the tsunami that washed over the Java and Sumatran coasts killed 36,000 people. The volcanic eruption was so powerful that for many nights, the sky shone with red lava dust.

15 June 1896: The "Sanriku Tsunami" struck Japan. The 23 m high giant tsunami that swept over masses of people gathered together for a religious festival cost the lives of 26,000 people.

17 December 1896: A tsunami destroyed part of the embankment of Santa Barbara in California, USA, and the main boulevard was flooded.

31 January 1906: The Pacific Ocean earthquake wave destroyed part of the city of Tumaco in Colombia, as well as all the houses on the coast between Rioverde in Ecuador and Micay in Colombia; 1,500 people died.

1 April 1946: The tsunami that destroyed the Aleutian Scotch Cap Lighthouse with its crew of five, proceeded to Hilo in Hawaii, killing 159 people.

22 May 1960: An 11-m high tsunami killed 1,000 people in Chile and 61 in Hawaii. The giant wave crossed to the opposite shore of the Pacific Ocean and rocked the Philippines and the Japanese island of Okinawa.

28 March 1964: The Alaskan "Good Friday" tsunami wiped three villages off the map with 107 people dead, and 15 in Oregon and California.

16 August 1976: A Pacific tsunami cost the lives of 5,000 people in the Moro Gulf in the Philippines.

17 July 1998: A tremor wave occurring in northern Papua New Guinea killed 2,313 people, destroyed 7 villages and left thousands homeless.

26 December 2004: The 8.9 earthquake and giant wave that struck six countries in Southeast Asia killed more than 156,000 people.

17 July 2006, South of Java Island: The 7.7 magnitude earthquake rocked the Indian Ocean seabed on July 17, 2006, 200 km south of Pangandaran, a beach famous to surfers for its perfect waves.

15 November 2006, Kuril Islands earthquake: An 8.3 magnitude earthquake occurred off the coast near the Kuril Islands. In spite of the quake's large 8.3 magnitude, a relatively small tsunami was generated. The small tsunami was recorded or observed in Japan and at distant locations throughout the Pacific.

2 April 2007, Solomon Islands earthquake: A powerful 8.1 (initially 7.6) magnitude earthquake hit the East Pacific region about 40 km (25 mi), south of the Ghizo Island in the western Solomon Islands at 7:39 a.m., resulting in a tsunami that was up to 12 m (36 feet) tall. The wave, which struck the coast of the Solomon Islands (mainly Choiseul, Ghizo Island, Ranongga, and Simbo), triggered region-wide tsunami warnings and watches extending from Japan to New Zealand to Hawaii and the eastern seaboard of Australia.

4 December 2007, British Columbia: A landslide entered Chehalis Lake in British Columbia, generating a large lake tsunami that destroyed campgrounds and vegetation many meters above the shoreline.

29 September 2009, Samoa earthquake and tsunami: At a magnitude of 8.1, it was the largest earthquake of 2009. A tsunami was generated which caused substantial damage and loss of life in Samoa, American Samoa, and Tonga. The Pacific Tsunami Warning Center recorded a 76 mm (3.0 in) rise in sea levels near the epicenter, and New Zealand scientists determined that the waves measured 14 m (46 ft) at their highest on the Samoan coast.

27 February 2010, Chile earthquake: An 8.8 earthquake offshore of Chile caused a tsunami which caused serious damage and loss of life; it also caused minor effects in other Pacific nations.

22 February 2011, Christchurch earthquake: A 6.3 magnitude earthquake hit the Canterbury Region of the South Island, New Zealand. Some 200 km (120 mi) away from the earthquake's epicenter, around 30 million tones of ice tumbled off the Tasman Glacier into Tasman Lake, producing a series of 3.5 m (11 ft) high tsunami waves, which hit tourist boats in the lake.

11 March 2011, Pacific coast of Japan Earthquake and Tsunami: On March 11, 2011, Japan was hit by an earthquake measuring 9 M_w. The most powerful earthquake to have hit the country, the Great East Japan Earthquake caused untold damages to life, property, and infrastructure. While the Japanese island of Honshu is believed to have shifted about eight feet to the east, the entire planet Earth is estimated to have shifted about four inches on its axis as a result of the earthquake.

6 February 2013, Solomon Islands earthquake: An earthquake measuring 8.0 on the Moment Magnitude scale struck the island nation of the Solomon Islands. This earthquake created tsunami waves up to around 1 meter high. The tsunami also affected some other islands like New Caledonia and Vanuatu.

16 September 2015, Chile earthquake: A major earthquake measuring 8.3 on the Moment Magnitude scale struck the west coast of Chile, causing a tsunami up to 16 feet (4.88 meters) high along the Chilean coast.

In all of the world's oceans and seas, damaging tsunamis have occurred. In the second half of the twentieth century, Pacific-wide, destructive tsunamis occurred in 1946, 1952, 1957, 1960, 1964, 2004, 2006, 2007, 2008, 2009, 2011 and 2013. Numerous tsunami occurred in inland seas around the periphery of the Pacific Ocean, which have been particularly destructive regionally and claimed thousands of lives. Such localized tsunamis took place in 1975, 1983, 1985, 1992, 1993, 1995, 1998, 1999 and 2001. The deadliest tsunami in recorded history was once the 2004 Indian Ocean tsunami, which killed nearly 230,000 humans in eleven countries. The 365 CE Mediterranean tsunami death toll may have been much greater [76].

2.7 Why aren't Tsunamis Seen at Sea or from the Air?

It is accepted reality for tsunami to have very low amplitudes and giant wavelengths whilst offshore, owing to the large quantity of surrounding water. A regular offshore tsunami can have an amplitude of 1 m and a wavelength of the order of 799.9 km. So the power related to offshore tsunami is additionally very low due to such low frequencies, or small amplitudes. Under these circumstances, it is truly impossible to see tsunami at sea or from the air [76].

2.8 Combination of Tsunami, Tide, Sea Level, and Storm Surge

Numerous people would raise a simple query about what will happen if the tsunami escorts excessive tide and powerful storm surge. The fabulous response ought to be that the water level will be extraordinary and turn out to be convoluted. In this understanding, if the tsunami strikes for the duration of a storm and excessive tide of about 2 m above the mean sea level, the inundation ranges will increase. Further, the tsunami has a great destructive energy along the coastal waters. Nevertheless, tsunamis cannot be occurred at times of storm surges and high tides due to a lower chance of these independent phenomena to combine together. Lastly, the water level must be retreated from the coastal zone before the second and third cycles of successive waves.

Chapter 3

Tsunami of Sumatra-Andaman Earthquake 26 December 2004

This chapter critically evaluates the comprehensive theories which indicate that more concepts are needed to bridge the gap found between 2004 tsunami and existing generation mechanisms. Thus, this chapter is devoted to the tsunami of 26 December 2004 that is known as Sumatra-Andaman earthquake. Consequently, this catastrophe is called the Asian tsunami in Asia region and additionally recognized as a boxing day within Australia, Canada, New Zealand, and the United Kingdom, attributable to the actual fact that it occurred on boxing day at 00:58:53 UT (07:58:53 neighborhood time) December 26, 2004.

3.1 Why Earthquakes and Tsunamis occur in the Sumatra Region

Indonesia is located between the Pacific ring of fireside on the north-eastern islands as well as Papua and Alpide belt that coincides with the south and west Sumatra, Java, Flores, and Timor.

Socquet et al. (2006) [78] expound that Sumatra is at the frontier between two tectonic plates. This plate boundary is named the Sumatran subduction trench. In this zone, the oceanic plate of Indian/Australian is slowly drizzled beneath the continental plate of Eurasian, eventually causing a subduction zone at the rate of 4.5 cm/year in the direction of 20º N. However, the Continental drift between the Australian and Sunda plates is occurred at a rate of 5 cm/year in direction of 8° N due to the impediment of the Himalayas. The subducting plate eventually sinks and goes into reverse into the earth's mantle, whereas a new ocean bottom is added to the plate zone at the mid-ocean ridge.

The seabed southwest of Sumatra is a portion of the Indian/Australian plate, whereas Sumatra and also the different islands of Indonesia and Kingdom of Thailand are fragments of the Eurasian plate (Figure 3.1). The two plates run into the ocean floor at the plate boundary (black line in Figure 3.1), that is 200 km (125 miles) off the western shore of Sumatra and about 5 km (3 miles) below the ocean surface [77].

In a subduction zone, the earthquake emphases generally intrigue a dipping plane at an angle of 33° to 60° which is identified as Benioff zone (Figure 3.2). Therefore, the Benioff Zone is well-defined as the energetic seismic zone in a subduction zone [85]. This feature is named after a US seismologist Hugo Benioff.

During Boxing day, the Benioff zone expanded to a depth of approximately 700 km (Figure 3.3). Figure 3.3 shows that Upper red star is 2004 Mw 9.0 mainshock and lower orange star is 2005 Mw 8.6 Nias Island earthquake. The thick black line (teeth on overriding plate) indicates the trench location, and gray triangles represent Holocene volcanoes [86].

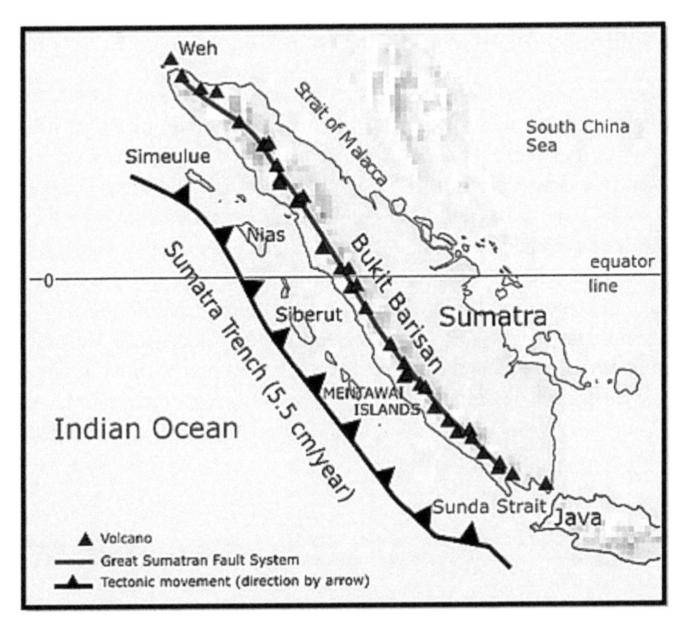

Figure 3.1 Subduction zone beneath Sumatra, Indonesia.

Black line= Benioff zone

Figure 3.2 Definition of Benioff zone.

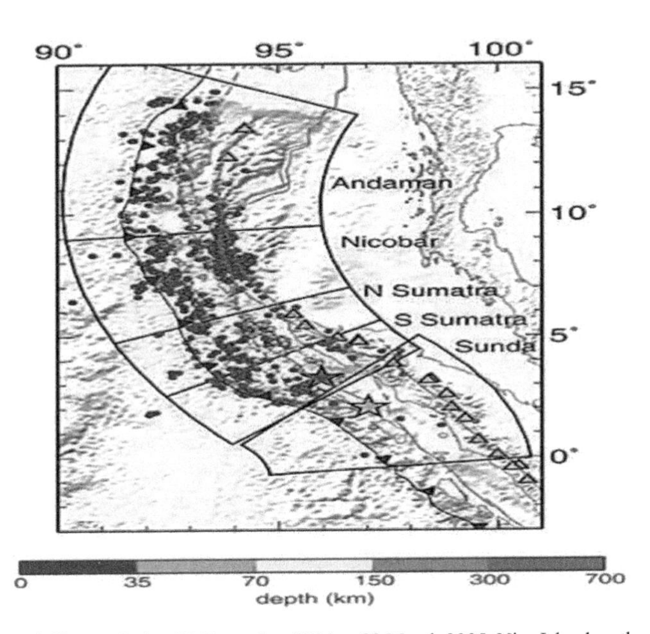

Figure 3.3 Benioff zone during 26 December 2004 to 28 March 2005, Nias Island earthquake [86].

3.2 Rupture of 2004 Earthquake and Tsunami

An undersea earthquake took place alongside a tectonic subduction zone in which the India Plate, an oceanic plate, is being subducted underneath the Burma micro-plate, phase of the larger Sunda plate. The earthquake's epicenter is off the west coast of Sumatra, Indonesia. In other words, the earthquake originated in the Indian Ocean just north of Simeulue island, off the western coast of northern Sumatra. The magnitude of the earthquake that was initially recorded as M_w 9.0 but has been elevated from 9.1 to 9.3 M_w, was a consequence of strain accumulated in the Indian/Sunda junction, some of which had no longer experienced a massive earthquake for the previous 150 years or so [79]. Consistent with Ioualalen et al. (2007) [79], there was unbalanced partition of previous earthquake magnitudes and recurrence instances between the two plate boundaries suggest that larger strains had accrued in the Indian/Sunda boundary prior to the 26 December 2004 event, and explains both the epicenter vicinity at the junction between the subducting Indian and Australian plates and the overriding Eurasian plate (Burma and Sunda subplates) and the northward rupture propagation (Figure 3.4), the place most of the aftershocks had been recorded alongside a ~ 1300 km arc of the Andaman trench.

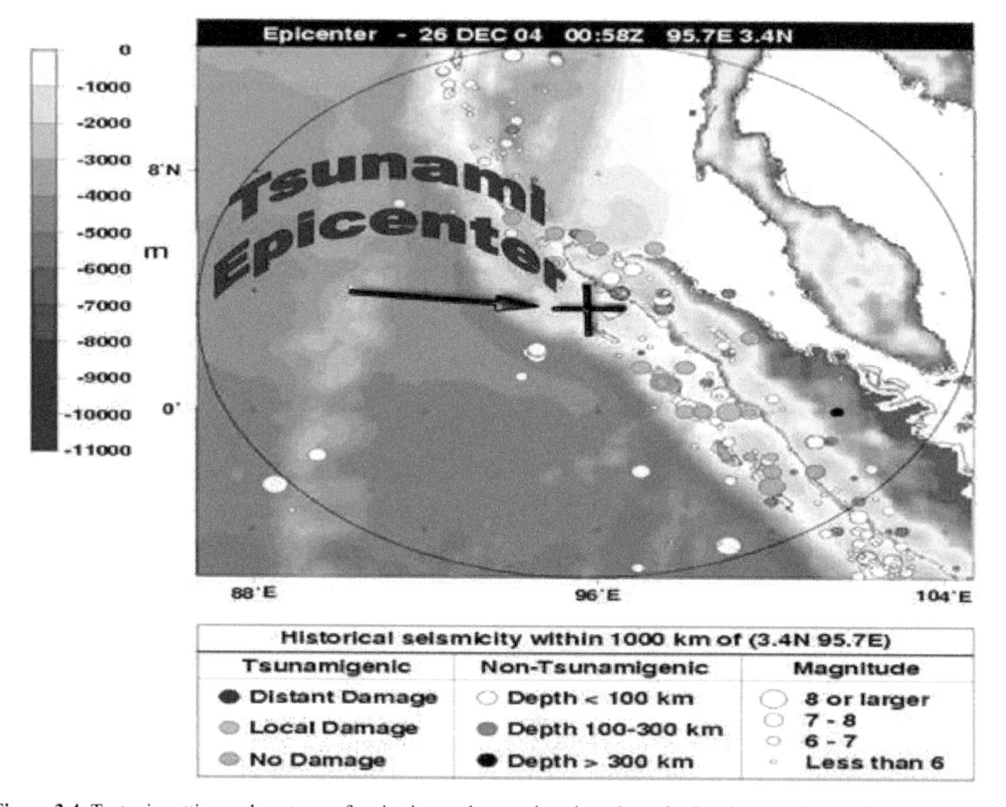

Figure 3.4 Tectonic setting and ruptures of major interpolate earthquakes along the Sunda megathrust. The yellow patches are estimated rupture areas of known large subduction events between 1797 and 2004 [http://www.noaanews.noaa.gov/stories2004/images/tsunami-map122604-0058z2.jpg] [free to use, share, commercial free too].

3.3 How Earthquakes occur in the Sumatra Region?

The mechanisms of Sumatran Earthquake can be explained as follows: (i) The two plates become stuck, or locked (Figure 3.5). (ii) As the lower plate slowly descends, it pulls the upper plate down with it, deforming the land surface above and building up stress. The island close to the trench slowly subsides,

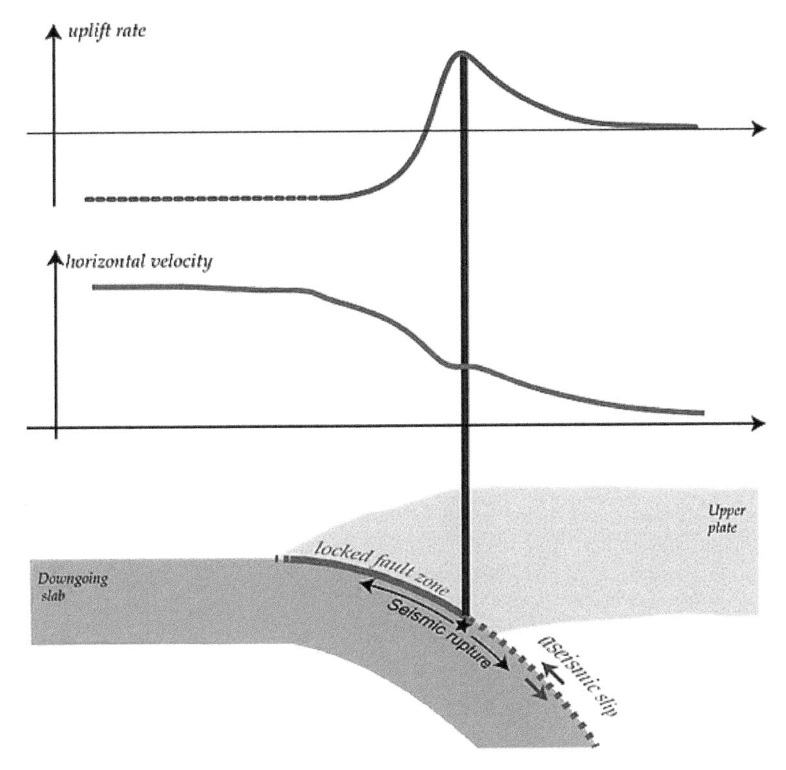

Figure 3.5 Geometry of locked fault zone.

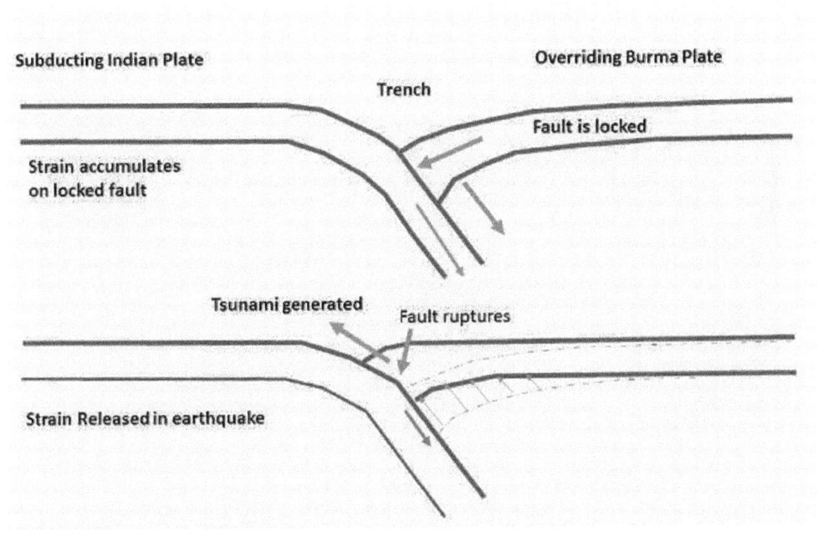

Figure 3.6 Tsunami due to strain accumulations.

while further inland the land slowly uplifts. (iii) When the stress buildup overwhelms the locking friction, the plates suddenly break free from each other and the upper plate slips back. This sudden slip generates an earthquake. The island pops back up, and the land to the right suddenly subsides. This quick displacement of the ocean floor can cause a tsunami (Figure 3.6).

3.4 Mechanisms of Sumatran Earthquake and Tsunami

The huge and sudden upward motion of the sea floor when the locking between two plates is ruptured displaces water (Figure 3.6). The great tsunami, that occurred after the December 2004 Sumatran quake, was due to the large vertical displacement. The quakes of 2005 and 2007 did not cause any major tsunami due to the fact the areas of sudden uplift primarily corresponded to areas of the Nias, Simeulue, and Mentawai islands, consequently restricting the quantity of water that was displaced.

During Boxing Day of 2004 tsunami, an estimated 1,600 km (994 mi) of fault line was slipped almost 15 m (50 ft) along the subduction zone where the India Plate bent beneath the Burma Plate (Figure 3.7). The slip did not manifest at once; it happened in two stages within countless minutes. Seismographic and acoustic information endorse that the first phase concerned a rupture of about 400 km (250 mi) long and 100 km (60 mi) wide, located 30 km (19 mi) below the seabed where the longest rupture to have ever been recognized was triggered by an earthquake. Beginning off the coast of Aceh, the rupture receded at a velocity of about 2.8 km/s (1.7 mi/s) or 10,000 km/h (6,300 mph) and propagated northwesterly over a duration of about 100 seconds. A pause of roughly 100 seconds passed off before the rupture resumed northwards closer to the Andaman and Nicobar Islands [82].

Nevertheless, the northern and western Andaman Islands rose at approximately 2.1 km/s (1.3 mi/s) or 7,600 km/h (4,700 mph), whereas the southern and eastern portions of the islands are subsided. This leads to transfer the fault from subduction zone to uplift region at the north of the Andaman and Nicobar Islands. In this view, the two plate boundaries push past one another in contrary directions [83]. This diminished the rapidity of the water motion, therefore reducing the magnitude of the tsunami which striked the northern portion of the Indian Ocean. The India Plate is the section of the extraordinary Indo-Australian Plate which underlies the Indian Ocean and Bay of Bengal, and is drifting northeast at an average movement of 6 cm/year (2 inches per year) [80].

Besides the sideways movement between the plates, the seabed is estimated to have risen through numerous metres, shifting an estimated of 30 km^3 (7 cu mi) of water and triggering devastating tsunami waves. The waves were not created from a point source, and were inaccurately depicted in some illustrations of their paths of travel, but alternatively radiated outwards along the entire 1,600 km (994 mi) length of the rupture (acting as a line source). This quickly expanded the geographical location over which the waves were observed, accomplishing and affected Mexico, Chile and the Arctic. The elevation of the seabed

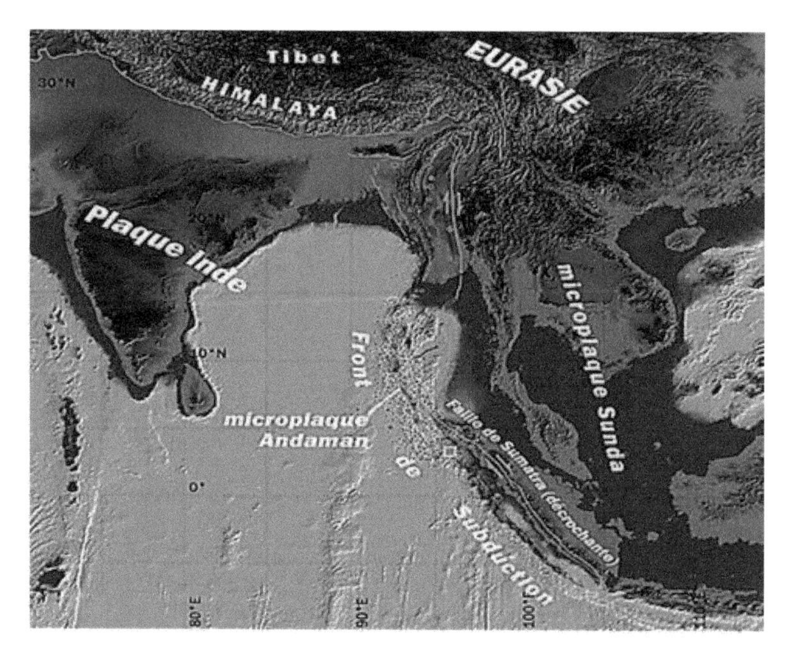

Figure 3.7 Location of the epicenter of the 2004 Indian Ocean earthquake.

considerably reduced the capacity of the Indian Ocean, producing an everlasting upward push in the world sea level through an estimated of 0.1 mm [81].

3.5 Physical Characteristics of the 2004 Earthquake

2004 earthquake is the second greatest earthquake ever recorded on a seismograph. This earthquake was additionally pronounced to be the longest period of faulting ever observed, lasting between 500 and 600 seconds (8.3 to 10 minutes), and it was estimated that it triggered the entire planet to shift as much as half an inch, or over a centimetre. Further, the Sumatran part of the Sunda megathrust generated extraordinary earthquakes south of the 2004 event in 1797, 1833 and 1861; however, there is no historical record of massive earthquakes in the north, between Sumatra and Myanmar (Figure 3.8). It had the highest speed of propagation, i.e., 750 km/hr and tsunami waves were up to 30 m.

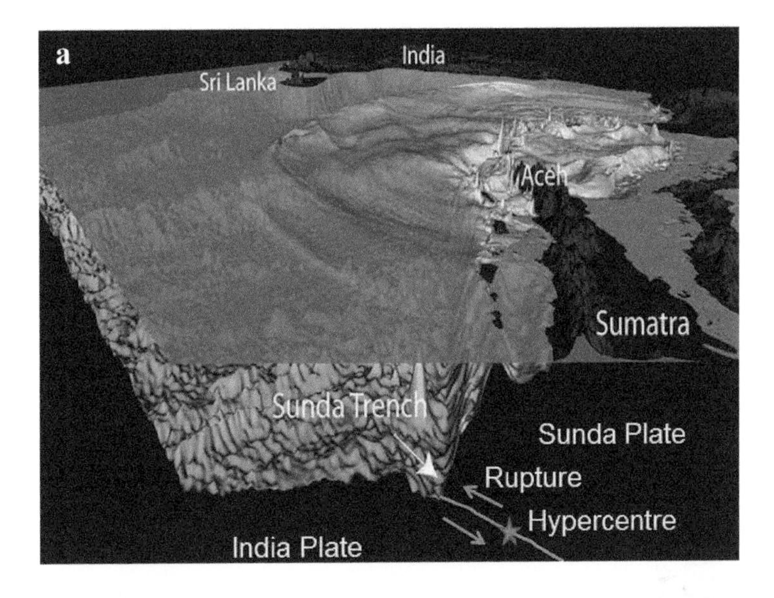

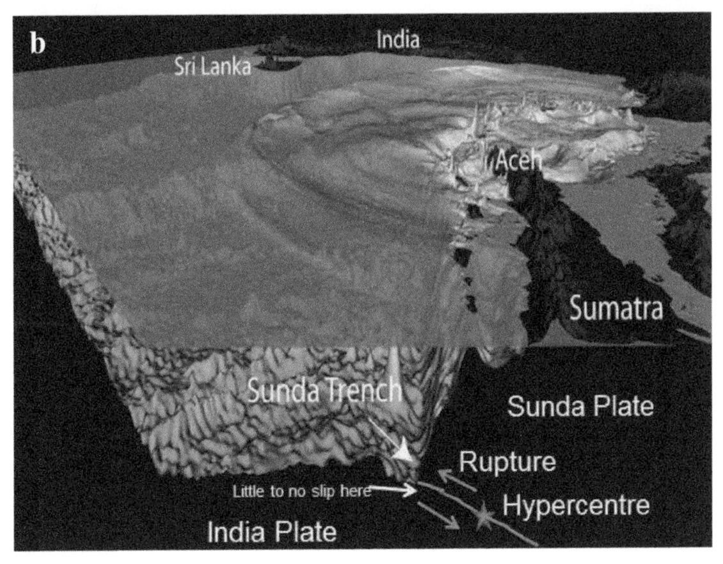

Figure 3.8 Location of Hypocenter of (a) 2004 and (b) 2005 earthquakes.

Of all the seismic moments released through earthquakes in the 100 years from 1906 through 2005, roughly one-eighth was due to the Sumatra-Andaman event. The earthquake was initially mentioned as moment magnitude, M_w 9.0 (note that this is not the Richter scale or local magnitude scale, Ml, which is recognized to saturate at greater magnitudes). On February 2005, scientists revised the estimate of the magnitude to M_w 9.3. Figure 3.8a suggests the hypocenter of the foremost earthquake which was at, approximately, 160 km (100 mi) west of Sumatra, at a depth of 30 km (19 mi) below mean sea level (initially mentioned as 10 km). The earthquake itself (apart from the tsunami) was felt as far away as Bangladesh, India, Malaysia, Myanmar, Thailand, Singapore and the Maldives. Three months after the devastating 2004 Indian Ocean tsunami, a 8.6 M_w earthquake happened offshore northern Sumatra (Figure 3.8b). Although there used to be excessively robust floor shaking and heavy damage associated with this earthquake, the tsunami was much less than expected. We can apprehend the elements that influence tsunami severity comparing this incident with the 2004 Sumatra-Andaman earthquake and other earthquakes alongside the Sunda subduction zone.

Generally, the magnitude of 2004 Indian Ocean tsunami is consistent with the measurement of local tsunamis generated by other earthquakes of comparable magnitude, for instance, the 1964 Great Alaska earthquake and tsunami. The 2005 northern Sumatra earthquake and, to some extent, the 2007 southern Sumatra earthquake (8.5 M_w) are deficient in terms of the local tsunamis produced. In contrast, the 2006 Java earthquake produced a tsunami greater than that was predicted and is a type of earthquake known as tsunami earthquakes.

3.6 2004 Tsunami Beaming

Figure 3.9 shows the tsunami beaming which refers to the higher tsunami amplitudes in a direction perpendicular to fault orientation during open-ocean propagation. Although tsunami waves are often described as waves spreading out in all directions (like when you throw a pebble into a pond), for long

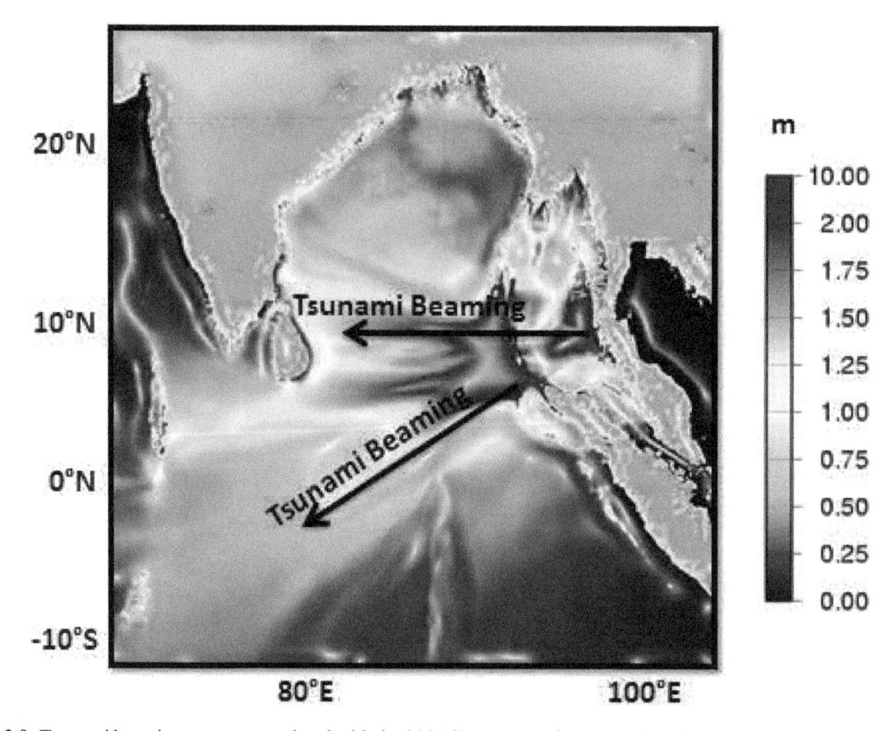

Figure 3.9 Tsunami beaming pattern associated with the 2004 Sumatra-Andaman earthquake. Lighter colors represent higher open-ocean tsunami amplitudes.

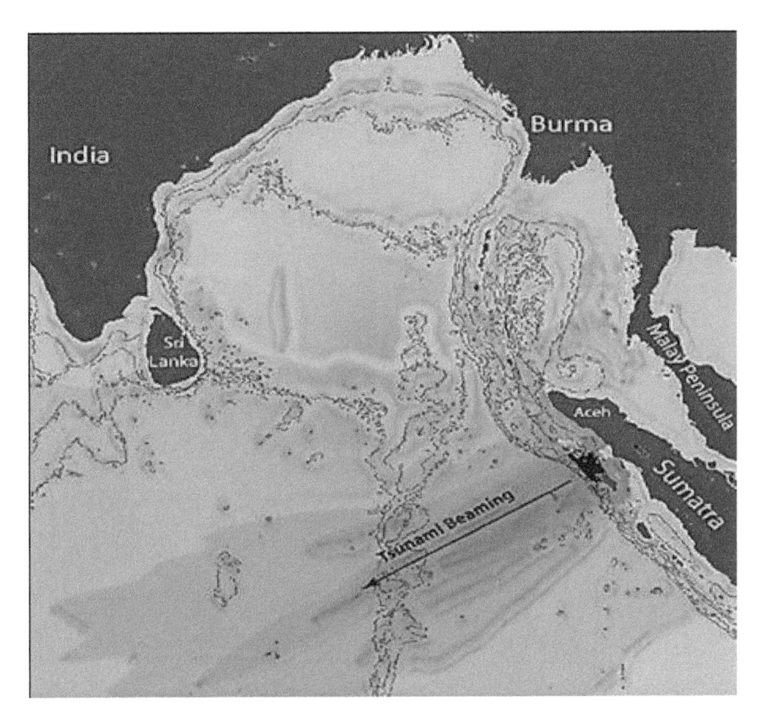

Figure 3.10 Tsunami beaming pattern associated with the 2005 northern Sumatra earthquake. Lighter colors represent higher open-ocean tsunami amplitudes.

earthquake ruptures, tsunami amplitudes are greater along the azimuth of tsunami beaming. Indeed, oceanic trench of subduction zones marks the orientation of inter-plate thrust, the tsunami beaming azimuth is also perpendicular to the orientation of the trench. Complexity of the earthquake source and refraction/scattering during propagation will affect the tsunami beaming pattern. It is interesting to find that the tsunami beaming plots are constructed by determining the maximum tsunami amplitude over four hours of propagation time. For the 2004 Sumatra-Andaman earthquake, the tsunami beaming pattern is particularly important in explaining high tsunami run-up in Sri Lanka and the western coast of the Malay Peninsula (Figure 3.9). However, for the 2005 northern Sumatra earthquake, tsunami beaming is directed south of Sri Lanka and essentially blocked by the island of Sumatra toward the Malay Peninsula (Figure 3.10).

3.7 Energy of the Earthquake and its Effects

Consistent with the U.S. Geological Survey (USGS), the Sumatra-Andaman earthquake was calculated to have released, approximately, the energy of 23,000 Hiroshima-type atomic bombs. In other words, the December 26, 2004 Indian Ocean tsunami was induced by an earthquake that is thought to have had the power of 23,000 atomic bombs. The complete power released through the 2004 Indian Ocean earthquake was estimated early on to be as much as 3.35 exajoules (3.35×10^{18} joules). This is equivalent to over 930 terawatt hours, 0.8 gigatons of TNT, or about as much energy as is used in the United States in 11 days. A new seismic data are released on September 30, 2005 stated that the earthquake was released approximately the energy of 1.1×10^{18} joules, which is equivalent to about 250 megatons of TNT. It was recorded that the earthquake generated seismic oscillation of the Earth's surface of up to 20–30 cm (8–12 in), equivalent to the impact of the tidal forces triggered through the Sun and Moon.

The shock waves of the earthquake were felt throughout the planet, as far away as the U.S. state of Oklahoma, where the uplift movements of 3 mm was occurred because of the fault deformation of 0.12 in. Because of its massive power release and shallow rupture depth, the earthquake generated tremendous seismic ground motions around the globe, particularly due to massive Rayleigh (surface) elastic waves that surpassed 1 cm in vertical amplitude everywhere on Earth. The report area plot beneath displays vertical displacements of the Earth's surface, recorded through seismometers from the IRIS/USGS Global Seismographic Network, plotted with recognizing time (since the earthquake initiation) on the horizontal axis, and vertical displacements of the Earth on the vertical axis (note the 1 cm scale bar at the bottom for scale). The seismograms are arranged vertically by way of distance from the epicenter in degrees.

The earliest, lower amplitude, signal is that of the compression (P) wave, which takes about 22 minutes to attain the other aspect of the planet (the antipode; in this case close to Ecuador). The largest amplitude indicators are seismic surface waves that attain the antipode after about 100 minutes. The surface waves can be truly considered to make stronger earthquake close to the antipode (with the closest seismic stations in Ecuador), and to consequently encircle the planet to return to the epicentral location after about 200 minutes. A foremost aftershock (magnitude 7.1) can be considered at the closest stations beginning simply after the 200 minute mark. This aftershock would be viewed a principal earthquake beneath ordinary circumstances; however, it is dwarfed by the great shock.

It can possibly be that the earth rotation was very slightly altered due to the shift of mass and the large launch of energy. The precise amount is not yet known; however, theoretical models inform that the earthquake shortened the size of a day by 2.68 microseconds (2.68 μs, or about one billionth of the length of a day), due to a limit in the oblateness of the Earth. It also triggered the Earth to minutely "wobble" on its axis by up to 2.5 cm (1 in) in the route of 145° east longitude, or perhaps up to 5 or 6 cm (2.0 to 2.4 in). However, due to the tidal effects of the Moon, the size of a day increases at an average of 15 μs per year, so any rotational exchange due to the earthquake will be occurred quickly. Similarly, the natural Chandler wobble of the Earth, which in some instances can be up to 15 m (50 ft), will subsequently offset the minor wobble produced through the earthquake.

More specifically, there was 10 m (33 ft) movement laterally, and 4–5 m (13–16 ft) vertically, alongside the fault line. Early speculation was that some of the smaller islands south-west of Sumatra, which is on the Burma Plate (the southern regions are on the Sunda Plate), may have moved south-west up to 36 m (118 ft); however, accurate information released more than a month after the earthquake located the movement to be about 20 cm (7.9 in). Since movement was vertical as well as lateral, some coastal areas might also have been moved beneath the sea level. The Andaman and Nicobar Islands show to have been shifted south-west by around 1.25 m (4.1 ft) and to have sunk by 1 m (3.28 ft) [84]. On February 2005, the Royal Navy vessel HMS "Scott" surveyed the seabed around the earthquake zone, which varies in depth between 1,000 m and 5,000 m (3,300 ft and 16,500 ft). The survey conducted the usage of a high-resolution, multi-beam sonar system and revealed that the earthquake had made a huge effect on the topography of the seabed. 1,500-meter (5,000 ft) high thrust ridges created through the preceding geologic activity along the fault had collapsed, generating landslides several kilometers wide. One such landslide consisted of a single block of rock some 100 m high and 2 km long (300 ft by 1.25 mi) [84].

The momentum of the water displaced with the aid of tectonic uplift had also dragged huge slabs of rock, each weighing millions of tons, as far as 10 km (7 mi) across the seabed. An oceanic trench several km wide was exposed in the earthquake area (Figure 3.11). Further, the 3-d sonar map of the ocean floor used to be captured using high-resolution multi-beam sonar from a UK Royal Navy survey ship, the HMS Scott. Marine geologists aboard the ship recognized aspects that bear testimony to the earthquake that wrenched the ocean bed, consisting of slabs of rock dragged up to 10 km along the seabed by using the pressure of the displaced water. The images also exhibit mountainous ridges 1500 meters tall and an oceanic trench countless kilometers wide, created over much greater intervals of time through recreation alongside the fault [87].

Further, Gaia [84] stated that a massive crack disturbance occurred due to the shifting of plates about 1200 km of fault line (Figure 3.12). In this regard, lines of strata snapped 4.4 km below sea-level are evidence of the massive upheaval in the seabed (Figure 3.12), which lifted one layer above the other [84].

Figure 3.11 The sonar image reveals ridges 1500 meters tall and a massive trench.

Figure 3.12 Massive upheaval seabed of Sunda trench.

3.8 Propagation of 2004 Tsunami

The greatest power of the tsunami waves was in an east-west direction. Because the 1,600 km (994 mi) of fault line affected by way of the earthquake was once in an almost north-south orientation, the Indian kingdom of Kerala was hit by the tsunami regardless of being on the western coast of India, and the western coast of Sri Lanka also suffered greatly. Also, distance alone is no guarantee of safety; Somalia was hit tougher than Bangladesh regardless of being a lot further away. Because of the distances involved, the tsunami took anywhere from fifteen minutes to seven hours (for Somalia) to hit the variety of coastlines (Figure 3.13).

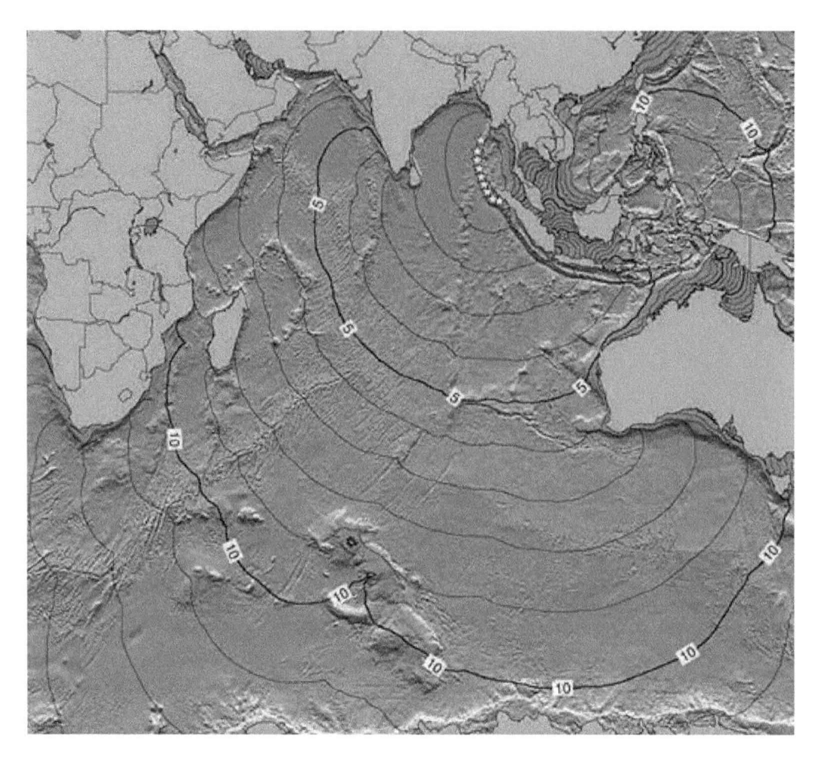

Figure 3.13 Tsunami propagation crosses the Indian Ocean.

Figure 3.13 suggests that, generally, the tsunami arrived at Phuket and Sri Lanka coasts in two hours after the earthquake, and the African coast in 8–11 hours. The tsunami propagation is also animated (up to 5 hours) from a 1200 km fault. The darker shade means that the water surface is greater than normal, while the bright color means lower wave height. It indicates that an initial tsunami to the east (e.g., Phuket) began with receding wave while to the west (e.g., Sri Lanka) massive wave abruptly struck the coastal water of Sri Lanka. The darker the color, the larger the amplitude. The tsunami had been larger in the east and west directions [88]. Consequently, the northern regions of the Indonesian island of Sumatra had been hit very quickly, whilst Sri Lanka and the east coast of India had been hit roughly 90 minutes to two hours later. Because the tsunami travelled more slowly in the shallow Andaman Sea off its western coast, Thailand was also struck about two hours later, despite being closer to the epicenter. The tsunami was observed as far as Struisbaai in South Africa, some 8,500 km (5,300 mi) away, where a 1.5 m (5 ft) excessive tide surged on shore about 16 hours after the earthquake. It took quite a while to reach this spot at the southernmost point of Africa, probably due to the vast continental shelf off South Africa and because the tsunami would have observed the South African coast from east to west [88].

The tsunami also reached Antarctica, where tidal gauges at Japan's Showa Base recorded oscillations of up to a meter, with disturbances lasting a couple of days (Figure 3.14). Some of the tsunami's strength escaped into the Pacific Ocean, where it produced small but measurable tsunamis alongside the western coasts of North and South America, usually around 20 to 40 cm (7.9 to 15.7 in). At Manzanillo, Mexico, a 2.6 m (8.5 ft) crest-to-trough tsunami was measured. The tsunami was also massive enough to be detected in Vancouver, Canada. This puzzled many scientists as the tsunamis measured in some parts of South America were larger than those measured in some parts of the Indian Ocean. It has been theorized that the tsunamis had been centered and directed at lengthy stages through the mid-ocean ridges which run along the margins of the continental plates.

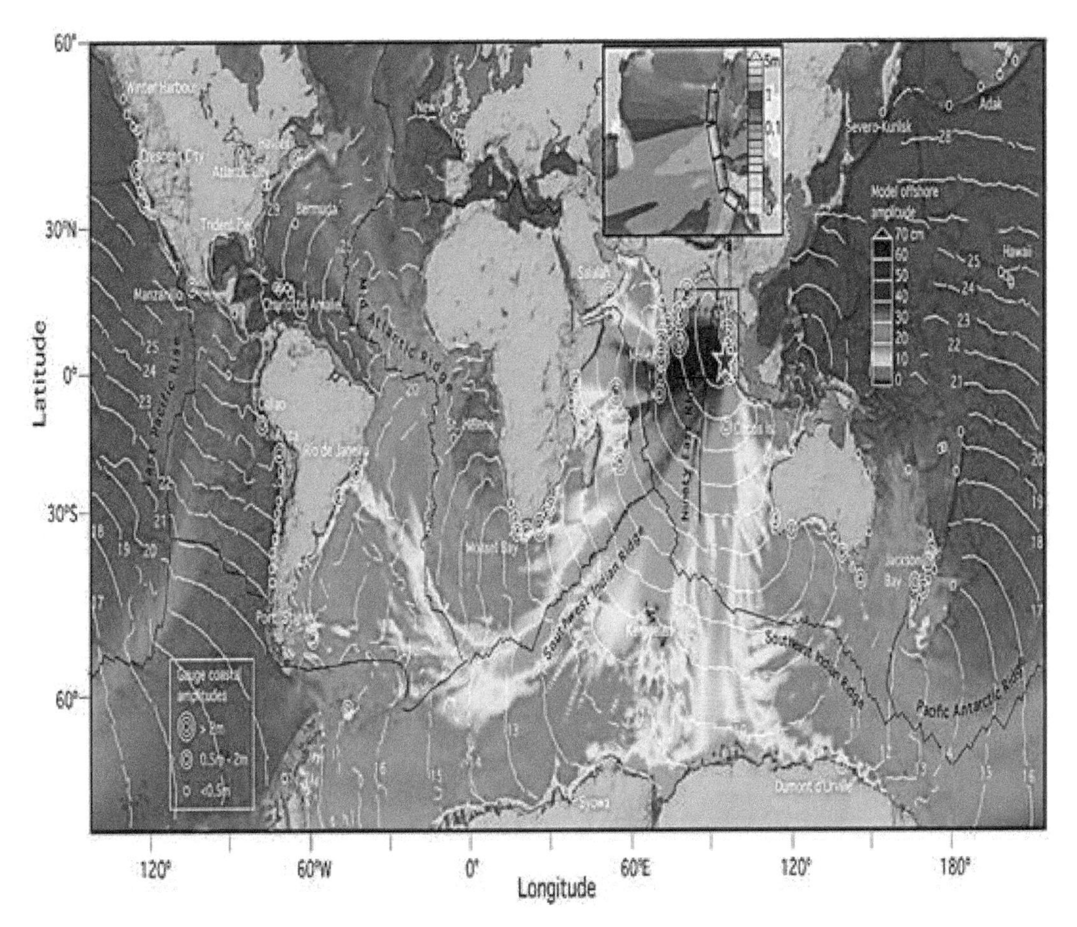

Figure 3.14 Massive 2004 tsunami wave propagations across the Atlantic Ocean [88].

3.9 Paths of Tsunami along Andaman Sea

The tsunami was caused by movement along a fault line running through the seabed of the Indian Ocean. As the fault runs north-south, the waves travelled out across the ocean in mainly easterly and westerly direction with duration of 7 hours that shook the world. At 00.58 GMT, an undersea earthquake measuring M_w 9.3 on the Richter Scale occurred off Indonesia. At +15 minutes later, the Indonesian Island of Sumatra, close to the epicenter of the quake, was hit by the full force of a tsunami. Many towns and villages in Aceh province on the western tip of the island were completely washed away, and the capital, Banda Aceh, was destroyed (Figure 3.15).

The remote Andaman, which lies only 100 km from the epicenter of the earthquake, was struck within +30 minutes later. An hour after it hit Sumatra, the tsunami reached Thailand. It had lessened slightly in height and power but still struck the Thai coast with incredible force and the sea surged out for about 200 m. One of the worst-hit places was Sri-Lanka, which lay almost directly west of the earthquake's epicenter. The tsunami wave reached Sri Lanka within +2.00 hours later (Figure 3.16). The tsunami height recorded in Sri Lanka ranged between 5 to 10 m (Table 3.1).

With no continental shelf to lift the tsunami as they near shore, the Maldives gets off relatively lightly within +3.30 hours. Finally tsunami waves strike African coast within +7 hours. The wave height along the Kenyan coast was between 2 and 3 m (Table 3.1).

Figure 3.15 Damages occurred along the coast of Banda Aceh.

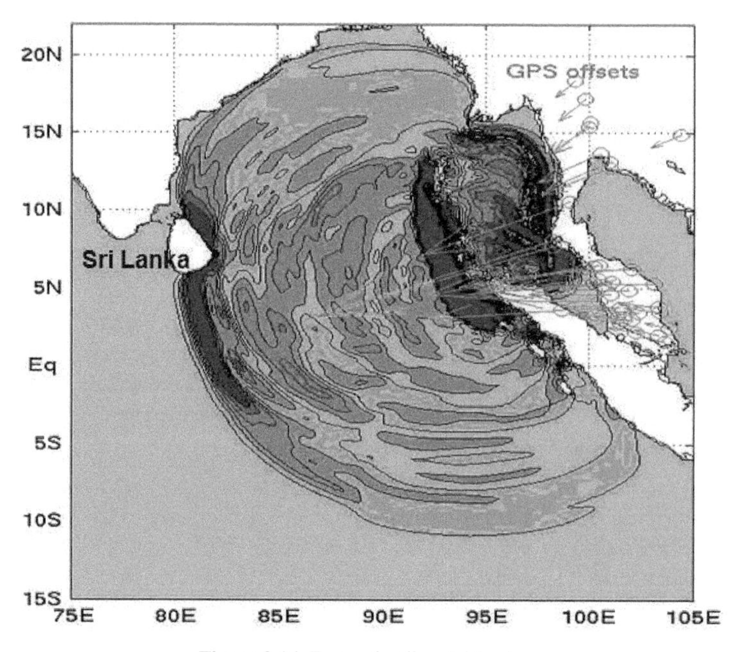

Figure 3.16 Tsunami strikes Sri Lanka.

Table 3.1 No trustworthy measured tsunami wave heights from the affected areas are available so far. According to mass-media reports, the waves reached the following height.

Locations	Tsunami Height
Sumatra, north-west coast	10–24 m
Sri Lanka, east coast	5–10 m
India, east coast	5–6 m
Andaman Island	> 5 m
Thailand, Phuket	3–5 m
Kenya	2–3 m

3.10 Retreat and Rise Cycle

The boxing day tsunami had unique characteristics that are known as a retreat and rise cycle. The tsunami was a succession of various waves, occurring in the throwback and upward thrust cycles with a duration of over 30 minutes between each peak. The third wave was the most effective and reached highest, taking place about an hour and a half after the first wave. Smaller tsunamis continued to manifest for the rest of the day. The first run-up has lower top whereas the second run-up starting to retreat, the Kata Noi Beach in Thailand at 10:17 a.m. (Figure 3.17). Three minutes later, the water recedes, inflicting turbulent moves (Figure 3.18).

Third tsunami reached the coast at 11:00 a.m. (Figure 3.19) with a stronger turbulent flow as compared to first tsunami cycle. It is clear that the time interval between second and third tsunami wave is 40 minutes. It might be because it is travelling from the epicenter to coastal water of Thailand.

Figure 3.17 Second tsunami run-up.

Figure 3.18 Receding waters after the second tsunami.

Figure 3.19 Third tsunami.

Figure 3.20 15 m wave height of 2004 tsunami.

Finally, why was the Sumatra tsunami so devastating? The Sumatra-Andaman Islands Earthquake on December 26, 2004 occurred in an area with high population densities along the coasts where it struck with maximal wave height of 15 m (Figure 3.20).

The worst exaggerated zones had two things in common: they had neither experienced hurricanes nor tides. Thus, human beings lived dangerously close to the water and regularly lived on ships that would no longer be seaworthy in areas where the tropical cyclones are common destructive natural phenomena. But from scientific observations, we are aware these days that tides and hurricanes are capable of regulating the coast in ways that may additionally minimize tsunami damage. Combined with waves, and particularly

Figure 3.21 Damages in Aceh due 2004 tsunami run-up.

those from storms, tides can erode the shoreline out, till a shore profile results that resist tsunamis. In such a coastal zone, a tsunami might not be capable of running inland as much distance as it did in Aceh (Figure 3.21), the region placed on the northern tip of the island of Sumatra, where about 170,000 Indonesians had been killed or went missing in the disaster, and about 500,000 had been left homeless [89].

Chapter 4

Novel Theories of Tsunami Generation Mechanisms

It is not easy to understand the nature of tsunamis because there are many circumstances which affect them. One approach to explain them is as ordinary water waves with lengths of hundreds of kilometers. In such an explanation, this depiction may legitimately be fit for tidal waves. Instead of considering them as ordinary, conventional water waves, they may be more similar to the edges of water or solitary waves. The major difficulty is that scientists are exploring not one-dimensional shallow waves, but two-dimensional waves of deep oceans. There is presently continuous research, both hypothetically and experimentally, to comprehend precisely the tsunami generation mechanisms and their behavior. In that sense, tsunami still remains unresolved. This chapter is devoted to deliver modern findings and theories beyond tsunami generation mechanisms.

4.1 5,000 Years of Tsunamis

The seaside cave is located along the northwestern coast of Aceh Province near the village of Lhong, 35 km south of Banda Aceh. Figure 4.1 is used to reveal historical tsunami [246]. Further, Figure 4.2 indicates the 2004 tsunami level at the top of the coastal cave.

Recently, Rubin et al. (2017) [246] utilized cave data in the Banda Aceh to measure tsunami recurrence intervals. Researchers reconnoitered a coastal cave containing layers of sand deposited by 11 ancient

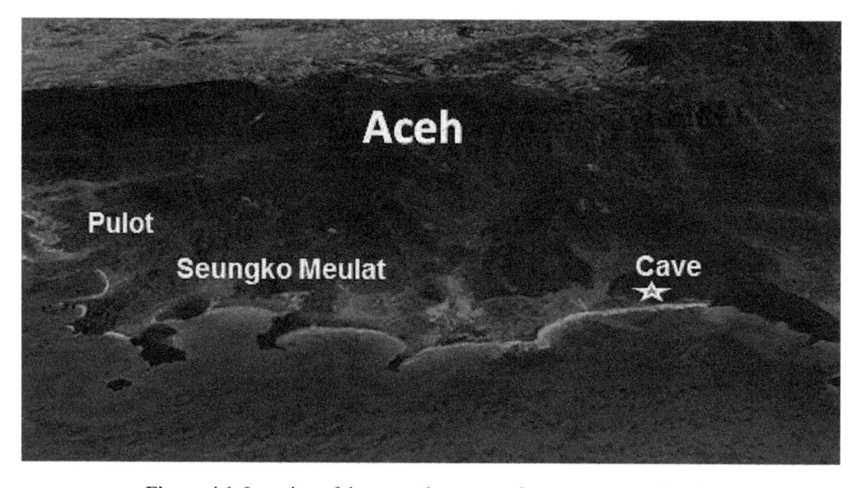

Figure 4.1 Location of the coastal cave, northwestern coast of Aceh.

Figure 4.2 Tsunami height hitting the coastal cave.

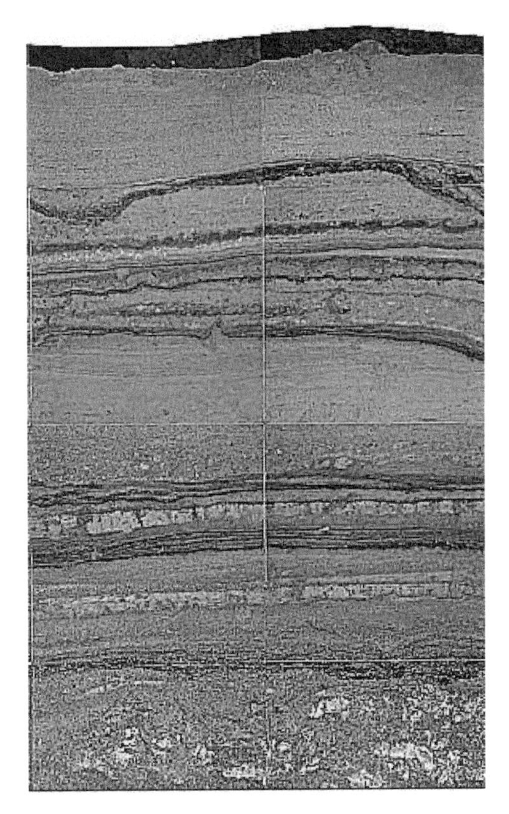

Figure 4.3 Sediment layers used for estimating tsunami ages in Aceh.

tsunamis and revealed that the time period between massive waves is extremely capricious (Figure 4.3). In this spirit, the massive waves were not regular in time. In other words, the relaxation periods of massive waves ranged from millennia to merely decades. In this understanding, over 5000 years, 11 tsunamis had swept over the Aceh.

Scientists implemented the laboratory radiocarbon measurements for depositing materials that were collected from six trenches at the rear of the 120-meter-long coastal cave. Thus, the approximate date of every tsunami and the probable age of each of the eleven buried layers of sand can be estimated by exploiting the radiocarbon measurements.

Rubin et al. (2017) [246] retrieved that the eleven sand layers bridged approximately 4,500 years, from about 7,400 to 2,900 years ago. Nevertheless, the guano-crusted 12th and the upmost sand layer—which restricted shreds of arraying, suggesting that it used to be deposited until very recently—differed from the stack of alternating deposits below it: its bottom face was jagged and irregular, in contrast to the smooth boundaries between the deeper layers.

Furthermore, the scientists expected that this irregularity resulted from effective waves from the 2004 tsunami caused by way of the Sumatra-Andaman earthquake sweeping into the cave, scraping away earlier deposited material, and actually erasing the geological document laid down post 900 BCE. The older layers of sediment were probably never disordered by tsunami waves due to the fact they are positioned in a natural depression in the cave. In this context, the deposited ancient sand layers are laid down, and completely sheltered.

4.2 Tsunami Recurrence

It is worth mentioning that the time span between successive tsunamis is far from constant. In this understanding, the researchers proved that 2,000-year period was free of tsunamis and four tsunamis were investigated in the century that followed the 2,000 year. This delivers a new evidence that tsunami recurrence can be extremely inconstant.

Based on the cave deposited sand information, it is hypothesized that the thickness of each sand layer displays the magnitude of the tsunami-causing earthquake because a huge earthquake would produce a great tsunami and consequently, plausibly transport greater sand into the cave. Along with this theory, the thickest sand layer, measuring approximately 25 centimetres, ought to correspond to the strongest earthquake that occurred within the almost 5,000 years of history recorded in the cave [246].

Moreover, the scientists concluded that no tsunamis took place for more than 2,100 years post this thickest layer of sand was probably packed down. This extremely lengthy interseismic gap is consistent with a period of reduced stress along faults—and therefore a lower probability of some other quake—after a huge temblor released a massive amount of energy. Conversely, the researchers discovered that the four sand layers corresponding to the four tsunamis that occurred inside one hundred years of each other had been all thin (less than 10 centimetres). They argued that quick interseismic intervals are steady with weaker earthquakes [245].

4.3 Can Tsunami Cause Marine Landslide?

Consistent with the above perception, the tsunami can accumulate massive sediments in the ocean bottom. In this view, the prevalence of earthquake can move the properly packed and accrued sediments. This displacement would possibly produce an underwater landslide. In fact, novel research discovered that massive earthquakes can trigger underwater landslides thousands of miles away, weeks or months post the quake strikes. Johnson et al. (2017) [247] revealed that recent great earthquakes and tsunamis around the world have heightened awareness of the inevitability of similar events occurring within the Cascadia Subduction Zone of the Pacific Northwest. They analyzed seafloor temperature, pressure, and seismic signals, and video stills of sediment-enveloped instruments recorded during the 2011–2015 Cascadia Initiative experiment, and seafloor morphology. They stated that thick accretionary prism sediments amplified and extended seismic wave durations from the 11 April 2012 Mw 8.6 Indian Ocean earthquake, located more than 13,500 km away. These waves triggered a sequence of small slope failures on the Cascadia margin that led to sediment gravity flows culminating in turbidity currents.

The triggering of sediment-laden gravity flows and turbidite deposition are related to local earthquakes, but this is the first study in which the originating seismic event is extremely distant (> 10,000 km). The possibility of remotely triggered slope failures that generate sediment-laden gravity flows should be considered in inferences of recurrence intervals of past great Cascadia earthquakes from turbidite sequences. Future similar studies may provide new understanding of submarine slope failures and turbidity currents and the hazards they pose to seafloor infrastructure and tsunami generation in regions both with and without local earthquakes.

The new findings could complicate sediment information used to estimate earthquake risk. If underwater landslides ought to be precipitated by using earthquakes, some distance away, not simply ones close by, scientists may additionally have to think about whether a neighborhood or a far away earthquake generated the deposits earlier than the usage of them to date local events and estimate earthquake risk.

The submarine landslides discovered in the study are smaller and more localized than huge landslides generated by a superb earthquake at once on the Cascadia margin itself; however, these underwater landslides generated by distant earthquakes may nonetheless be capable of generating nearby tsunamis and adversely affect underwater communication cables.

4.3.1 Mechanisms of Earthquake Causing Landslides

Scientists revealed that the steep underwater slopes hundreds of feet high line of the Cascadia margin. Then, sediment accumulates on the pinnacle of these steep slopes. When the seismic waves from the Indian Ocean earthquake reached these steep underwater slopes, they jostled the thick sediments piled on the pinnacle of the slopes. This shaking triggered areas of sediment to damage off and slide down the slope, creating a cascade of landslides all along the slope. The sediment did not fall as soon as the slope collapsed so the landslides befell for up to four months after the earthquake (Figure 4.4) [247].

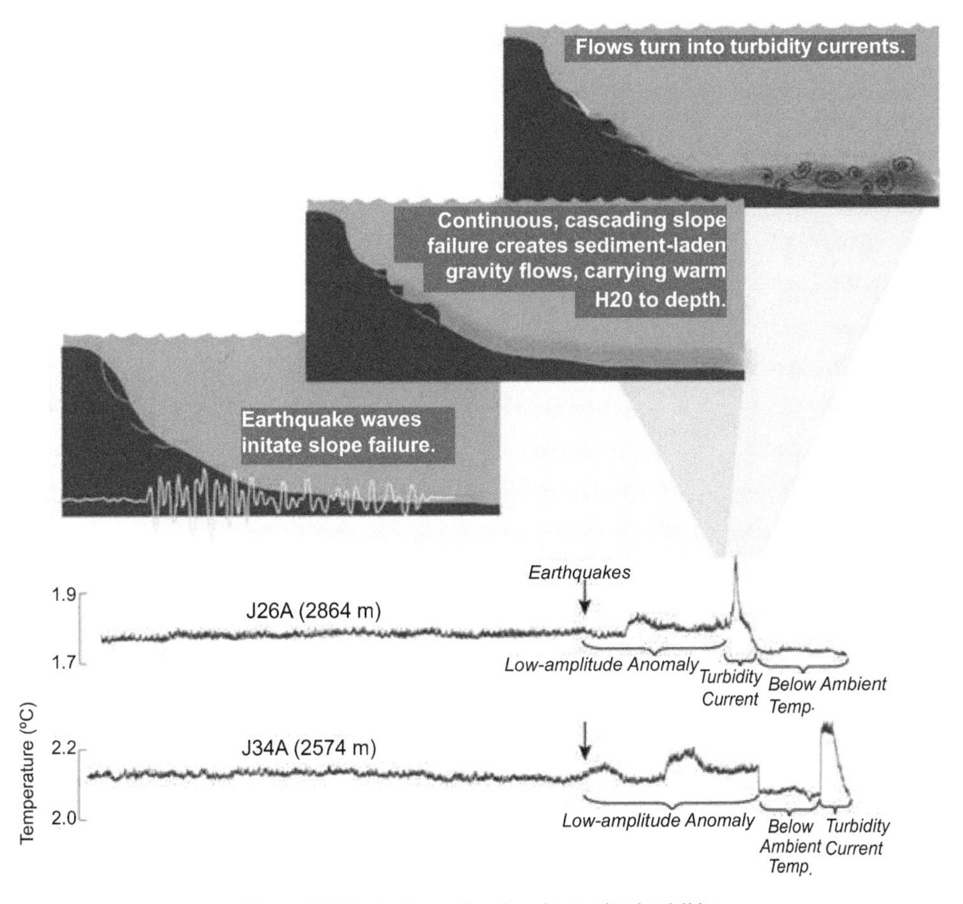

Figure 4.4 Mechanisms of earthquake causing landslides.

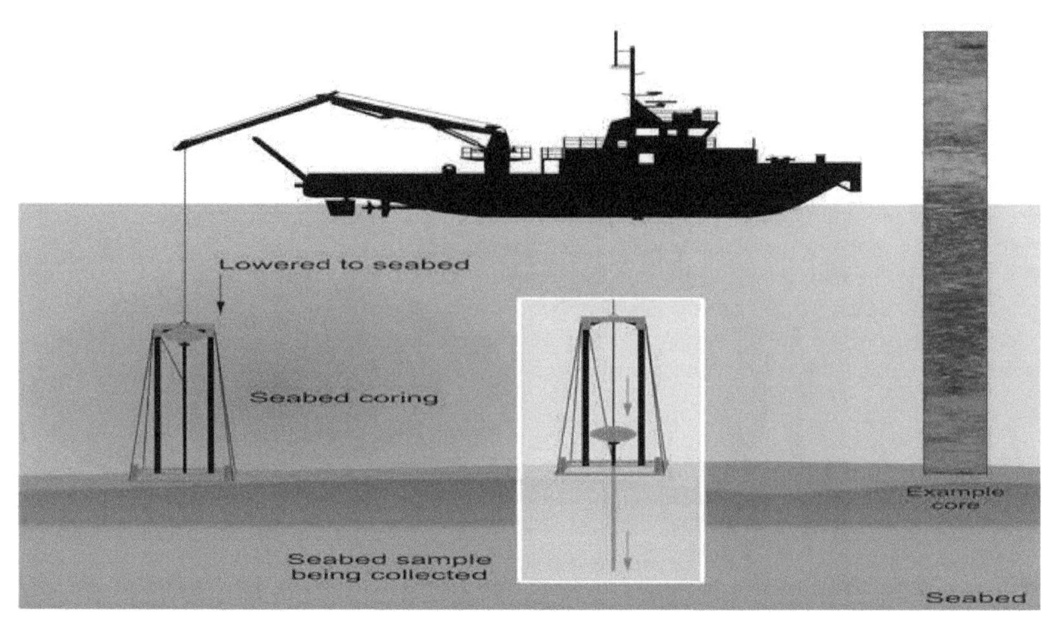

Figure 4.5 Submarine core sampling.

The downslope drift of sediments entrains heat seawater from the shallow ocean, producing temperature anomalies as the sediment flows before the ocean bottom seismometer. Low amplitude anomalies are the smaller slope disasters that precede the massive temperature spikes from the major turbidity currents, comparable to what is discovered with terrestrial landslides. Consequently, the changes in temperature could only be signs of multiple underwater landslides that shed sediments into the water. These landslides caused warm, shallow water to become denser and flow downhill, which caused the temperature spikes [247].

Submarine landslides push ocean water out of the way when they occur, which could spark a tsunami on the local coast [247]. It is worth mentioning that the landslide deposits can be used to deliver the timeliness of past earthquakes. This can be used to predict how often an earthquake might occur in the region in the future and how intense it could be. The new study, however, stipulates that a distant earthquake might only result in landslides up to 20 or 30 km wide. That means when scientists take sediment cores (Figure 4.5) to define how frequent local earthquakes occur, they may not be able to tell if the sediment layers arrived on the seafloor as a result of a distant or local earthquake. In this regard, huge records of core sampling of margine would be required to guess the precise earthquake risk.

4.4 Slow Slip and Tsunami

Uchida et al. (2016) [249] introduced a novel theory which is known as slow-slip. They reported that slow-slip or "aseismic" events can manifest at any plate boundary; however, they are discovered most certainly in subduction zones (Figure 4.6). For instance, aseismic events exist where the Pacific Plate, an oceanic slab of the Earth's crust, creeps beneath the continental Okhotsk Plate. Such aseismic motions happen over weeks or months, in contrast to earthquakes, which occur in seconds.

Additionally, on 25 October 2010, Mentawai islands in western Indonesia, offshore, were struck by the moment magnitude 7.8 earthquake (Figure 4.7). However, less than a magnitude of 8 earthquake created a massive tsunami which resulted in more than 400 human causalities. Therefore, this earthquake is considered as a rare slow-source tsunami earthquake based on (i) excessively large tsunami; (ii) disproportionate rupture duration near 125s; (iii) mainly shallow, near trench slip determined through finite-fault modeling; and (iv) shortages in energy-to-moment and energy-to-duration-diced ratios, the former in near-real time. This event can re-occur any time at any place of the Sunda trench in the future [251].

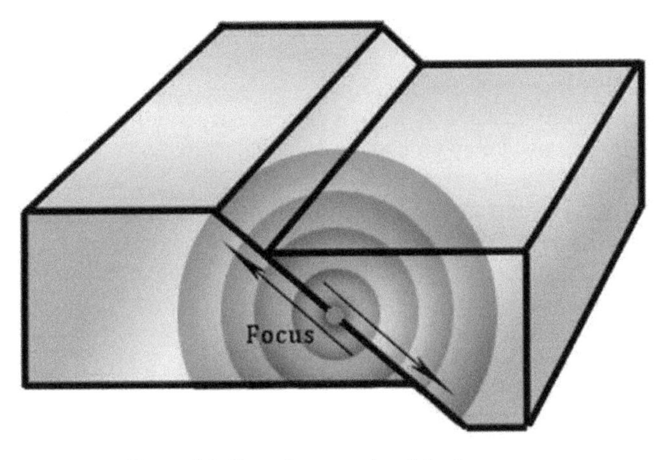

Figure 4.6 Slow-slip occurs in subduction zones.

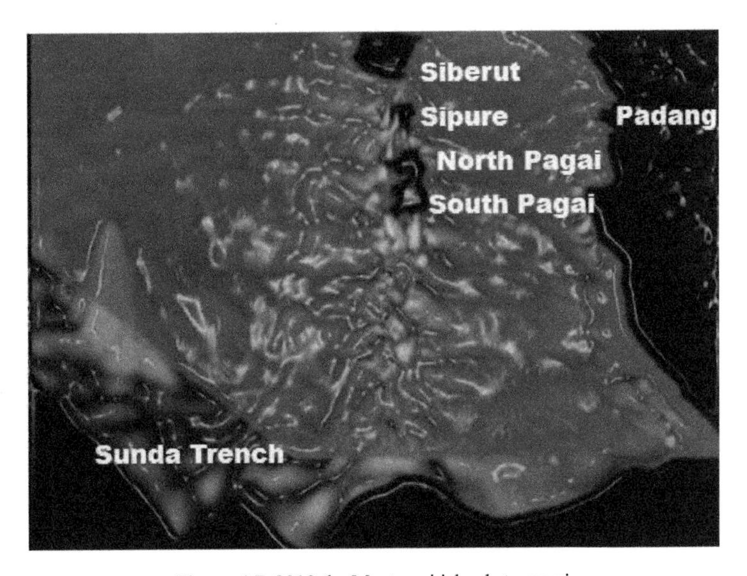

Figure 4.7 2010 the Mentawai islands tsunami.

The slow slip does not create seismic waves, so seismometers cannot record it or, maybe, researchers track the little, reasonable tendencies through recording the movement of GPS collectors protected on the ground. Seismologists, consequently, distinguish average slip events about once at normal intervals. These events occur so gradually that what may properly be known as an extent 6.5 event can appear throughout a month—without anybody feeling it. In this view, seismologists have watched reasonable slip events in the days following large quakes. They have been considering the fact that the past presumed an association between those events and tremors; however, in the event that occurred and how the two are related has remained a riddle.

The researchers evaluated over 6000 fault ruptures and the slow slip that caused them, and located a comparatively stable regularity of slow-slip events—they occur each 1–6 years in some places and each 2–3 years in alternative places of the subduction zone. The researchers then checked out the incidence of huge earthquakes—magnitude 5 and higher in slow slip plate and found that once the speed of slow-slip events magnified, earthquakes were far more probable to occur. They even found a swarm of slow-slip events on the fault preceding the Tohoku earthquake. The correlation between higher rates of slow-slip events and huge earthquakes can hopefully give well understanding of the development of additional precise and shorter-term earthquake forecasts.

However, the slow-slip events were inferred from repeaters, rather than from remote GPS measurements. Slow-slip events in subduction zones depart from periodicity in other subduction zones—this location of the slow-slip plate may be unique. The periodicity is not perfect, and it is difficult to quickly assimilate [248].

In other words, over the past decade, several subdivisions of the megathrust have ruptured consecutively, inflicting a sequence of earthquakes alongside the western coast of Sumatra (Figure 4.8). Therefore, this tsunami propagated widely towards Antarctica and South Africa within 12 hrs. In addition, it moved towards Penang, Malaysia within 3 hrs (Figure 4.9). In this regard, there is a possibility of tsunami occurrence in the future.

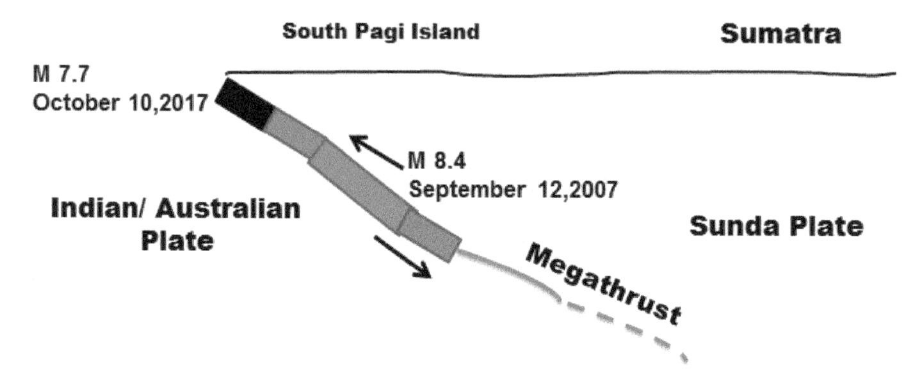

Figure 4.8 2010 Tsunami occurs along the Sunda trench.

Figure 4.9 2010 the Mentawai islands tsunami propagation across the Indian Ocean [https://nctr.pmel.noaa.gov/indonesia20101025/MentawaiMaxAmpCrop.png].

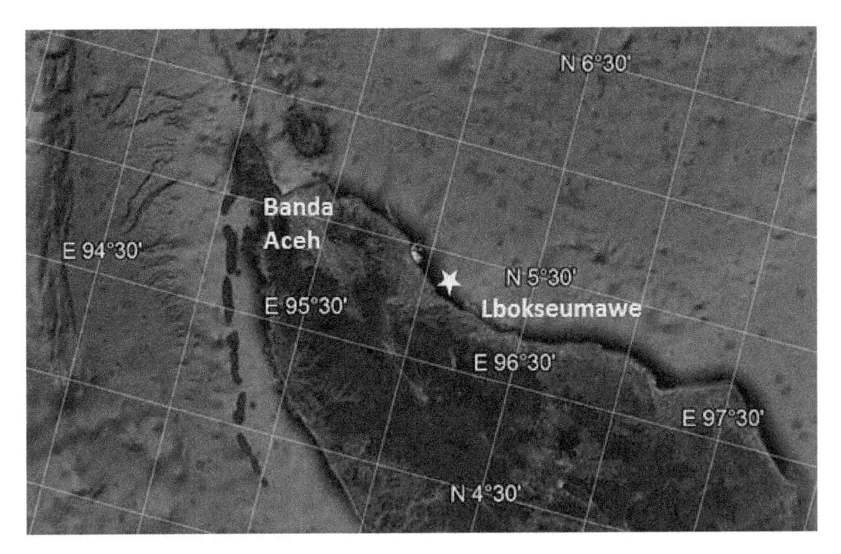

Figure 4.10 Earthquake without tsunami along northern Sumatra.

In this view, scientists agree that the superior quake can be produced by sequence of earthquakes and having the same dimension of the one that triggered the Indian Ocean tsunami in 2004. It could happen "in 30 minutes or in 30 years." For instance, a robust and shallow earthquake was recorded through the USGS as M6.5 struck northern Sumatra, Indonesia at 22:03 UTC on December 6, 2016 (Figure 4.10). This earthquake caused secondary hazards such as landslides that might have contributed to losses [250].

4.5 Low-frequency Earthquake Event

Consistent with Nakano et al. (2016) [177], low-frequency earthquakes reveal crustal deformation in the subduction zone. On the other hand, the generation mechanism of these extraordinary earthquakes has not been explicated yet because no ordinary earthquakes happen in the sedimentary wedge. In fact, low-frequency earthquakes are the solitary advantageous signs of the deformation. Revealing the source process of these earthquakes as a different earthquake activity will enhance our grasp of plate movement at subduction zones where megathrust earthquakes occur.

4.5.1 Characteristics of Low-frequency Earthquakes

Presently, scientists have spotted relaxing earthquakes and crustal distortion with periods continuing from 1 second to as long as 1 year. Further, like usual earthquakes, sluggish earthquakes are triggered by shear slip on a fault plane; nonetheless, the slip rate is slow and continues for an extended length. Rupture of normal earthquakes continues at the most for a few minutes. Additionally, shallow low-frequency earthquakes are indicated by means of the ascendancy of seismic signals longer than 10 seconds, unlike regular earthquakes of similar ground velocity of 5×10^{-5} m/s. In this view, low-frequency earthquakes are peaked between 1 and 3 Hz (Figure 4.11).

They are also created beneath the seafloor. Nonetheless, observation near the source is challenging to acquire. In this view, their impacts are restricted to partial zone. Generally, low-frequency earthquakes are generated due to the shear failure in the accretionary prism or plate boundary [250]. In accretionary prism or edge, sediments are the top layer of material on a tectonic plate, which accrue and distort where oceanic and continental plates crash. These sediments are scraped off the top of the downgoing oceanic crustal plate and are affixed to the facet of the continental plate (Figure 4.12). Therefore, the magnitudes

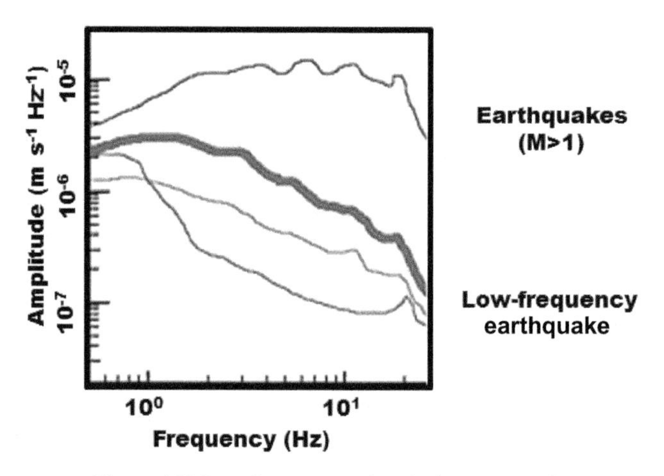

Figure 4.11 Low-frequency earthquake frequency peak.

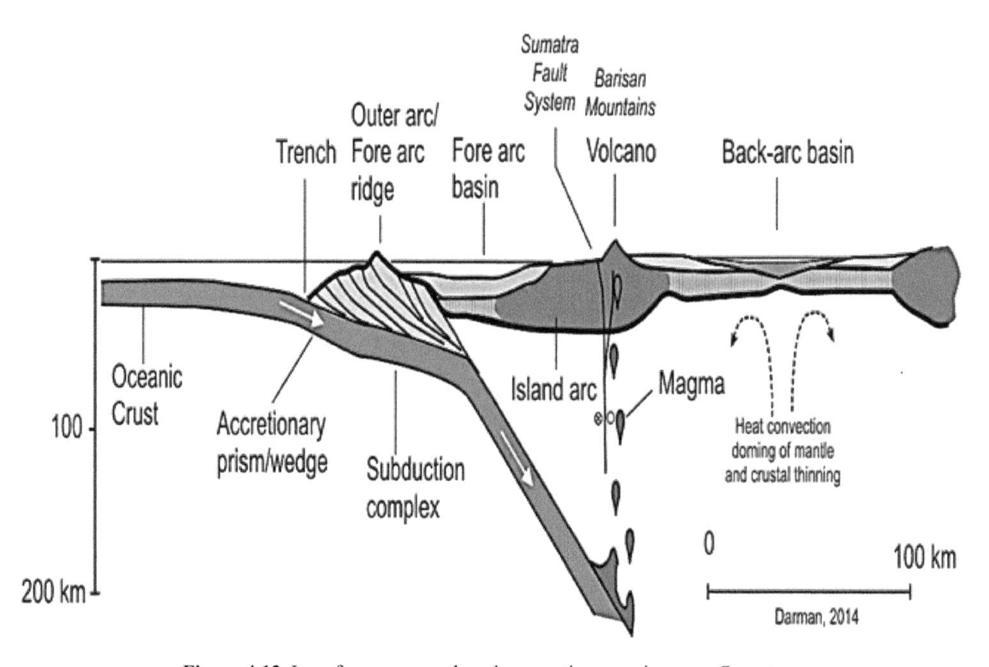

Figure 4.12 Low frequency earthquake accretionary prism near Sumatra.

of the low-frequency earthquakes were less than 4 and occurred in the accretionary prism or the oceanic crust at depths shallower than 10 km. These magnitudes are considered small compared to the regular earthquakes where the magnitudes are higher than 5.

At some point or another, like regular earthquakes, slow earthquakes are caused by shear slip on a fault plane, but the slip velocity is slow and lasts for a longer duration. Rupture of regular earthquakes continues at most for several minutes. In a tsunami caused by submarine earthquake, the rupture propagates alongside the fault more slowly than usual; however, the power blast-off arises on a time scale comparable to different earthquakes.

4.6 New Tsunami Generation Mechanisms and Models

Scientists have agreed that the tsunami is just a long wave. In this context, the tsunami models are established totally based on the conventional shallow water wave equations which are non-frequency dispersive long wave equations. In this view, it is required to develop a new approach of dispersive wave model to imitate the beaming of the tsunami. In spite of there being a wave-current interaction, the scientists ignore this concept and never consider that stream flow can be created by earthquakes. In this understanding, a tsunami wave cannot precisely describe either a long wave or a short wave. Conversely, it can be described as an intermediate wave. Under this circumstance, the physical parameters of the ocean and seafloor can alter the tsunami beaming across the water body. In this spirit, the equilibrium theory is applied by Rivera [252].

Let us assume that the force which is generated by seabed displacement in the case of the submarine earthquake is much higher than the hydrostatic pressure that exists above the submarine fault deformation. Under this circumstance, strong current can be generated along the fault and its direction and magnitude are the function of strength of fault deformation. The seismic movements due to fault deformation will be vertical. However, this seismic vertical movement generates cutting-edge shear flow in its opposite direction. In this understanding, the sequence of waves emit away from the epicenter as a feature of physics of sudden momentum of horizontal shear flow such as consequent convergence and divergence.

We assume that the weight of the column of water is balanced by the difference in pressure forces (*P*) on the bottom (*z*) and top faces prior to Sunda plate vibration. This balance is termed as the hydrostatic balance:

$$0 = -\frac{\partial P}{\partial z} - \rho(T,S,P)g \tag{4.1}$$

where the density of seawater (ρ) is a nonlinear function of salinity (S), temperature (*T*), and pressure (*P*). Seawater is slightly compressible (its volume decreases under pressure), thus its density increases with pressure. Following the Newton's second law, the net force applied to the parcel of water equals the mass of parcel of water column multiplied by its acceleration. Therefore, the equilibrium force acts on the water column, then can be given by

$$\rho p \frac{D^2\Delta}{Dt^2} = -\frac{\partial P}{\partial z} - \rho p g \tag{4.2}$$

For the water parcel acceleration to approach the equilibrium state, it must equal the water density gradient changes through the water column as:

$$\frac{D^2\Delta}{Dt^2} = N^2 = \frac{g}{\rho}\frac{\partial p}{\partial z} \tag{4.3}$$

where N is the buoyancy frequency or frequency of Brunt-Väisälä. In other words, in steady state of $-\frac{\partial p(z)}{\partial \rho}$ the mathematical expression of N is expressed by

$$N = \sqrt{-\frac{g}{\rho}\frac{\partial p(z)}{\partial z}} \tag{4.4}$$

We consider the simple harmonic oscillator of the water buoyancy as:

$$\frac{d^2 z}{Dt^2} + N^2 z = 0 \tag{4.5}$$

Equation 4.5 has two solutions:

In case $N^2 < 0$ then

$$z(t) = Ae^{\sqrt{-N^2 t}} + Be^{\sqrt{-N^2 t}} \tag{4.6}$$

In case $N^2 > 0$

$$z(t) = A\cos(Nt) + B\sin(Nt) \tag{4.7}$$

Equation 4.7 is implemented prior to Sunda plate movement, as the bottom density above the epicenter (ρ_c) is larger than the picnocline (ρ_i). In this view, the buoyancy frequency N is positive and produces more stability of the water density through the water column. During the seabed deformation, N is displaced vertically upwards and causes the turbulent shear flow, which is governed by R_i Richardson number

$$U_s = \sqrt{R_i N^2} \tag{4.8}$$

$$U_s = \sqrt{\left(\frac{\partial u}{\partial z}\right)^2 + \left(\frac{\partial v}{\partial z}\right)^2} \tag{4.8.1}$$

$$U_s = \sqrt{\left(\frac{\partial u}{\partial z}\right)^2 + \left(\frac{\partial v}{\partial z}\right)^2 / R_i N^2} \tag{4.9}$$

Then velocity shear (du/dz) must be large enough to generate turbulence along the fault slope. During the Sunda plate movement, two current movements are generated: one is turbulent flow above the epicenter and the second is vertical displacement due to upward movement of N (Figure 4.13). The ocean currents $u(x, y, z, t)$ created by the submarine earthquake can be estimated by [252]:

$$u(x,y,z,t) = \kappa\left[0.25\left[\frac{10^M g}{\rho Z v}\right]^{0.33}\right] f^{-1} U_S \beta_s \sin(\omega t) + V_s f^{-1} \tag{4.10}$$

where κ is non-dimensional tuning parameter, f is Coriolis force, $\beta_s = \frac{\partial d}{\partial s}$ is the bathymetric slope of the disturbed seabed, d is the water depth, ∂s is slope deformation, v is the kinematic viscosity of seawater and M is the moment magnitude of the earthquake. Further, ω is a local current whose frequency is given by

$$\omega = \frac{2\pi}{2\pi N^{-1}} = \frac{2\pi}{T_c} \tag{4.10.1}$$

Moreover, U_s represents a turbulent flow, which is based on Equation 4.9 and the right side of Equation 4.10 V_s represents the horizontal shear flow which equals

$$V_s = vA\sqrt{\left(\frac{\partial u}{\partial x}\right)^2 + \left(\frac{\partial v}{\partial y}\right)^2 + 0.5\left(\frac{\partial u}{\partial x} + \frac{\partial v}{\partial y}\right)^2} \tag{4.10.2}$$

where v is eddy viscosity constant which ranges between 0.01 to 0.5 and A is the area of a grid element. Finally, Z equals the total of the focal depth of the earthquake (z_c) and water depth through a water column h. This can be expressed as

$$Z = z_c + h = (z_v^2 + z_h^2)^{0.5} + h \tag{4.10.3}$$

Figure 4.13 The current generated due to Sunda fault displacement.

where z_v and z_h are the actual focal depth directly beneath the epicenter and the horizontal distance from the epicenter to the ocean water, respectively. Commonly, Equation 4.10 indicates that the large bathymetric slope created the strong current which interfered with the seismic energy and induced peculiarly massive tsunami.

Generally, Sunda plate deformation created a complex shear turbulent flow, which caused instability of water column. In this regard, irregular patterns of currents and wave were developed and oscillated continuously even after cessation of the fault line rupture until dissipated by friction. Therefore, the fastest growing modes to be associated with local minima of the gradient Richardson number were consistent with observations of internal waves. This leads to a greater current speed along the fault compared to the upper layer of the water column. It means that the current logarithmically increased upwards because the earthquake applied force generated due to seabed deformation. In this view, a dead zone appeared along the epicenter west of Sumatra as a strong proof that current is not created by the wave but by the seabed displacement.

4.7 Molecular Hydrodynamic Tsunami Generation

This theory was introduced by Burkov and Melker (2012) [253]. In this spirit, the tsunami generation mechanism was developed based on the molecular hydrodynamic theory. The scientific explanation of molecular dynamics can mathematically be written as:

$$m_i \frac{\partial^2 r_i}{\partial t^2} = F_i + m_{ig} = \sum_{j=1, j\neq 1}^{n} F_{ij} g \tag{4.11}$$

where r_i is the radius-vector of particle i, m its mass and $\boldsymbol{F}$ is the force acting on a particle which is the sum of interaction forces between particles and the gravity force $g\,m$. This operates as a natural condition owing to the particles which divert the distributed space and form a free surface.

The force, which models the generation of particles was integrated into the water, subsequently accomplishing the thermal equilibrium. Initially, the disturbance creates the wave of giant amplitude which transmits with a fast speed and a cavity at the site of disturbance (t = 5–12.5 ps). Once the wave reaches shallow water, the amplitude and velocity reduce and the wave alters its form (t = 32.5–67.5 ps) (Figure 4.14). This is in keeping with the observations in areas with a large angle of the coastal zone. It is worth mentioning that the wave density decreases, and foam formation occurs as the coastal zone is approached. Then, the rarefied water penetrates deeper during the wave propagation. Thus, this theory is able to acquire a better insight into the nature of tsunami waves.

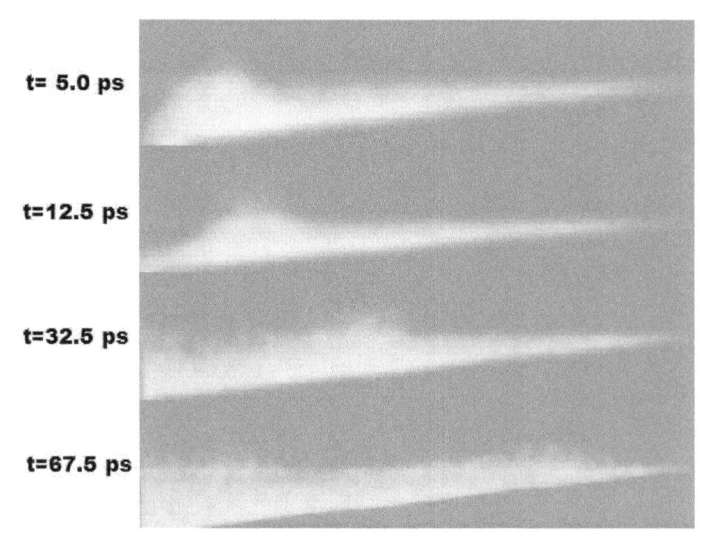

Figure 4.14 Molecular hydrodynamic theory of tsunami propagation with different times.

4.8 Can Gravity Cause Tsunami?

Earthquakes are some of the most devastating and unexpectable forces on the planet. It seems, however, some of the world's largest earthquakes may be following a pattern after all—they seem to occur at times around the full or new moon. These included the massive Indian Ocean shock in 2004 and the one that shook off the coast of Chile in 2010. In fact, the 2004 Boxing day earthquake took place within days of the maximum tidal stress. Tidal stress, therefore, is triggered by the pull of gravity on our planet as it orbits around our solar system. The gravity of the moon causes the oceans to bulge and relax as it orbits round the Earth, developing the tides. At a full moon, the moon is aligned on the opposite side of our planet to the sun, causing perhaps the biggest tides, distinguished as a spring tide [254].

Correspondingly, spring tides also arise once the moon is allied on the same direction as the sun, creating a new moon where its face is in shadow. Under both circumstances, the gravitational pull of the sun is combined with the gravitational pull of the moon, producing the oceans to bulge more and generating higher tides.

This is when the gravitational pull of the moon and the sun on the Earth are at their greatest and it could trigger fault lines into slipping. These seem to have happened as trifling failures in the fault lines poured into huge ruptures owing to the strain caused on the Earth's crust by the moon and the sun. Furthermore, seismic gravity alters since earthquakes shift adjoining blocks of mass and create permanent alterations in the gravity field around the epicenter. For instance, the Earth's gravity is approximately 9.8×10^8 microgals, so the largest relative change recorded was just 1 part in two billion.

In this understanding, the Earth's crust is very thin, relatively speaking. If it is shrunk to the size of a chicken's egg, the crust would be 3 to 4 times thinner than the egg shell. The moon, obviously, pulls on liquid rock just below the thin crust just as it does on sea water [254]. However, geologists have suspected that a similar stress is placed on the rocks in the Earth's crust; nevertheless, there has been little evidence to support it till date.

Generally, information about the tidal stress state in seismic regions can be used to improve probabilistic earthquake forecasting, especially for extremely large earthquakes.

4.8.1 Why Would Gravity and Topography be Related to Seismic Activity?

The frictional behavior of the fault is keystone to understanding the earthquake mechanisms. Once two plates rub up in opposition to each other, the friction between the plates makes it more difficult for them to slide. If the friction is great enough, the plates will stick. Over long periods of time, as the stack plates push towards each other, they may deform, creating spatial variants in topography and gravity.

Friction triggers stress between plates can generate additional deformations. Once extraordinary stress accumulates, the plates will unexpectedly leap and circulate the strain in the violent shaking of an earthquake. If there have been no friction between the plates, they would just slide right by way of each other smoothly, barring bending or creating the strain that ultimately effects on earthquakes. In this view, in subduction zones, areas beneath excessive stress probably have greater gravity and topography anomalies, and are additionally more susceptible to earthquakes. Though this concern delivers a fundamental clarification for an alternatively intricate and intuitive phenomenon of tsunami generation mechanism.

The gravity anomalies occur over a long time. However, they have tiny change over timescales up to at least 1 million years. Therefore, earthquakes can cause the small variations of the gravity field, about 4×10^{-4} m/s^2, as compared with the long-term anomalies.

Since topography and gravity fluctuations persevere over periods of time much longer than the usual periods between earthquakes, which is ranged from 100 to 1,000 years, great earthquakes would be consistently absent from zones with large positive gravity anomalies. Consequently, short-term seismic activity is strongly correlated with long-term tectonic behavior, thereby affording a comprehensive understanding of earthquake dynamics. Although large earthquakes occur where gravity and topography are low, there are low-gravity areas in subduction zones with no seismic activity.

Finally, the mechanisms are ambiguous. Conversely, the moon's attraction generates tidal fluctuations which are sequences of magnitude lower than those experienced in an earthquake. In addition, not every variation in tide originates with a resultant earthquake. In this view, part of the difficulty is that scientists cannot determine precisely what produces a foremost earthquake. It is worth mentioning that not all great earthquakes are triggered by the moon's attraction.

4.9 Did Himalayan Mountain Cause 2004 Tsunami?

Recently, the research team from the university of Southampton and Colorado School of Mines conducted the ocean drilling expedition along the Sumatra region in 2016. The team collected the sediments and rock samples from the ocean Sunda tectonic plate, approximately 1.5 km beneath the seabed. The aim of this expedition was to find out what are the main causes of the massive destructive 2004 tsunami and compare them with other regions that have similar geological characteristics.

Further, they investigated the sediment composition, chemical, and physical properties and then modeled how the sediments and rock had transported 250 km to the east towards the subduction zone. Moreover, the team investigated how the sediment and rock that were buried deeply, approached maximum temperature [263].

The new theory claims that the ocean floor contains sediments which eroded from the Himalayan mountain variety and Tibetan Plateau. This can lead to a critical question which is that how did these sediments flow thousands of kilometers to settle beneath the plate tectonic? They were carried out by rivers, and drainage networks on land and in the ocean (Figure 4.15). Further, Figure 4.16 confirms the existence of countless of drainage networks and two large rivers of Ganges and B' putra rivers, which are connected to the Himalayan mountain by Sunda plate tectonic through the Bay of Bengal. In this view, dehydration process took place prior to the sediments approaching the subduction zone. This creates a surprisingly robust material, permitting earthquake slip on the subduction fault surface to shallower depths and over a bigger fault region—inflicting the exceedingly strong earthquake as seen in 2004 [262].

Furthermore, the team determined that the thickness of the sediment on the subducting plate could be a cause for the great magnitude of an earthquake and tsunami. In this understanding, the thickest sediments can be extremely compressed and warmer too. Under these circumstances, the strongest sediments are formed because of heat and temperature. As this occurred in the shallowest region of the subduction zone, the slipping phase can occur during an earthquake. Consequently, massive volumes of water are displaced due to high pressure that was a result of the seabed moving horizontally and vertically [263].

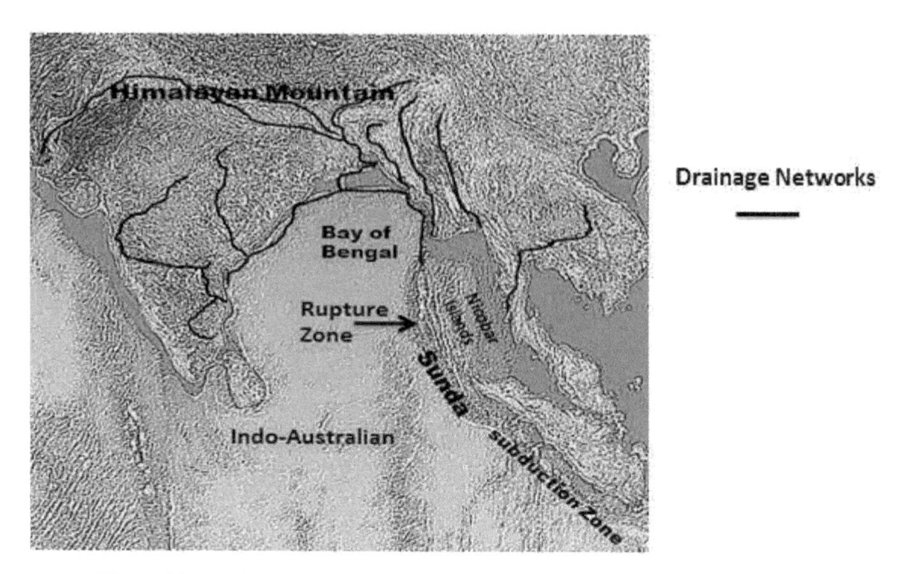

Figure 4.15 Drainage networks carried eroded sediments from Himalayan mountain.

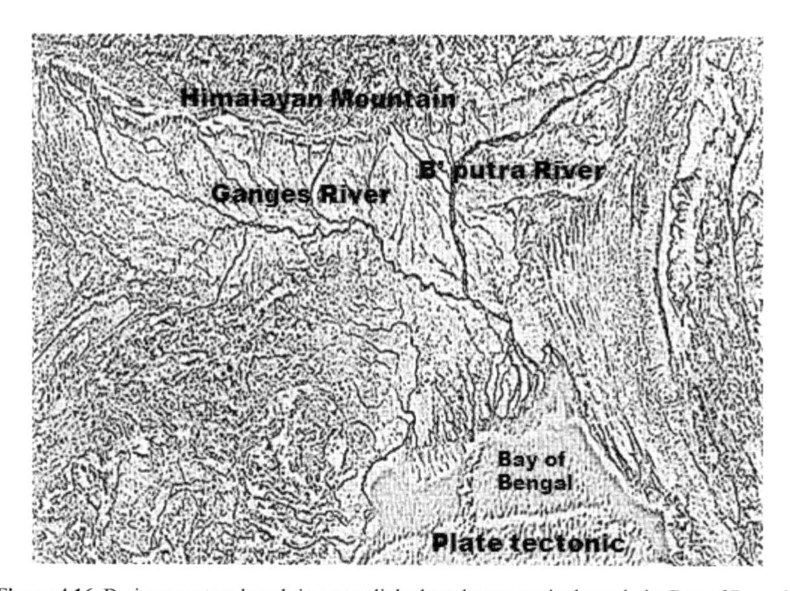

Figure 4.16 Drainage network and rivers are linked to plate tectonic through the Bay of Bengal.

This suggests that a similar earthquake and tsunami of 2004 can occur in the Western Indian Ocean. In this regard, the coastlines of India, Pakistan, Iran and Oman could be struck by a future tsunami. This requires further investigation of pre-historic earthquakes, compiled with the tsunami.

4.10 Did Deep Heat Spawn the 2004 Tsunami?

Over tens of millions of years, warmness constrained water out of submerged minerals inside the sediment close to the Sunda trench. Under this circumstance, the rock became super brittle, permitting a superior earthquake to occur. Consequently, the minerals in the sediment reduced the salinity in the sediments and released a layer of fresh water. Then the minerals created the crystal structures on the top of plate tectonic with a blanket of thick sediments.

In this view, the huge heat created by thick sediment chemically interacts with minerals. It may suggest that thermal process links to siliciclastic deposition dehydration which is driven by rapid temperature increases in the past of 9 million years. As the siliciclastic deposition is buried deeply, it becomes compacted and cemented, forming sedimentary rock. Consequently, if the sediment attribution begins to sustain the deposition under high-conductivity, the siliciclastic depositions abide by the earlier low-conductivity carbonates that can imitate the heat flow pattern albeit the same deposition level.

In this regard, the extraordinary heat triggered a chemical transformation within the sediment, driving water out of the mineral crystals and into tiny pores between the grains. This causes the sediment dehydration and then gets buried under plate tectonic within a few kilometers.

Over millions of years, the accumulation of completed dehydrated sediments became closer to the subduction zone. In this context, rough and unstable sediments puts more force on the hydrostatic pressure above the fault. Under this circumstance, a mega-quake dramatically occurs.

This mechanism requires more evidence to argue that it can cause future mega-quake from other faults. However, the Himalayan and heat theories have confirmed the cause of the 2004 tsunami.

4.11 Can Nuclear Bomb Create a Tsunami?

The primary conspiracy in terms of the 2004 Indian ocean tsunami is that it occurred because of the underground nuclear tests done through the United States. Underground nuclear tests are commonplace practised in the great region and this can result in destructive consequences for land masses.

In fact, the tsunami bomb became a conflict at some stage in the global struggle, i.e., World War II, to invent a tectonic weapon that might create massive destructive tsunamis. The venture started after US navy officer E.A. Gibson observed small waves generated by using explosions used to clean coral reefs. The concept was evolved by the United States and New Zealand navy in a prospective législation-named project seal. The nuclear weapon (fission bomb or thermonuclear bomb) opinion was deemed viable; nevertheless, the weapons themselves have, by no means, completely evolved or been used. In this regard, the bouncing bomb was developed and used in World War II, to be dropped into water as a method to damage German dams, instigating loss of commercial potential and big flooding.

Consequently, if a nuclear weapon is exploded in a vacuum, i.e., within the space, the high energy of neutron and gamma rays will be emitted within picoseconds of the explosion and travel miles. In a vacuum, they will travel even faster because they do not have to push air molecules aside. So, the hot gasses from the explosion will travel outward at thousands of miles per hour, slowing down as they go, because they are losing heat. Can the nuclear bomb trigger earthquake? The maximum direct cause-impact relationship is that the passage of the seismic waves, created with the aid of the thermonuclear explosion, through the epicentral area in Afghanistan by some means caused the earthquake. Therefore, the opposite nuclear tests took place 2 to 20 days prior to the earthquake. In this view, the elastic strains were triggered in the epicentral location by the passage of the seismic wavefield, generated by using the most important of the nuclear checks. A 40 kilotons bomb for instance, releases 100 times less energy than the traces brought about by using the earth's semi-diurnal (12 hour) tides which are produced by way of the gravitational fields of the moon and the sun. If small nuclear assessments should cause an earthquake at a distance of 1000 km, equal-sized earthquakes, which occur globally at a charge of the numerous of tests consistent with a day, could also be expected to cause earthquakes. As a result, there may be no proof of a causal connection between the nuclear tests and the huge earthquake in Afghanistan and it's only natural twist of fate that they passed off close to in time and vicinity. Because the geology of the Sunda Trench is already risky because of massive-scale fracturing as a result of preceding tests, similar critical landslides are probable. Such landslides within the past have given an upward push to tsunamis inflicting coastal damage. They can also release radioactive material into the ocean, with catastrophic consequences at the food chain in a place.

Nevertheless, the impression that nuclear tests results in tsunamis are new and under-researched. No atomic bomb became at the time robust enough to create a tsunami of that stature. Further, no bomb is robust enough at the present to generate a tsunami wave. The electricity required would be huge. Yet, it's plausible that it might cause an earthquake which might cause a tsunami.

It is far more believable that a nuclear explosion can cause an earthquake which could bring about a tsunami. The problem with this principle, and with most of the conspiracy theories, is that there may be, without a doubt, no evidence whatsoever to suggest that the United States was testing nuclear bomb at the time. And if there was, then a hyperlink between the nuclear tests, the earthquake and the tsunami would have to be hooked up. This principle has no documentary evidence associating it with the 2004 tsunami, no testimonies, no cause, no prior report, and no purpose. As it stands, in reality, it is conjectured and joins billions of different theories that "ought to" be authentic, but until it comes up with a few proofs, it is considered as false information.

4.12 Can HAARP Technology Create a Tsunami?

HAARP—which stands for High Frequency Active Auroral Research Program, was designed by the US government in 1993, and then implemented by the US Air Force and US Navy. In this regard, USA is using it as a weapon for controlling everything from weather to earthquake. On the contrary, HAARP can certainly send radiation into the ground to detect underground munitions, minerals and tunnels. As the ionosphere is opaque in HAARP's frequency band, it absorbs nearly all of the incident power. In other words, HAARP is an Ionospheric heater. It changes the shape of the ionosphere, allowing for beamed energy to be concentrated on a known spot.

HAARP is based on the phased array of antennas whose phases (timing) and amplitudes are exclusively modifiable. In other words, a great variety of various combinations can be strained out, synthesizing novel and complex complete wave configurations. Therefore, phasing permits the waveforms to be directed in diverse routes without having to transfer the antennas themselves. Under these circumstances, phased arrays generate an entirely altered sort of radiation, which cannot be sensed with standard electromagnetic devices. In this regard, the HAARP uses high frequency radio waves ranging from 2.8 to 10 megahertz. Further, these radio waves as a function on their sort and frequency can affect human emotion and biology, influence the weather, alter tectonic dynamics, and even manipulate our surrounding hyper-dimensional environment to permit regulated invisible and interdimensional manoeuvring of military personnel and vehicles. In essence, everything that regular electromagnetic waves cannot really do.

In this regard, the magnetic field is disrupted and a massive earthquake can be triggered. In fact, the recent earthquakes have blamed HAARP. In this understanding, the HAARP microwave signal can penetrate a depth of 10 km which is appropriate location for faults such as the Sunda Trench with a depth of 6 km.

On April 27, 2008, both the cyclone that hit Burma and the earthquake that struck China strictly corresponded to the period of the HAARP activities. In this view, an area of deep convection continued near a low-level circulation in the Bay of Bengal, approximately 1150 km east-southeast of Chennai, India.

On May 12, 2008, a destructive earthquake struck China. However, the Chinese military experts confirmed that there were no geological shocks in the southeast Asia after the earthquake. Consequently, they found that this quake was caused by an underground nuclear explosion. In other words, the energy released was equal to an underground nuclear explosion.

Chapter 5

Modification of the Earth's Rotation by 2004 Earthquake

Humans are taught that the Earth's day is 24 hours or 86,400 seconds. Scientists have found that there are inequalities on the millisecond level, which might affected predicting weather and earthquakes. The earth day of 24 hours is attributable to the rotation of the Earth around its axis. On the other hand, it takes 3 hours, 56 minutes and 4.0916 seconds for the Earth to turn once on its axis which is identified as sidereal days. The key question is why this seems different post 2004 tsunami?

5.1 Earth Rotation

Consistent with Cain (2015) [91], the Earth orbiting the Sun takes a full 365 days, 5 hours, 48 minutes and 46 seconds to complete the entire journey. Simultaneously, the Earth is spinning on its axis (Figure 5.1). The Earth would not rotate easily round its axis, instead, the poles meander in rough circles about 10 meters in diameter.

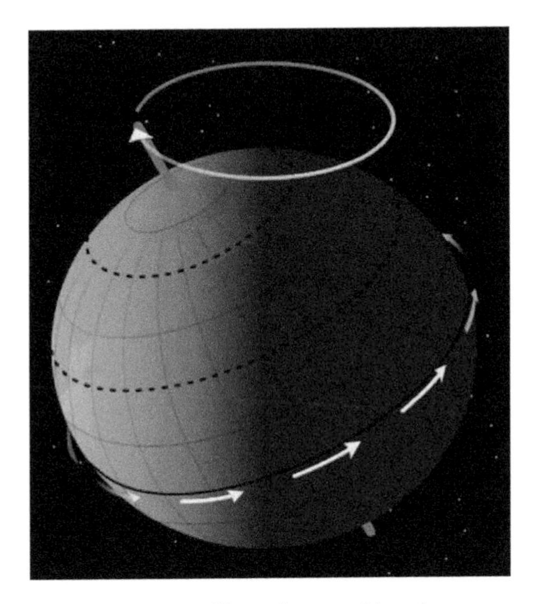

Figure 5.1 Earth's rotation around its axis.

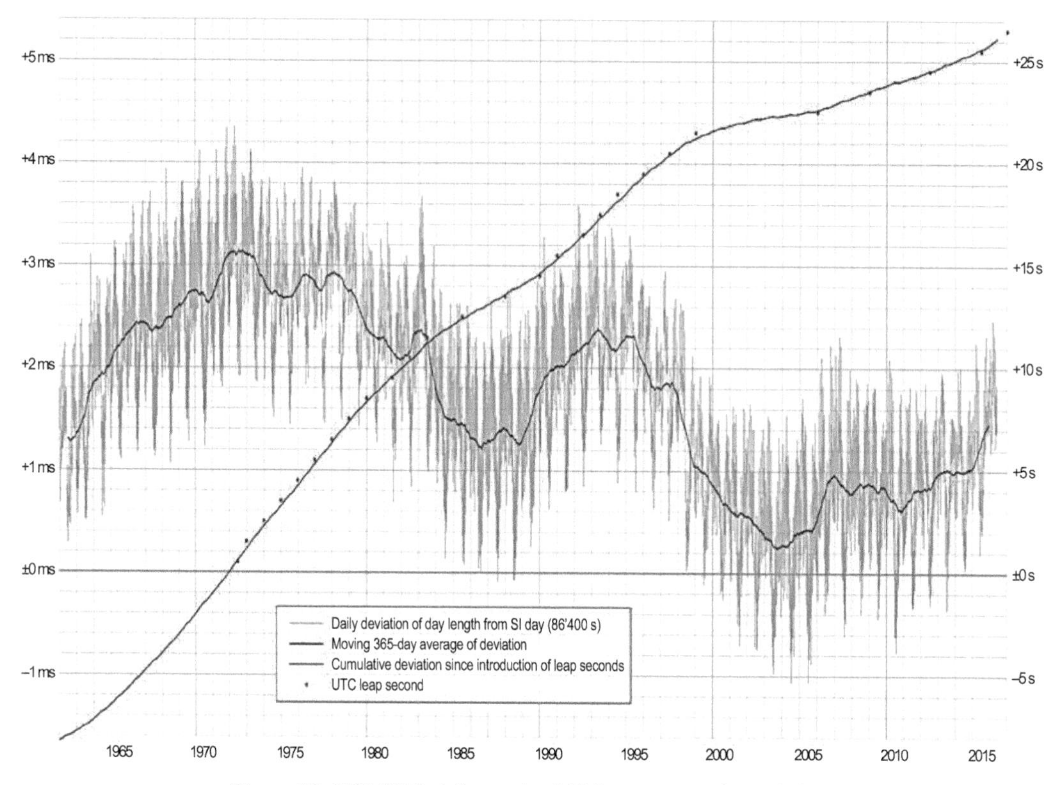

Figure 5.2 1962–2015: daily, moving 365-day average and cumulative.

In other words, latitude changes slightly, i.e., 0.7 seconds of arc over a cycle of nearly a year. Each day that goes by, the Earth requires to revolve a little more for the Sun to reappearance to the same place in the sky. In this regard, the required additional time is approximately 4 minutes. If we tend to solely measure sidereal days, the position of the Sun slips back daily. For half of the year, the Sun is up between 12 am and 12 pm. Additionally, for the opposite 0.5 degree, it might be between 12 pm and 12 am. There is be no connection between what time it is, and whether or not the Sun is within the sky.

Therefore, Figure 5.2 suggests that the average of 50 years is about 86,400.002. The yearly average over that period has ranged between about 86,400 and 86,400.003, whilst the size of individual days has varied between about 86,399.999 and 86,400.004 seconds [92].

5.2 Forces Affecting the Length of the Earth's Day

5.2.1 Tidal Forces and Earthquakes

Tidal forces can play tremendous role in changing the length of an Earth day. As stated by Cain [91], tidal interactions with the Moon over the last one century have increased the length of a day on the Earth approximately by 1.7 milliseconds. This brings us to the critical question of how could earthquakes change the length of a day on Earth? Powerful earthquakes can modify the Earth's rotational time by a few microseconds depending on how the tectonic plates shove around. Even as the glaciers melt, the rotation speed slows down a little more.

5.2.2 Wind Force

The Earth's rotation oscillates over several completely different time scales. For example, the earth slows down in January and February. It, therefore, turns out strongly throughout the northern hemisphere winter because the winds are preponderant to the west to east. In other words, the fluctuation of Earth's rotation is seasonal. This is often a result of the angular momentum of the atmosphere, doubled by additional forceful winds. In this regard, the whole system concerns the revolving earth and whirling atmosphere. This contributes to modify the stormy winter months by decelerating down the solid Earth's spinning. This means the time becomes longer by a few thousandths of a second. In this respect, the winter in the Southern Hemisphere does not yield similar growth in the wind because the extremity of the Earth is commonly ocean and therefore the temperature fluctuations are, consequently, not as important [91].

5.2.3 Madden-Julian Cycle

Other than the regular impacts, Madden-Julian cycle plays significant role to slow down the Earth's spinning. As indicated by Roland and Paul (1994) [93], the Madden-Julian oscillation (MJO) is the largest phase of the intra-seasonal (30-to 90-day) variability in the tropical atmosphere. It is a huge-scale connection between the climatic route and tropical profound convection (Figure 5.3). To cite a frequent example, like the El Niño Southern Oscillation (ENSO), the Madden-Julian oscillation is a voyaging layout that flows eastbound at round 4 to 8 m/s (14 to 29 km/h, 9 to 18 mph), through the air overhead in the warm parts of the Indian and Pacific seas. This regular dissemination layout indicates itself most obviously as peculiar precipitation.

Generally, the Madden-Julian cycle variation lasts about 30 to 60 days, which correlates to deviations in the Earth's rotation. Through an El Nino, the length of the day increases slightly which occurs on longer ranges.

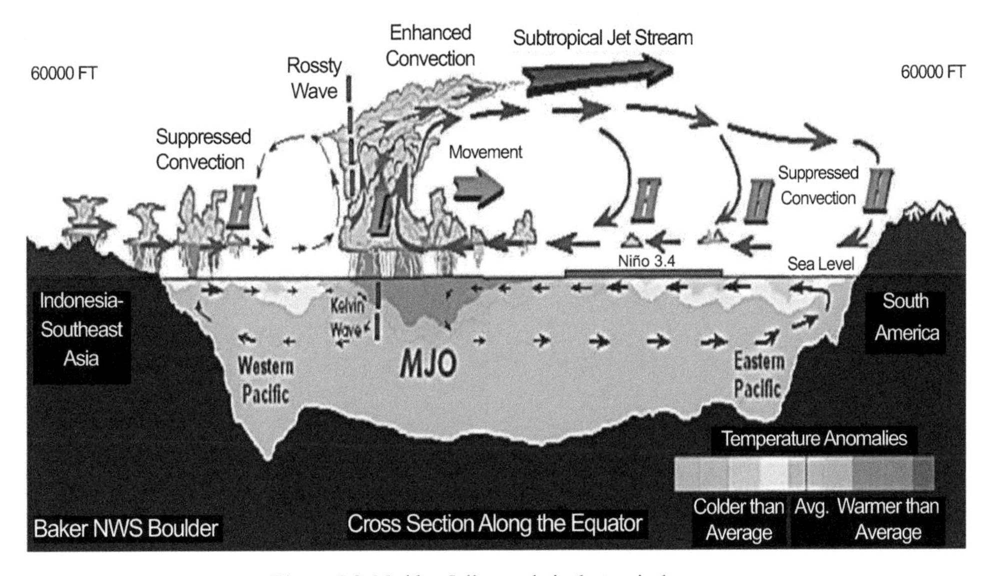

Figure 5.3 Madden-Julian cycle in the tropical zone.

5.2.4 Climate Changes

The global climate changes are a result of a shift in the Northern Hemisphere jet stream, which is dominated by a high-altitude and fast-moving wind current. In fact, it was compiled by the two- to four-year cycle in Pacific Ocean temperatures which drove a global climate change. Some scientists have recommended that global warming can slow down the Earth's spinning. This can occur due to the faster warming of the poles than the equator. In this regard, huge water masses from the poles will move and generate huge jet stream volumes, which slow down the rotation of the Earth. Generally, the Earth slowing down may drive a change in wind patterns, which causes another alteration to the rotation, which then kicks up more wind [90].

5.3 2004 Tsunami's Effects on Earth's Rotation

Some scientists declare that December's tsunami, which was induced by the Sumatra earthquake, triggered in a shortening of the day length through 2.68 millionths of a second. Likewise, the Moon and the Earth are slowly drifting away from each other. To preserve angular momentum, the Earth is slowing down, and the day is lengthening at a rate of approximately a millisecond per century.

5.3.1 Chandler Wobble or Variation of Latitude

The Chandler Wobble or variant of latitude is a minor deviation in the Earth's axis of rotation compared to the solid earth (Figure 5.4). It amounts to change of about 9 m (30 ft) in the point at which the axis intersects the Earth's surface and has a period of 433 days. This wobble, which is a swaying combines with some other wobble with a period of one year, so that the complete polar movement varies with a period of about 7 years [94].

The Chandler wobble is an illustration of the range of movement that can occur in a revolving object which is no longer a sphere. Consequently, this is also acknowledged as free nutation. Somewhat confusingly, the route of the Earth's spin axis is relative to the stars which varies with exclusive periods, and these motions—formed through the tidal forces of the Moon and Sun—are furthermore acclaimed

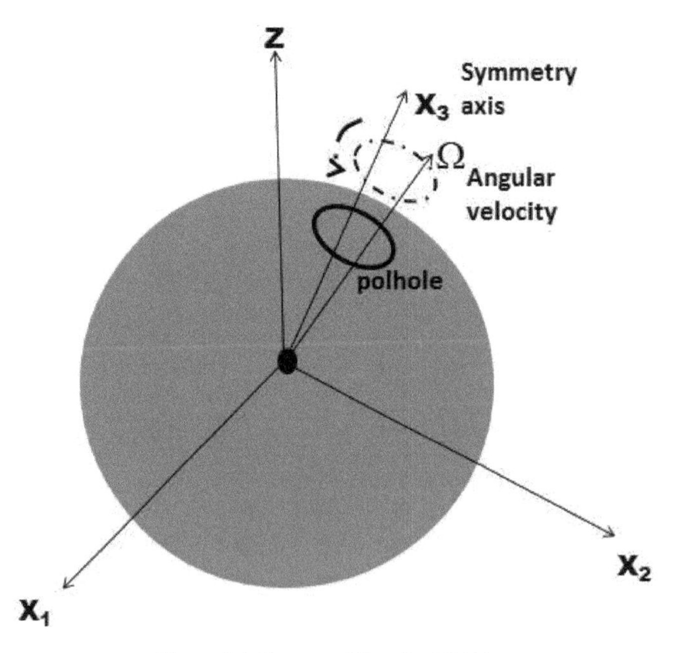

Figure 5.4 Concept of Chandler Wobble.

as notations, apart from for the slow-moving of the Earth's spinning, which are precessions of the equinoxes [95].

Gross (2000) [95] implemented angular momentum models of the atmosphere and the oceans in computer simulations to exhibit that from 1985 to 1996, the Chandler wobble used to be excited through an aggregate of atmospheric and oceanic processes, with the dominant excitation mechanism being ocean-bottom strain fluctuations. Gross [95] determined that two-thirds of the "wobble" was once triggered through fluctuating strain on the seabed, which, in turn, is triggered through modifications in the circulation of the oceans induced with the aid of variations in temperature, salinity, and wind. The ultimate third is due to atmospheric fluctuations. Some scientists, consequently, forecast the last wobble to have befallen about 4000 to 7500 years ago, all through a time when there had been dramatic modifications in the pattern of monsoons throughout northern Africa. It is declared that it has distorted the Great Saharan Desert into an incredibly greener region. Additionally, Gross [95] assumes the wobble to appear once in about 20000 years, which implies that the subsequent wobble can be predicted after a duration of 16000 years.

5.3.2 How Chandler Wobble is Impacted by Earthquakes?

During boxing day, a giant portion of the northern Indian ocean leaned northward, roughly 10 to 20 meters, compared to Asia. Consequently, it was inclined various meters into the mantle and flattened and elevated northern Sumatra by quite a few meters. It also probably triggered massive submarine landslides that caused the tsunamis. Afterward, an affordable amount of mass became decentralized. Thus, the question that arises is what impact did this 2004 earthquake have on the rotation of the earth? It modified the length of the day by –2.676 microseconds. Additionally, the horizontal and vertical polar motions are excitation approximately –0.670 milliarcseconds and 0.475 milliarcseconds, respectively (Figure 5.5). This suggests that tiny motion excitation being approximately 0.82 milliarcsecond in amplitude which is difficult to be detected. In fact, the 2004 earthquake was situated close to the equator. In other words, the little changes within the length of the day caused by 2004 earthquakes are extraordinarily tiny to be determined.

How precisely did the earthquake modify the earth's spinning velocity? It is due to the fact that the earthquake convoluted plate convergence, and effectively diminished the Earth's equatorial circumference by a few millimeters whereas forceful denser material sinks into the Earth's equatorial bulge. In this regard, it also decreased the Earth's equatorial radius by a fraction of a millimeter. The usual fault slip used to be approximately 10 to 20 meters, despite the fact that some of that was pointed north-south; consequently, the east-west compression was smaller. The impact would be shrunk of the day length. In fact, the crust

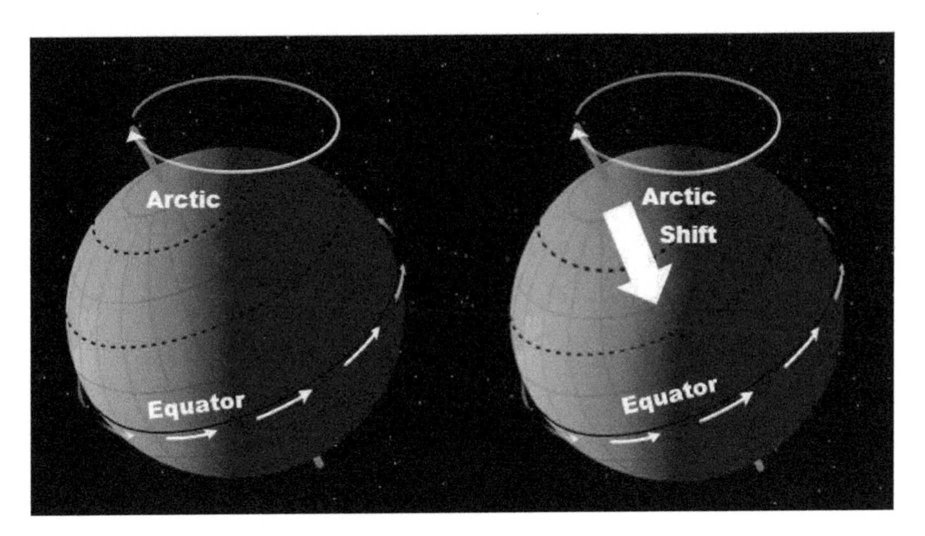

Figure 5.5 Pole shift hypotheses.

above the fault zone would have to impact the Earth's bulge, resulting in a bulge that would incredibly offset the reduction in circumference. Not all the nearby compression would influence the equator.

The Earth's rotational properties are pronounced by its moment of inertia (Figure 5.6), which is proportional to the square of its equatorial radius. The calculated change in rotation rate is about 3×10^{-11} of the length of the day, implying a similar change of 3×10^{-11} in the earth's equatorial radius (about 0.4 millimeters) or about 2.4 millimeters (1/10 inch) shortening of the equatorial circumference. For a rigid (i.e., non-bendy) object, the angular momentum is also proportional to the rotation rate of the object: if we double the speed of rotation (rotations/second) of a wheel, we double its angular momentum. If a wheel has no rotation, it has an angular momentum of zero.

Generally, angular momentum is a conserved quantity. In order to make an object spinning, we also have to expose an equal and contrary angular momentum to another object. This leads to some incredibly immaculate conclusions. For instance, every time the wheels of a bicycle are spinning, the angular momentum imparted to the wheels has been countered by using an equal and contrary angular momentum imparted to the Earth itself! Gratefully, the Earth is so massive and large (i.e., it has an excessive second of inertia) that the real rotation conveyed to the Earth is negligible.

Figure 5.6 presents the angular momentum of the bicycle, which is presented as coming out of the page, whereas the angular momentum of the Earth is shown going into the page. The angular momentum of a rigid body is mathematically described as having a "direction" that is perpendicular to the plane of rotation, and follows a right-hand rule [98] (Figure 5.7): curl the fingers of your right hand in the direction of rotation, and your thumb points in the defined direction of angular momentum. The complete angular momentum of a system of rotating bodies simply lines the arrows of the individual objects tip to tail. The net combined arrow represents the whole angular momentum. In the bicycle/Earth picture, there is no internet angular momentum, as the two arrows are equal and opposite.

In this understanding, angular momentum turns out to be clear. However, a few researchers at first imagined that it is incomprehensible for a feline to turn over in freefall. On the off chance that the feline is discharged with no underlying revolution, so the thinking goes, its precise force is zero. In freefall, there are no forces following up on the feline that can give it a net angular momentum. Since angular momentum energy is relative to the revolution speed, there is no chance for a feline to finish a 180 degree turn. So their (inaccurate) thinking goes!

In this view, Hopkin (2004) [96] claimed that the devastating earthquake that struck the Indian Ocean on 26 December was so effective that it has accelerated the Earth's rotation. Further, the shockwave shortened

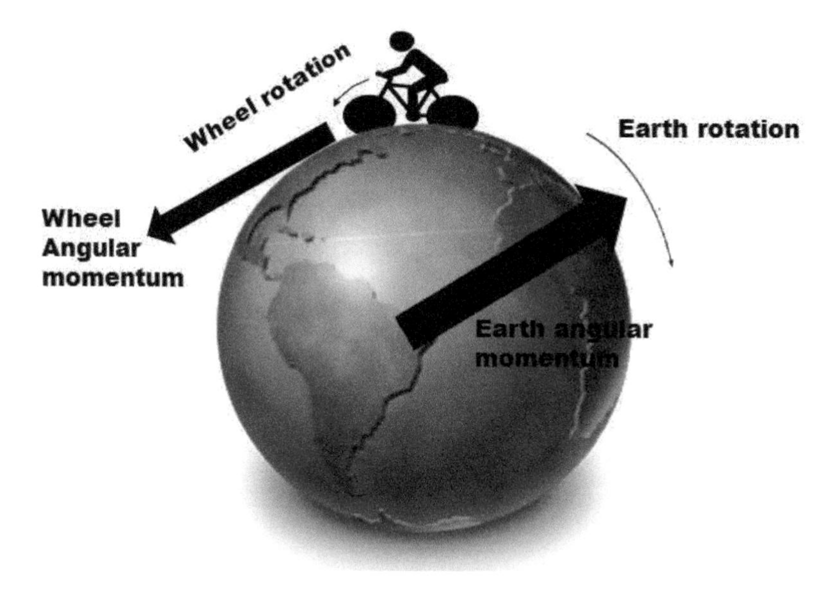

Figure 5.6 Earth's moment of inertia concept.

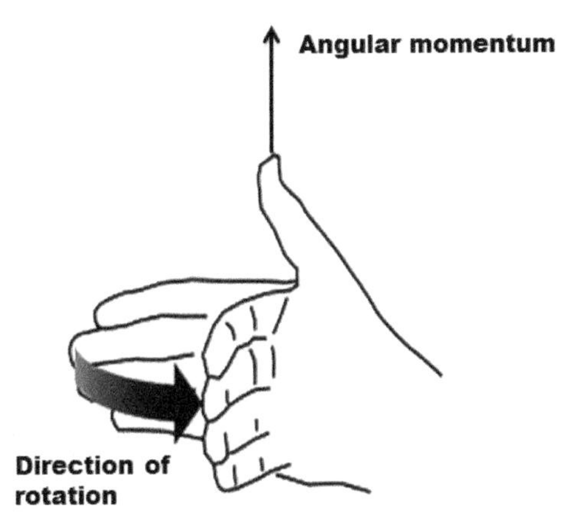

Figure 5.7 Right-hand rule.

the duration of our planet's rotation by some three microseconds [96]. Consequently, he mentioned that the alternate of the Earth's rotation was induced through a shift of mass towards the planet's center, as the Indian Ocean's heavy tectonic plate lurched beneath Indonesian one. This precipitated the Earth to rotate faster. The blast literally rocked the Earth on its axis, which made the Earth tilt an extra 2.5 centimetres in the wake of the jolt [96].

Lastly, Stevenson (2005) [276] stated that there would be no net tendency for great earthquakes to spin up Earth if Earth's great earthquake zones were randomly distributed. The zones, conversely, are not arbitrarily dispersed and their gravitational influences denote that Earth is not accidentally angled. According to the theory of true polar wander (Euler's equations with a small amount of dissipation), Earth always drifts toward the state in which the axis of maximum moment of inertia coincides with the Earth's revolution axis. Consequently, the growing impact of many large earthquakes causes the earth to spin up. On the contrary, the research done by convection between earthquakes retains everything in balance.

Comparative changes to the Earth's mass conveyance were computed from GPS information acquired amid the 2004 Sumatra quake and the 2010 Chile seismic tremor. On account of Sumatra, the alteration in the length of the day was bigger which was about 6.8 microseconds. However, for the Japan quake, it may be that the alteration in the Earth's wobble was more than twice as sprawling as those ascertained for the 2004 and 2010 events.

Generally, the length of day is shorten due to the effects of the massive destructive tsunami on the Earth's bulge. On the contrary, different sorts of seismic tremors, for instance, even strike-slip shudders, in which two plates slide on a level plane past each other, do not influence Earth's revolution.

Chapter 6

Principles of Optical Remote Sensing for Tsunami Observation

6.1 Introduction to Remote Sensing

Remote sensing technologies play a remarkable role in comprehending the mechanisms and characteristics of the 2004 Boxing Day tsunami. This was accomplished through exploitation of both optical and microwave remote sensing technologies. The main aim of this chapter is to clarify how remote sensing technologies are used to investigate the tsunami and its damage. Therefore, both techniques are mechanically connected to techniques that determine and reckon electromagnetic energy which has interacted with tsunami wave propagation, the atmosphere and materials that are made through tsunami influences. Understanding of electromagnetic (EM) spectrum and EM radiation, of that light, radar, and radio waves are based totally on the photoelectrical theory.

6.2 Electromagnetic Spectrum

The electromagnetic spectrum definitely refers to the wavelengths of light. In this regard, the electromagnetic spectrum is the time period used to describe the all-inclusive range of light that is existent. However, most of the light in a universe from radio waves to gamma rays, is vague to us! Light is a wave of irregular electric and magnetic fields. The transmission of light is not much exceptional than wave propagation in an ocean. A vital descriptive characteristic of a waveform is its wavelength or distance between succeeding peaks or troughs. In remote sensing, the wavelength is most frequently measured in micrometres, each of which equals one-millionth of a meter. The variation in wavelength of electromagnetic radiation is so significant that it is usually proven on a logarithmic scale (Figure 6.1) [98, 100].

Physically, the Earth is brightened through electromagnetic radiation from the Sun. The peak solar power is in the wavelength variation of visible light (between 0.4 and 0.7 μm) (Figure 6.1). Additional large components of incoming solar energy are in the configuration of invisible ultraviolet and infrared radiation. Merely tiny portions of solar radiation encompass the microwave spectrum. Imaging radar systems used in remote sensing create and transmit microwaves, and then measure the component of the signal that has backscattered to the antenna from the Earth's surface [99, 100]. The electromagnetic spectrum ordinarily breaks up into seven regions, in order of decreasing wavelength and growing energy and frequency: radio waves, microwaves, infrared, visible light, ultraviolet, X-rays and gamma rays.

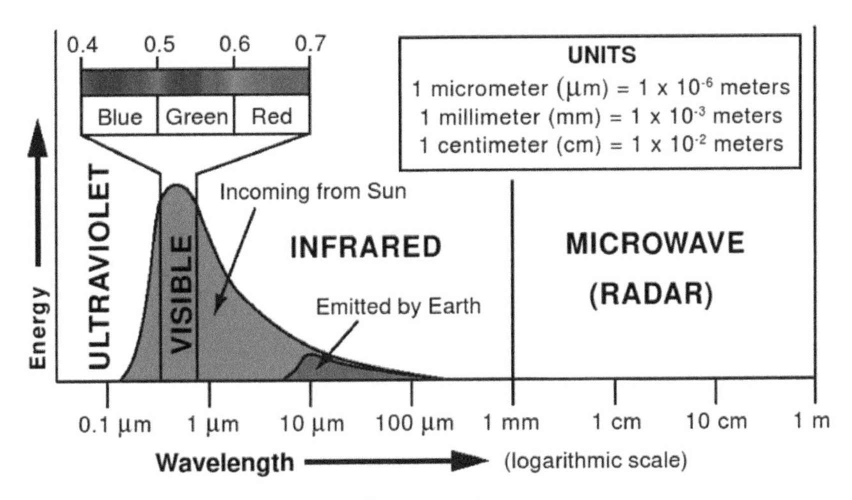

Figure 6.1 Electromagnetic wave spectra.

6.2.1 Radio Waves

Radio waves are at the lower range of the EM spectrum, with frequencies of up to about 30 billion hertz, or 30 gigahertz (GHz). These correspond to the wavelengths as low as 30 cm and as high as 1000 m. Radio waves are used specifically for communications, including voice, data and leisure media.

6.2.2 Microwaves

Microwaves fall in the range of the EM spectrum between radio and IR. They have frequencies from about 3 GHz up to about 30 trillion hertz, or 30 terahertz (THz), and wavelengths of about 10 mm (0.4 inches) to 100 micrometers (μm), or 0.004 inches. Microwaves are used for high-bandwidth communications, radar and as a source of warmth for microwave ovens and industrial applications.

6.2.3 Infrared

Infrared is in the range of the EM spectrum between microwaves and visible light. IR has frequencies from about 30 THz up to about 400 THz and wavelengths of about 100 μm (0.004 inches) to 740 nanometers (nm), or 0.00003 inches. IR light is invisible to human eyes, but we can feel it as heat if the intensity is sufficient.

6.2.4 Visible Light

Visible light is found in the middle of the EM spectrum, between the IR and UV. It has frequencies of about 400 THz to 800 THz and wavelengths of about 740 nm (0.00003 inches) to 380 nm (0.000015 inches). More generally, visible light is defined as the wavelengths that are visible to most human eyes.

6.2.5 Ultraviolet

Ultraviolet light is in the range of the EM spectrum between visible light and X-rays. It has frequencies of about 8×10^{14} to 3×10^{16} Hz and wavelengths of about 380 nm (0.000015 inches) to about 10 nm (0.0000004 inches). UV light is a component of sunlight; however, it is invisible to the human eye. It has numerous medical and industrial applications, but it can damage living tissue.

6.2.6 X-beams

X-beams are generally arranged into two sorts: soft X-beams and hard X-beams. The soft X-beams have frequencies of around 3×10^{16} to around 10^{18} Hz and wavelengths of approximately 10 nm (4×10^{-7} inches) to around 100 picometers (pm), or 4×10^{-8} inches. Hard X-beams possess an indistinguishable area of the EM of gamma beams. The only difference between them is their source: X-beams are delivered with the guide of quickening electrons, while gamma beams are created by nuclear cores (Table 6.1).

Table 6.1 Summary of principle of electromagnetic spectrum.

Spectra Wavelength	Description and Usages
Gamma rays	Gamma rays
X-rays	X-rays
Ultraviolet (UV) region 0.30 μm–0.38 μm (1 μm = 10^{-6} m)	This band is the violet portion of the visible wavelength, and hence its name. Approximately earth's surface material, primarily rocks and minerals, emanate visible UV emission. Nevertheless, UV emission is generally scattered by the earth's atmosphere and hence, not used in the field of remote sensing.
Visible Spectrum 0.4 μm–0.7 μm Violet 0.4 μm–0.446 μm Blue 0.446 μm–0.5 μm Green 0.5 μm–0.578 μm Yellow 0.578 μm–0.592 μm Orange 0.592 μm–0.62 μm Red 0.62 μm–0.7 μm	This is merely the portion of the spectrum that can be correlated with the concept of color. The color of an object is identified through the color of the reflected light.
Infrared (IR) Spectrum 0.7 μm–100 μm	Remote sensing is used by the reflected IR (0.7 μm–3.0 μm). Thermal IR (3 μm–35 μm) is the radiation radiated from the Earth's surface in the form of heat.
Microwave Region 1 mm–1 m	This is the longest wavelength used in remote sensing, which has the potential to penetrate through clouds.
Radio Waves (> 1 m)	This is the longest portion of the spectrum regularly used for meteorology and commercial broadcast.

6.2.7 Gamma-rays

Gamma-rays are in the range of the spectrum above soft X-rays. Gamma-rays have frequencies greater than about 10^{18} Hz and wavelengths of less than 100 pm (4×10^{-9} inches). Gamma radiation causes damage to living tissue, which makes it useful for killing cancer cells when applied in carefully measured doses to small regions. Uncontrolled exposure, though, is extremely dangerous to humans (Table 6.1).

Generally, the electromagnetic spectrum is composed of the low frequencies used for contemporary radio communication to gamma radiation at the short-wavelength (high-frequency) end, thereby covering wavelengths from thousands of kilometers down to a fraction of the dimension of an atom. Visible light lies towards the shorter end, with wavelengths from 400 to 700 nm. The restriction for long wavelengths is the dimension of the universe itself, while it is understood that the short wavelength constraint is in the vicinity of the Planck length which is a unit of length and equal to $1.616229(38) \times 10^{-35}$ meters. Until the middle of the 20th century, it used to be believed by most physicists that this spectrum was countless and continuous [98].

Most waves are both longitudinal and transverse. For instance, sound waves are longitudinal. Nonetheless, all electromagnetic waves are transverse. Moreover, EM is created through the movement of electrically charged particles. Consequently, EM can travel in a "vacuum" (they do NOT want a medium) at the speed of light, i.e., 300,000 km/sec in space [99, 100].

6.3 Energy in Electromagnetic Waves

The wave theory of EM radiation is adopted based on the concept of particle waves. Consequently, EM can behave like a wave or like a particle whereas a "particle" of light is called a photon. In other words, the insufficiency of the wave theory preceded a revival of the idea that EM radiation might better be thought of as particles, dubbed photons. The energy of a photon can mathematically be written as:

$$E = h * v \tag{6.1}$$

where E is energy, v is frequency for electromagnetic radiation, h is constant Planck's quantum of energy, and equals:

$$\begin{pmatrix} 6.626 \times 10^{-34} & \text{joule seconds} \\ 4.136 \times 10^{-15} & \text{eV seconds} \end{pmatrix}.$$

The electron volt (eV) is a convenient unit of energy related to the standard metric unit (joules, or J) by the relation 1 eV = 1.602×10^{-19} J. Photon energy E is determined by frequency of EM radiation: the higher the frequency, the higher the energy. Even though the photons propagate at the speed of light, they have zero rest mass, so the rules of special relativity are not ruined.
The relationship between wavelength (λ) and frequency (v) for electromagnetic radiation is formulated as:

$$\lambda * v = c \tag{6.2}$$

From these relationships, we can define the relationship between energy and wavelength as:

$$E = h * \frac{c}{\lambda} \tag{6.3}$$

or, rearranging as:

$$\lambda = h * \frac{c}{E} \tag{6.4}$$

The relationship between wavelength (λ) and momentum (m*v) for DeBroglie's "particle wave" is determined from:

$$\lambda = \frac{h}{\mathrm{m} * \mathrm{v}} \tag{6.5}$$

From the above relationships, we can calculate the relationship between energy (E) and momentum (m*v) as:

$$\frac{h}{\mathrm{m} * \mathrm{v}} = h * \frac{c}{E} \tag{6.6}$$

Simplify, and solve for E:

$$E = \mathrm{m} * \mathrm{v} * c \tag{6.7}$$

The highest velocity (v) attainable by matter is the speed of light (c), therefore, the maximum energy would seem to be:

$$E = \mathrm{m} * c * c \tag{6.8}$$

or

$$E = \mathrm{m} * c^2 \tag{6.9}$$

Equation 6.5 is the DeBroglie equation where photons behave like matter particles and extend it to all particles. Consequently, Equation 6.7 relates the wave like behavior of matter to its momentum. Figure 6.2 summarizes the concepts of EM energy spectra. For instance, visible-light wavelengths correspond to a wavelength range from 0.38–0.75 µm of 2–3 eV.

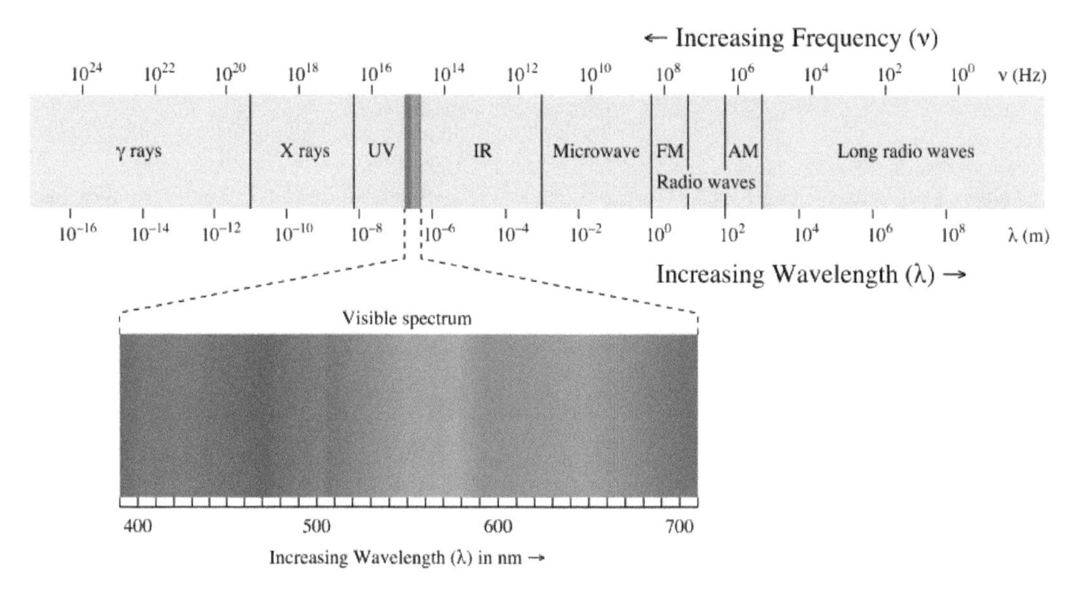

Figure 6.2 The spectrum of electromagnetic radiation.

6.4 Photoelectric Effect

The concept of energy in photons is very important for remote sensing technology. The photoelectric effect was the first example of a quantum phenomenon to be seen at the end of the 19th Century. Light or other forms of electromagnetic radiation shone onto metals release electrons; the energy supplied by the radiation frees the electrons from the metal. The way in which the numbers and energies of electrons released changes when the frequency and intensity of the radiation changes cannot be explained using the classical wave model of light. Increasing the intensity of the radiation does not increase the energy of the electrons, but releases more of them per second. Increasing the frequency of the light increases the energy of the electrons. Below a certain frequency of radiation, v_0, no electrons are emitted no matter how intense the radiation (Figure 6.3).

These facts are explained using the photon model of light. Light (and all EM radiation) is emitted and absorbed in little packets or quanta called photons. The energy of a photon is equal to its frequency multiplied by Planck's constant (Equation 6.1), $h = 6.63 \times 10^{-34}$ Js.

KE_{max} of electrons = $hv-\Phi$ where Φ is the energy needed to escape from the metal. This phenomenon introduces the **wave-particle duality** of nature: light behaves as a wave at times (e.g., Young's slits) and

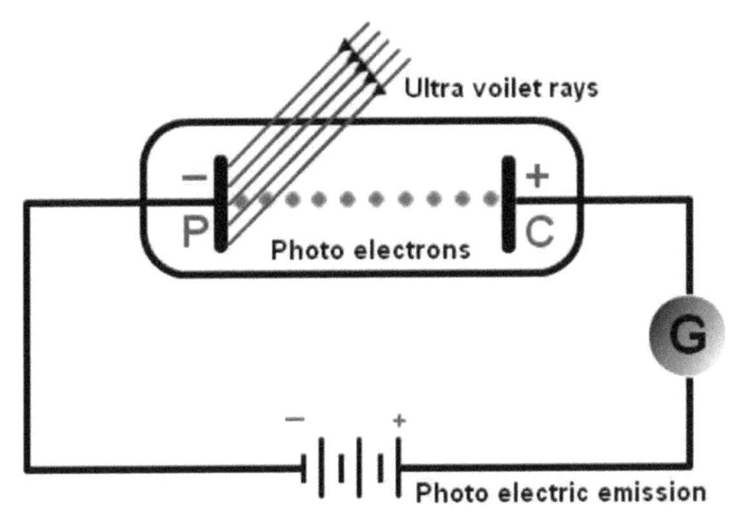

Figure 6.3 Photoelectric concept.

as a particle at times. This duality is central to the way quantum mechanics explains nature as it applies to everything.

6.5 Young's Slits

Thomas Young used this experiment to 'prove' that light was a wave at a time when light was thought to be a particle (Figure 6.4). The light going through two slits interferes and produces a pattern that is easy to explain using a wave model, but which cannot be explained if light acts like particles.

Figure 6.4 Young's slits.

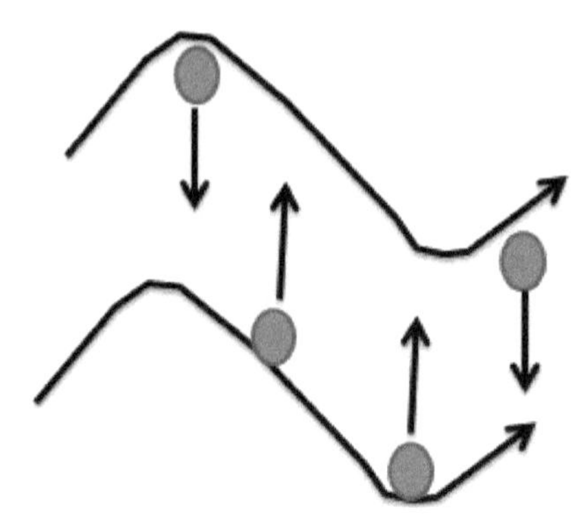

Figure 6.5 Sketch of photon behavior.

Wave particle duality: If light was just a wave, then the electrons would absorb some energy no matter what the frequency. If light was just a wave, then the emission of an electron would take longer when a lower intensity light was used, not instantaneously.

Nonetheless, light is not just a wave: It can also behave as though it was made of tiny energy packets or particles. We call these particles, **photons** (Figure 6.5). It is one of these photons that will hit one electron on the plate, the electron will absorb the energy and it will fly off the plate. So if the intensity is greater, i.e., there are more photons, then more electrons can be knocked off.

6.6 Electromagnetic-radiation-matter Interactions

The main question that arises is how tsunami waves and their effects have been imagined from space? To answer this question, we need to understand how remote sensing sensors can imagine this phenomenon which mainly is a function of the interaction of EM radiations with tsunami waves and the debris remains after the disaster. From the point view of optical remote sensing, two important issues must be well understood: (i) EM-radiation-matter interactions; and (ii) blackbody radiation [101–103].

Remote sensors determine electromagnetic (EM) radiation that has intermingled with the Earth's surface. Interactions with matter can exchange the bearing, intensity, wavelength content, and polarization of EM radiation. The nature of these changes is reliant on the chemical makeup and physical structure of the material exposed to the EM radiation. Changes in EM radiation resulting from its interactions with the Earth's surface therefore provide major clues to the characteristics of the surface materials. Upon striking matter, EM may be transmitted, reflected, scattered, or observed in proportion that depends upon: (i) the combinational and physical properties of the medium; the wavelength of the frequency of the incident radiation; and the angle at which the incident radiation strikes a surface. In other words, the total amount of radiant flux in specific wavelengths (λ) incident to the terrain (Φ_{i_λ}) must be accounted for by evaluating the amount of radiant flux reflected from the surface ($\Phi_{reflected\lambda}$), the amount of radiant flux absorbed by the surface ($\Phi_{absorbed\lambda}$), and the amount of radiant flux transmitted through the surface ($\Phi_{transmitted\lambda}$) [103]:

$$\Phi_{i\lambda} = \Phi_{reflected\lambda} + \Phi_{absorbed\lambda} + \Phi_{transmitted\lambda} \tag{6.10}$$

The radiation passes through a substance without significant attenuation. Transmission through material media of different densities (e.g., air, water) causes radiation to be reflected or deflected from a straight-line path with an accompanying change in its velocity and wavelength; the frequency always remains constant [102].

Let us take the incident beam of EM at an angle θ_i which is deflected toward the normal in going beam from a low density medium to a denser one at an angle θ_t. Emerging from the far side of the denser medium, the beam is refracted from the normal at an angle θ_r. The angle relationships in Figure 6.6 are

$\theta_i > \theta_t$ and $\theta_i = \theta_r$ (Figure 6.6). The change in EMR velocity is explained by index of reaction n, which is the ratio between the velocity of EMR in vacuum c and its velocity in a material medium V [102]:

$$n = \frac{c}{V} \tag{6.11}$$

The index of refraction of a vacuum (perfectly transparent medium) is equal to 1, or unity. V has never been greater than c, and n can never be less than 1 for any substance. Indices of refraction vary from 1.0002926 (for the earth's atmosphere) to 1.33 (for water) and 2.42 (for a diamond). The index of refraction leads to Snell's law [102]:

$$n_1 \sin\theta_i = n_2 \sin\theta_t \tag{6.12}$$

Figure 6.7 shows the EMR refraction behavior. Reflection is also known as specular reflection. It describes the process whereby incident radiation bounces off the surface of a substance in single predictable direction. The angle of reflection is always equal and opposite to angle of incidence $\theta_i = \theta_r$.

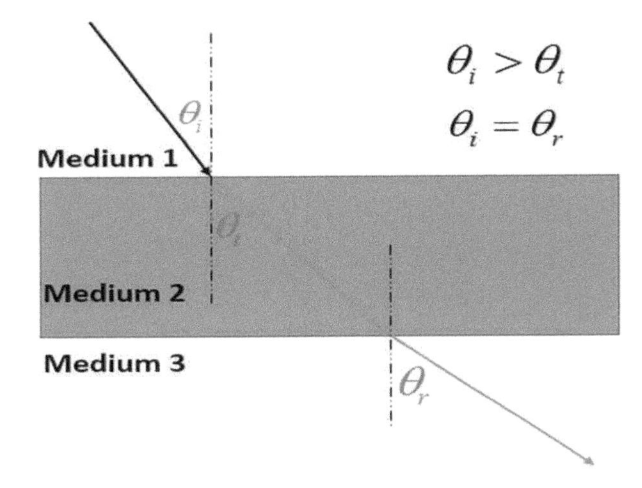

Figure 6.6 Transmission and refraction of EMR.

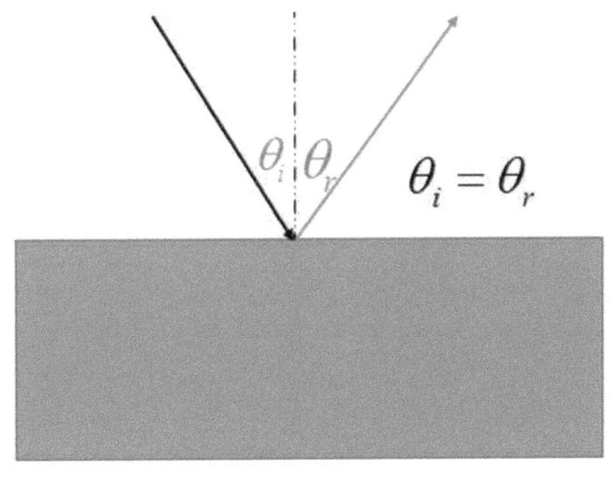

Figure 6.7 EMR reflection.

Reflection is caused by surfaces that are smooth to the wavelengths of incident radiation. These smooth, mirror-like surfaces are called specular reflectors. Therefore, specular reflection causes no change to either the EMR velocity or wavelength. According to Hecht (2001) [101], the theoretical amplitude reflectance of a dielectric interface can be derived from electromagnetic theory. In this regard, $\vec{E}$ polarized perpendicular to the plane of incidence:

$$r = \frac{n_1 \cos\theta_i - n_2 \cos\theta_r}{n_1 \cos\theta_i + n_2 \cos\theta_r}; \tag{6.13}$$

For $\vec{E}$ polarized parallel to the plane of incidence then:

$$r = \frac{n_2 \cos\theta_i - n_1 \cos\theta_r}{n_2 \cos\theta_i + n_1 \cos\theta_r}; \tag{6.14}$$

where, n_1, θ_i, n_2, and θ_r are the refractive indices and angles of incidences and refraction, respectively. Here r is the ratio of the amplitude of the reflected electric field to incident field. Consequently, the intensity of the reflected EMR is the square of this value [102].

Once electromagnetic radiation is created, it is transmitted through the earth's atmosphere almost at the speed of light in a vacuum. In other words, in a vacuum, electromagnetic radiation of short wavelengths travels as fast as radiation of long wavelengths. Nevertheless, the atmosphere might disturb not only the velocity of EMR but also its wavelength, intensity, spectral distribution, and/or direction. However, scatter differs from reflection in that the direction associated with scattering is unpredictable, whereas the direction of reflection is predictable. There are essentially three sorts of scattering: (i) Rayleigh; (ii) Mie; and (iii) Non-selective (Figure 6.8).

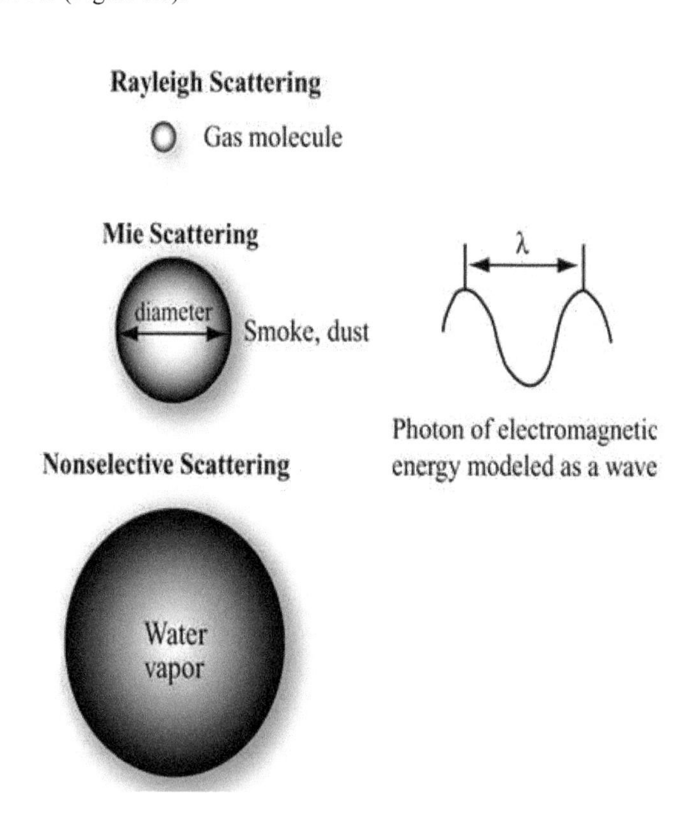

Figure 6.8 EMR scattering [104].

Rayleigh scattering arises when the diameter of the matter (usually air molecules) is many times smaller than the wavelength of the incident electromagnetic radiation. This type of scattering is named after the English physicist who offered the first coherent explanation for it. All scattering is accomplished through absorption and re-emission of radiation by atoms or molecules in the manner described in the discussion on radiation from atomic structures. It is impossible to predict the direction in which a specific atom or molecule will emit a photon, hence scattering.

The energy required to excite an atom is associated with short-wavelength, high frequency radiation. The amount of scattering is inversely related to the fourth power of the radiation's wavelength. For example, blue light (0.4 mm) is scattered 16 times more than near-infrared light (0.8 mm). The intensity of Rayleigh scattering varies inversely with the fourth power of the wavelength (λ^{-4}) [104].

Mie scattering takes place when there are essentially spherical particles present in the atmosphere with diameters approximately equal to the wavelength of radiation being considered. For visible light, water vapor, dust, and other particles ranging from a few tenths of a micrometer to several micrometers in diameter are the main scattering agents. The amount of scatter is greater than Rayleigh scattering and the wavelengths scattered are longer. Pollution also contributes to beautiful sunsets and sunrises. The greater the amount of smoke and dust particles in the atmospheric column, the more violet and blue light will be scattered away and only the longer orange and red wavelength light will reach our eyes [104].

Non-selective scattering is produced when there are particles in the atmosphere several times the diameter of the radiation being transmitted. This type of scattering is non-selective, i.e., all wavelengths of light are scattered, not just blue, green, or red. Thus, water droplets, which make up clouds and fog banks, scatter all wavelengths of visible light equally well, causing the cloud to appear white (a mixture of all colors of light in approximately equal quantities produces white). Scattering can severely reduce the information content of remotely sensed data to the point that the imagery loses contrast and it is difficult to differentiate one object from another. Non-selective scattering is a function of: (i) the wavelength of the incident radiant energy; and (ii) the size of the gas molecule, dust particle, and/or water vapor droplet encountered [104].

Absorption is the process by which radiant energy is absorbed and converted into other forms of energy. An absorption band is a range of wavelengths (or frequencies) in the electromagnetic spectrum within which radiant energy is absorbed by substances such as water (H_2O), carbon dioxide (CO_2), oxygen (O_2), ozone (O_3), and nitrous oxide (N_2O). The cumulative effect of the absorption of the various constituents can cause the atmosphere to close down in certain regions of the spectrum. This is bad for remote sensing because no energy is available to be sensed.

In certain parts of the spectrum, such as the visible region (0.4–0.7 mm), the atmosphere does not absorb all of the incident energy but transmits it effectively. Parts of the spectrum that transmit energy effectively are called "atmospheric windows".

Absorption occurs when the energy of the same frequency as the resonant frequency of an atom or molecule is absorbed, producing an excited state. If, instead of re-radiating a photon of the same wavelength, the energy is transformed into heat motion and is recruited at a longer wavelength, absorption occurs. When dealing with a medium like air, absorption and scattering are frequently combined into an extinction coefficient.

Transmission is inversely related to the extinction coefficient of the thickness of the layer. Certain wavelengths of radiation are affected far more by absorption than by scattering. This is particularly true of infrared and wavelengths shorter than visible light [104]. Following Jensen [104], the *absorption* of the Sun's incident electromagnetic energy in the region from 0.1 to 30 mm is caused by various atmospheric gases (Figure 6.9). The first four graphs depict the absorption characteristics of N_2O, O_2 and O_3, CO_2, and H_2O, while the final graph depicts the cumulative result of all these constituents being in the atmosphere at one time. The atmosphere essentially "closes down" in certain portions of the spectrum while "atmospheric windows" exist in other regions that transmit incident energy effectively to the ground. It is within these windows that remote sensing systems must function. The combined effects of atmospheric absorption, scattering, and reflectance reduces the amount of solar irradiance reaching the Earth's surface at sea level.

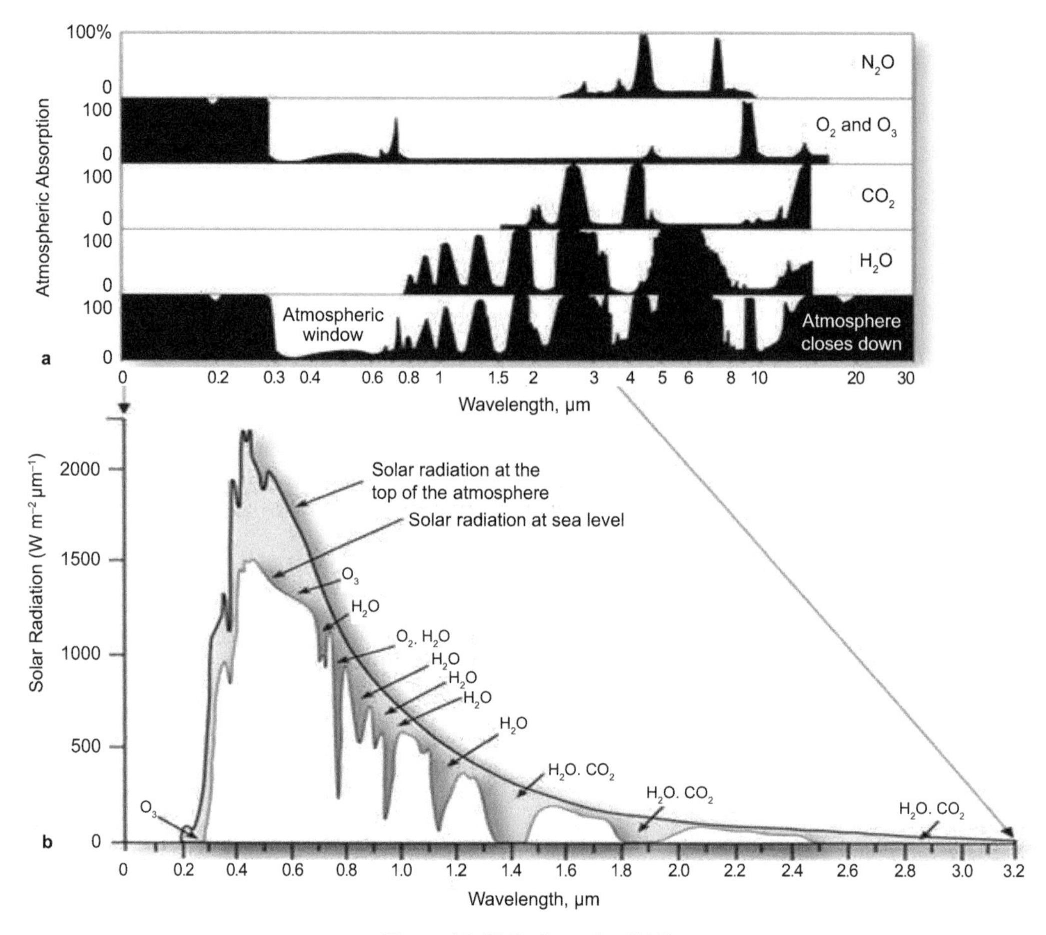

Figure 6.9 EMR absorption [104].

6.7 Interaction Processes on Remote Sensing

To understand how different interaction processes impact the acquisition of aerial and satellite images, let's analyze the reflected solar radiation that is measured at a satellite sensor. As sunlight initially enters the atmosphere, it encounters gas molecules, suspended dust particles, and aerosols. These materials tend to scatter a portion of the incoming radiation in all directions, with shorter wavelengths experiencing the strongest effect. The preferential scattering of blue light in comparison to green and red light accounts for the blue color of the daytime sky. Clouds appear opaque because of intense scattering of visible light by tiny water droplets. Although most of the remaining light is transmitted to the surface, some atmospheric gases are very effective at absorbing particular wavelengths. The absorption of dangerous ultraviolet radiation by ozone is a well-known example. As a result of these effects, the illumination reaching the surface is a combination of highly filtered solar radiation transmitted directly to the ground and a more diffused light scattered from all parts of the sky, which helps illuminate shadowed areas (Figure 6.10).

As this altered sunlight based radiation reaches the ground, it might experience soil, shaking surfaces, vegetation, or different materials that absorb a bit of the radiation. The measure of vitality assimilated fluctuates in wavelength for every material naturally, making a kind of spectral signature. The vast majority of the radiation not ingested is diffusely reflected (scattered) moving down into the environment, some of it toward the satellite. Consequently, the upwelling radiation experiences a further round of dissemination and assimilation as it goes through the climate before being distinguished and measured by the sensor.

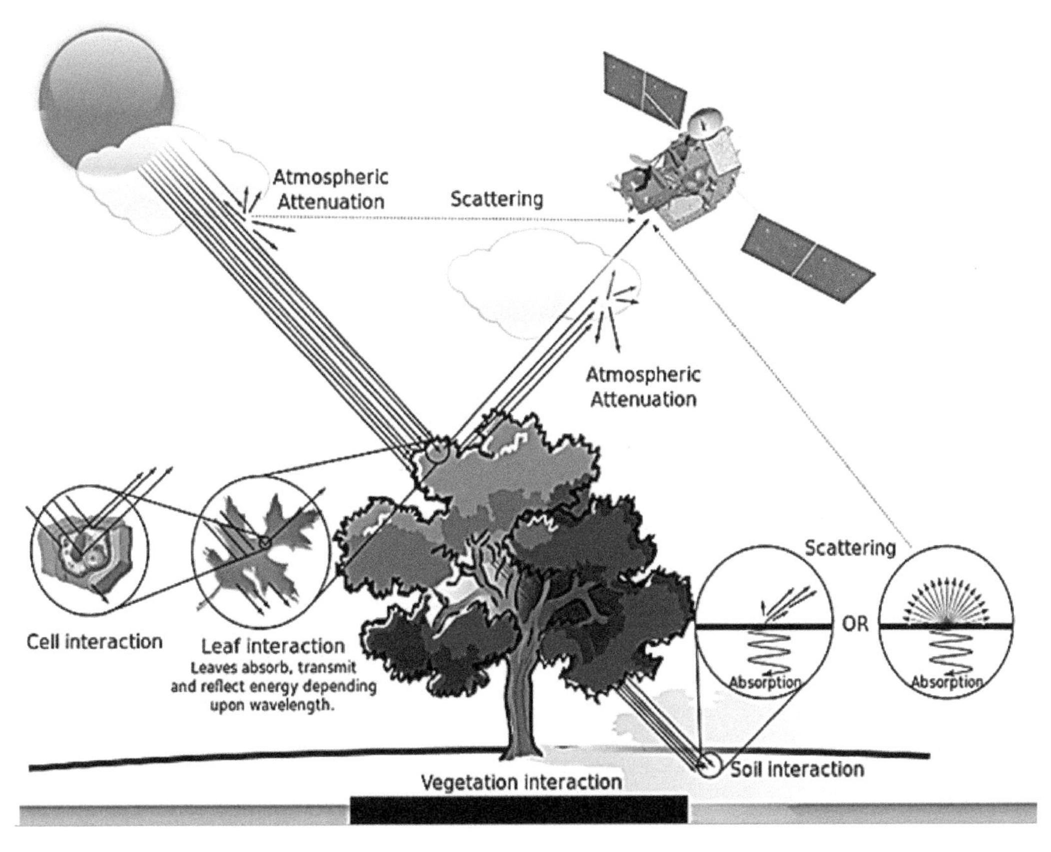

Figure 6.10 Typical EMR interactions in the atmosphere and at the Earth's surface.

In the event that the sensor is equipped for identifying thermal infrared radiation, it will likewise acquire radiation transmitted by surface materials as a result of solar heating.

6.8 Black Body Radiation

A black body is a sentimentalized physical body which retains all occurrences of electromagnetic beams, irrespective of frequency or angle of incidence. A white body is unified with a coarse surface which mirrors all occurrences beams absolutely and consistently in all paths. Black body radiations are transmitted by hot solids, fluid, or thick gases and has a persistent dispersion of emanated wavelength, as shown in Figure 6.11.

The curve in this figure gives the radiance L in the following dimensions:

$$\frac{Power}{unit\ \text{area.wavelength.solid angle}}; \tag{6.15}$$

Or units of watts/(m^2 μ ster). The radiance equation is

$$Radiance = L = \frac{2hc^2}{\lambda^5} \cdot \frac{1}{e^{\frac{hc}{\lambda kT}} - 1}; \tag{6.16}$$

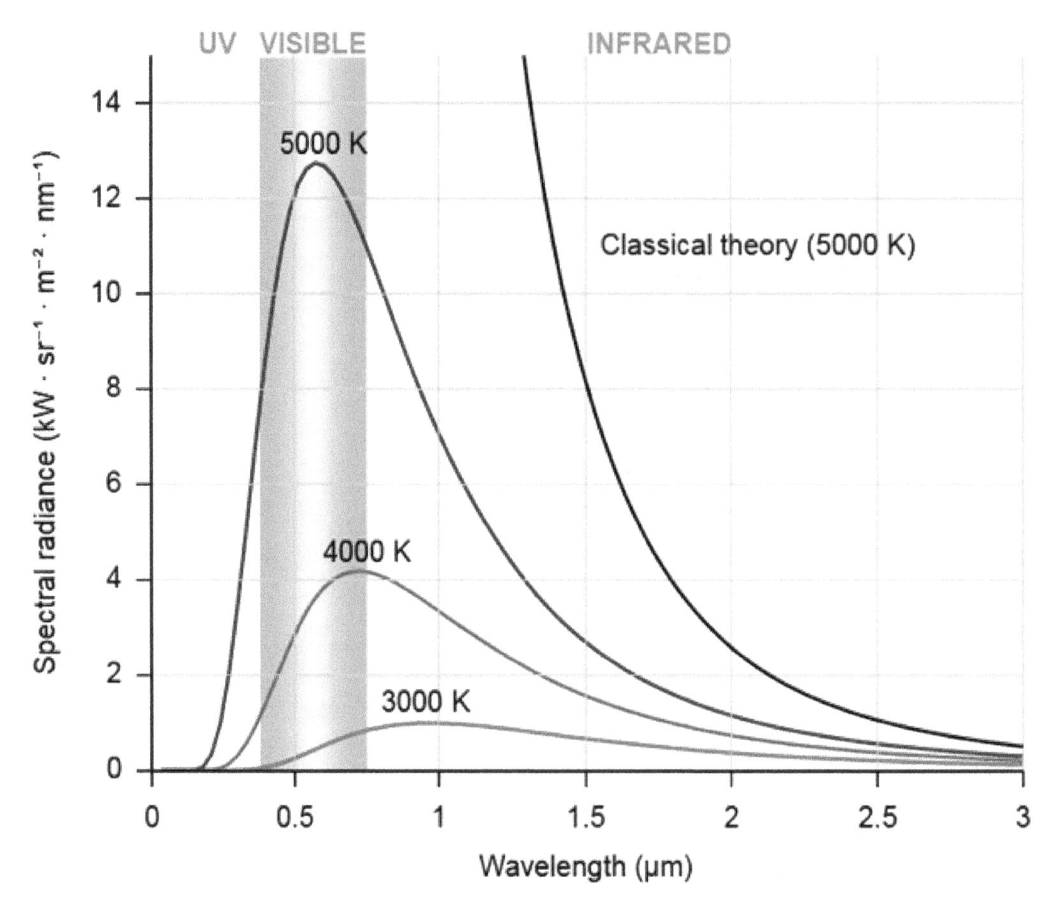

Figure 6.11 Temperature of a black body decreases, its intensity also decreases and its peak moves to longer wavelengths.

Here, $c = 3 \times 10^8$ m/s, $h = 6.626 \times 10^{-34}$ joules per second (J/s), and $k = 1.38 \times 10^{-23}$ joules per Kelvin (J/K). Real materials will differ from the idealized blackbody in their emission of radiation. The emissivity of the surface is a measure of the efficiency with which the surface absorbs (or radiates) energy and lies between 0 (for perfect reflector) and 1 (for a perfect absorber). A body that has $\varepsilon = 1$ is called a "black" body. In the infrared, many objects are nearly blackbodies-in particular, vegetation. Materials with $\varepsilon < 1$ are called gray bodies. Emissivity ε will vary with wavelength [102].

Another form of Planck's law can be given by

$$Radiant\ exitance = M = \frac{2\pi hc^2}{\lambda^5} \cdot \frac{1}{e^{\frac{hc}{\lambda kT}} - 1} \cdot \frac{watts}{m^2 \mu m}. \tag{6.17}$$

The difference is that the dependence on the angle of the emitted radiation has been removed by integrating of the solid angle. This can be done for blackbodies because they are "Lambertian" surface by definition-the emitted radiation does not rely on the angle, and $M = \pi L$ [102]. The power radiated is given by the Stefan-Boltzmann law:

$$R = \sigma \varepsilon T^4 \ \mathrm{W/m^2}, \tag{6.18}$$

where, R is the power radiated per square meter, ε is the emissivity, $\sigma = 5.67 \times 10^{-8}$ W/m^2 K^4 which is Stefan's constant, and T is the temperature of the radiator in K [102]. Wien's displacement law gives the wavelength at which the peak in radiation occurs:

$$\lambda_{max} = 2.898x10^{-3}\,(m/K)T^{-1} \tag{6.19}$$

Wien constant is $2.898x10^{-3}$ (m/K) for known temperature T in K, and λ_{max} is in meters.

6.9 Spectral Signatures

The spectral signatures produced by wavelength-dependent absorption provide the key to discriminating different materials into images of reflected solar energy. The property used to quantify these spectral signatures is called *spectral reflectance*: the ratio of reflected energy to incident energy as a function of wavelength. The spectral reflectance of different materials can be measured in the laboratory or in the field, providing reference data that can be used to interpret images. As an example, the illustration below shows contrasting spectral reflectance curves for three very common natural materials: dry soil, green vegetation, and water [103].

The reflectance of dry soil rises uniformly through the visible and near infrared wavelength ranges, peaking in the middle infrared range. It shows only minor dips in the middle infrared range due to absorption by clay minerals. Green vegetation has a very different spectrum. Reflectance is relatively low in the visible range, but is higher for green lighter than for red or blue, producing the green color we see. The reflectance pattern of green vegetation in the visible wavelengths is due to selective absorption by chlorophyll, the primary photosynthetic pigment in green plants [101]. The most noticeable feature of the vegetation spectrum is the dramatic rise in reflectance across the visible-near infrared boundary, and the maximum peak of near infrared reflectance. Infrared radiation penetrates plant leaves, and is intensely scattered by the leaves' complex internal structure, resulting in high reflectance [102]. The dips in the middle infrared portion of

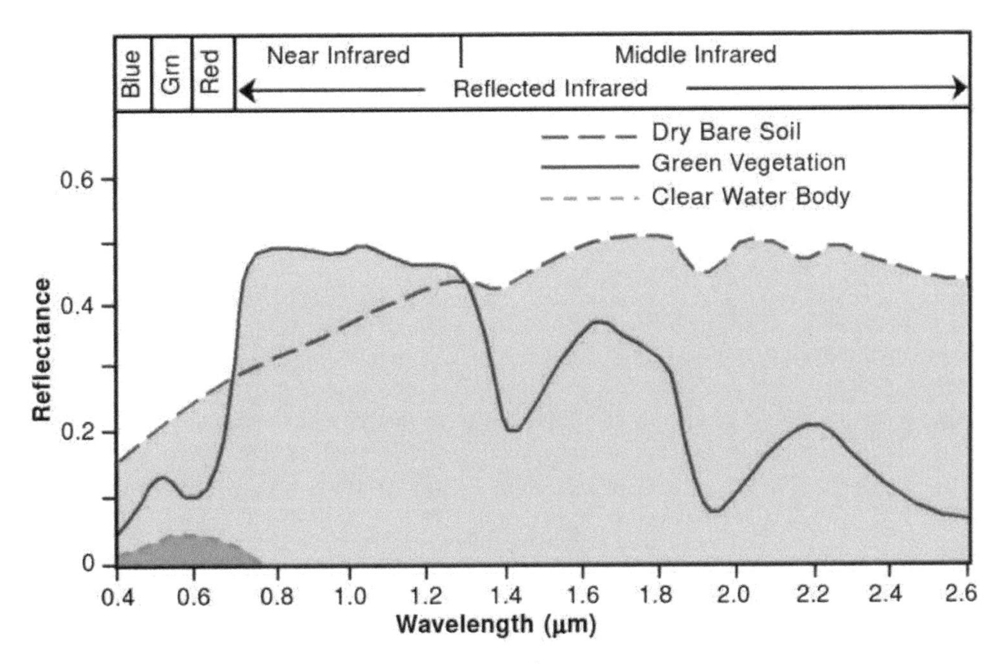

Figure 6.12 Spectra signature curve.

the plant spectrum are due to absorption by water. Deep clear water bodies effectively absorb all wavelengths longer than the visible range, which results in very low reflectivity for infrared radiation [100].

6.10 Spatial Dimension

The spatial dimension plays a great role for different satellite remote sensing to capture the clues about the 2004 Boxing Day tsunami, particularly, the three dimensions associated with remote sensing imagery: (i) spectral resolution; (ii) spatial resolution; and temporal resolution (Figure 6.13). In this regard, the three dimensions identify competing requirements for design and operation.

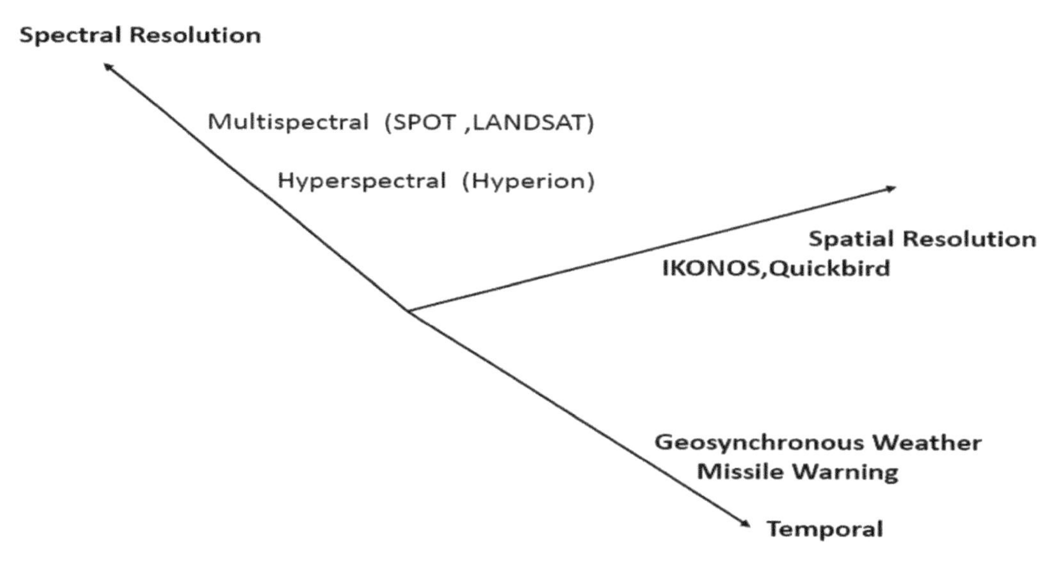

Figure 6.13 Three dimensions for remote sensing.

6.10.1 Spectral Resolution

The *spectral resolution* of a remote sensing system can be described as its capability to differentiate dissimilar parts of the range of determined wavelengths. In other words, spectral resolution describes the ability of a sensor to define fine wavelength intervals. The finer the spectral resolution, the narrower the wavelength range for a particular channel or band. An "image" produced by a sensor system can contain one very wide-ranging wavelength band, a few broad bands, or many narrow wavelength bands. The names usually used for these three image categories are *panchromatic*, *multispectral*, and *hyperspectral*, respectively [102].

Multispectral images determine even a finer variety of spectrum or wavelength that is required to replicate color. That is why the multispectral data have a greater spectral resolution than regular color data. Spectral resolution is the potential to unravel spectral features and bands into their separate components. The spectral resolution required through the analyst or researcher relies upon the utility involved. For instance, activity analysis for fundamental pattern identification generally requires low/medium resolution.

The multispectral bands are listed as follows:

- **Blue**, 450–515 and 520 nm, is used for atmosphere and deep water imaging, and can reach depths up to 150 feet (50 m) in clear water.
- **Green**, 515..520–590..600 nm, is used for imaging vegetation and deep water structures, up to 90 feet (30 m) with clear water.

- **Red**, 600..630–680..690 nm, is used for imaging man-made objects, in water up to 30 feet (9 m) deep, soil, and vegetation.
- **Near infrared**, 750–900 nm, is used primarily for imaging vegetation.
- **Mid-infrared**, 1550–1750 nm, is used for imaging vegetation, soil moisture content, and some forest fires.
- **Far-infrared**, 2080–2350 nm, is used for imaging soil, moisture, geological features, silicates, clays, and fires.
- **Thermal infrared**, 10400–12500 nm, uses emitted instead of reflected radiation to image geological structures, thermal differences in water currents, and fires, and for night studies.

The HRV sensor aboard the French SPOT (Système Probatoire d'Observation de la Terre) 1, 2, and 3 satellites (20 meter spatial resolution) has this design. Color-infrared film used in some aerial photography provides similar spectral coverage, with the red emulsion recording near infrared, the green emulsion recording red light, and the blue emulsion recording green light. The IKONOS satellite from Space Imaging (4-meter resolution) and the LISS II sensor on the Indian Research Satellites IRS-1A and 1B (36-meter resolution) add a blue band to provide complete coverage of the visible light range, and allow natural-color band composite images to be created. The Landsat Thematic Mapper (Landsat 4 and 5) and Enhanced Thematic Mapper Plus (Landsat 7) sensors add two bands in the middle infrared (MIR).

Like other spectral sensors, Hyperspectral sensor collects and processes data from throughout the electromagnetic spectrum (Figure 6.14). The conception of hyperspectral sensor is to attain the spectrum for every pixel in the data of a scene, with the persistence of determining objects, identifying materials, or designing an algorithm [105]. Even though the human eye realizes the color of visible light with commonly three bands (red, green, and blue), spectral imaging divides the spectrum into many extra bands. This method of dividing data into bands can be extended beyond the visible. In hyperspectral imaging, the recorded spectra have fine wavelength resolution and cover a wide range of wavelengths [106].

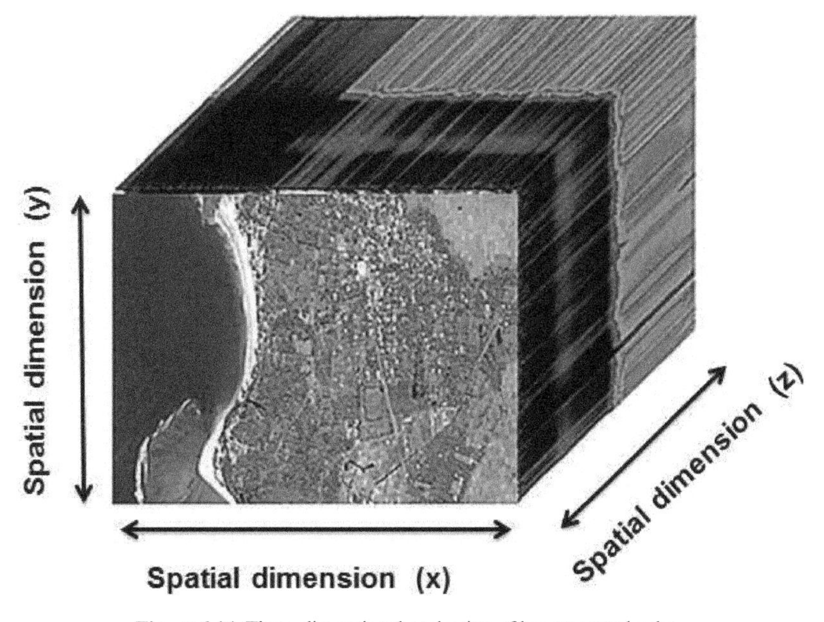

Figure 6.14 Three-dimensional projection of hyperspectral cube.

6.10.2 Spatial Resolution

Spatial Resolution pronounces how much information in a remote sensing data is seen with the human eye. The potential to resolve or separate small details are one way of describing the ground target and is known as spatial resolution. Spatial resolution is a measure of the spatial detail in an image, which is a feature of the design of the sensor and its operating altitude above the surface. Each of the detectors with a remote sensor measures energy acquired from a finite patch of the Earth's surface. The smaller these individual patches are, the greater the detail which can be inferred from the data. For digital images, spatial resolution is most frequently expressed as the ground dimensions of an image pixel.

Further, it relies mainly upon their Instantaneous Field of View (IFOV). The IFOV is the angular cone of visibility of the sensor (A) and determines the area on the Earth's surface which is "seen" from a given altitude at one unique moment in time (B). The dimension of the location considered is determined by multiplying the IFOV by means of the distance from the ground to the sensor (C) (Figure 6.15). This location on the ground is referred to as the resolution cell and determines a sensor's more spatial resolution. For a homogeneous feature to be detected, its dimension usually has to be equal to or larger than the resolution cell. If the feature is smaller than this, it may not be detectable as the common brightness of all features in that resolution cell will be recorded. Nevertheless, smaller features may also on occasion be detectable if their reflectance dominates within an articular resolution cell, permitting sub-pixel or resolution cell detection [100, 103].

Shape is one visual factor that we can use to recognize and identify objects in an image. Shape is usually discernible only if the object dimensions are several times larger than the cell dimensions. On the other hand, objects smaller than the image cell size may be detectable in an image. If such an object is sufficiently brighter or darker than its surroundings, it will dominate the average brightness of the image cell it falls within, and that cell will contrast in brightness with the adjacent cells.

Finally, there are three types of spatial resolution (i) low resolution (Figure 6.16); (ii) moderate spatial resolution satellite data (Figure 6.14); and (iii) high resolution (Figure 6.14). The low resolution ranges between 30 – > 1000 m while high resolution ranges from 0.41–4 m.

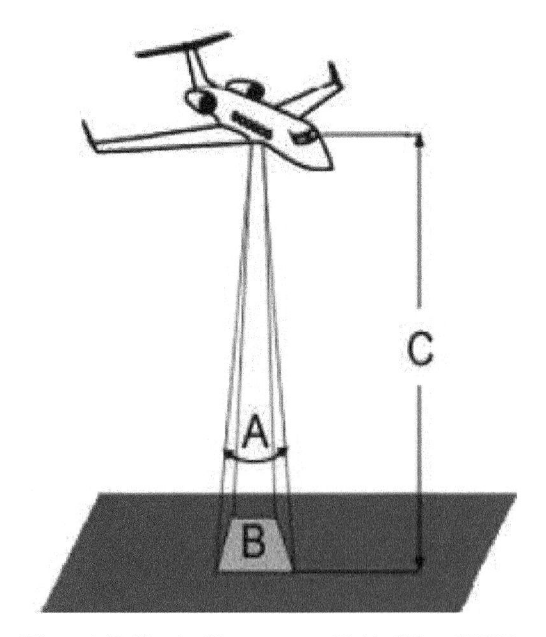

Figure 6.15 Sketch of Instantaneous Field of View (IFOV).

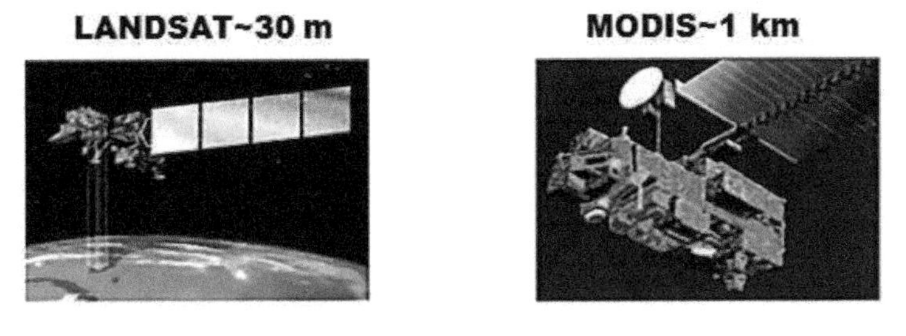

Figure 6.16 Example of low spatial resolution satellite sensors [http://www.satimagingcorp.com/satellite-sensors/].

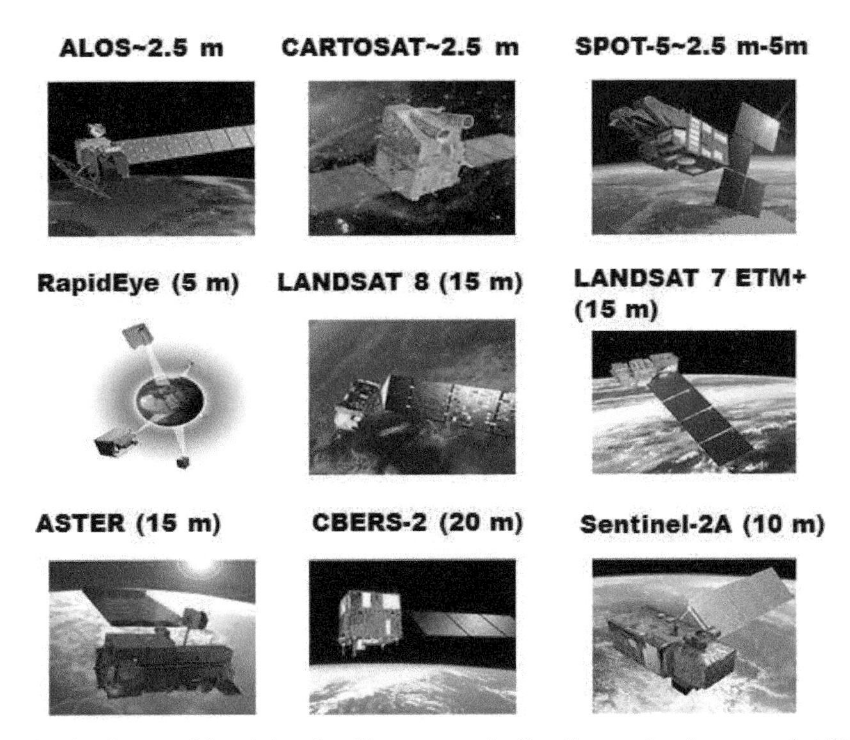

Figure 6.17 List of moderate spatial resolution of satellite remote sensing [http://www.satimagingcorp.com/satellite-sensors].

6.10.3 Temporal Resolution

Temporal resolution (TR) denotes the precision of a dimension as a function of time. In other words, the temporal resolution specifies the revisiting frequency of a satellite sensor for a particular location.

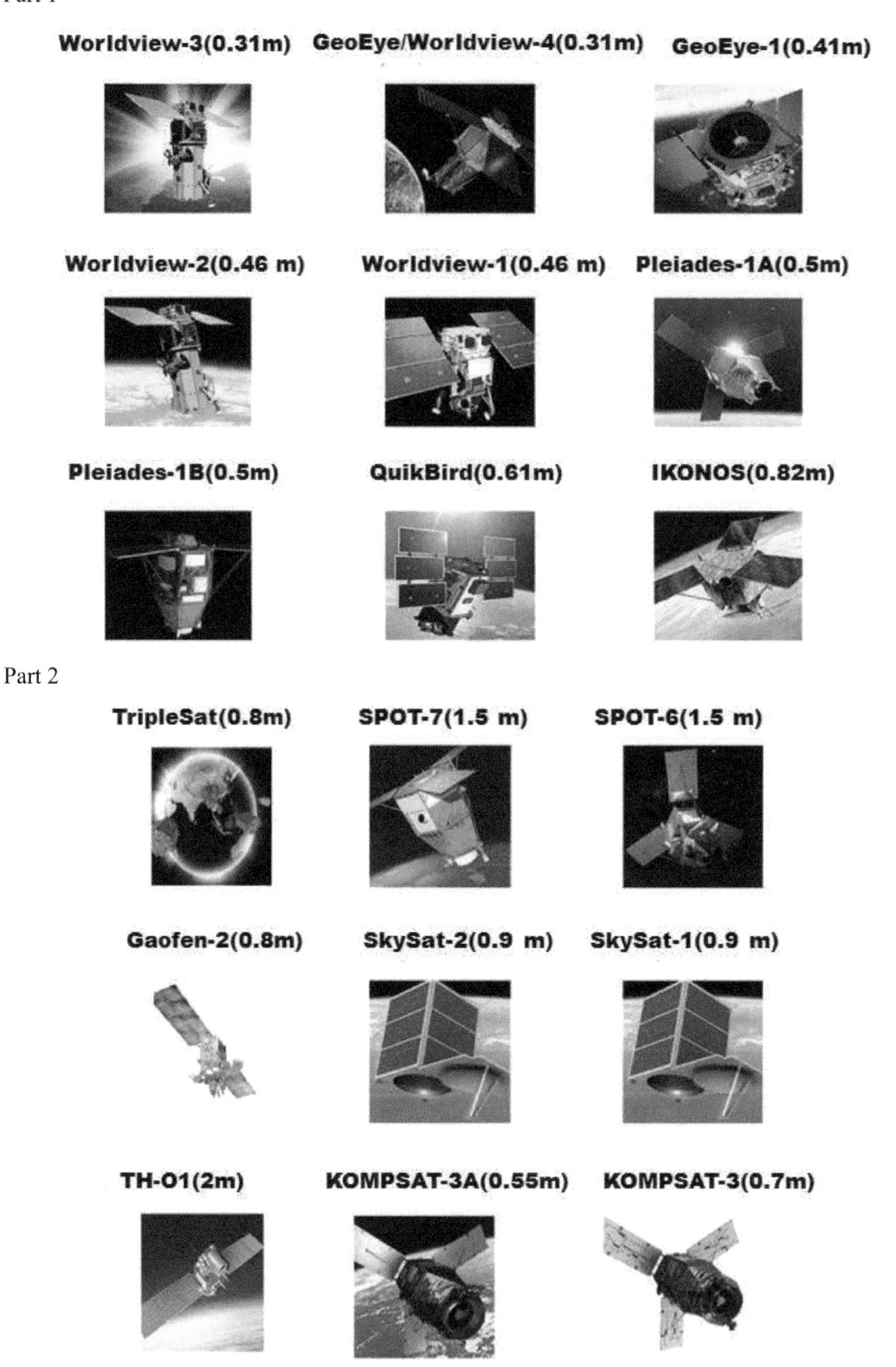

Figure 6.18 List of high resolution satellite remote sensing [http://www.satimagingcorp.com/satellite-sensors/].

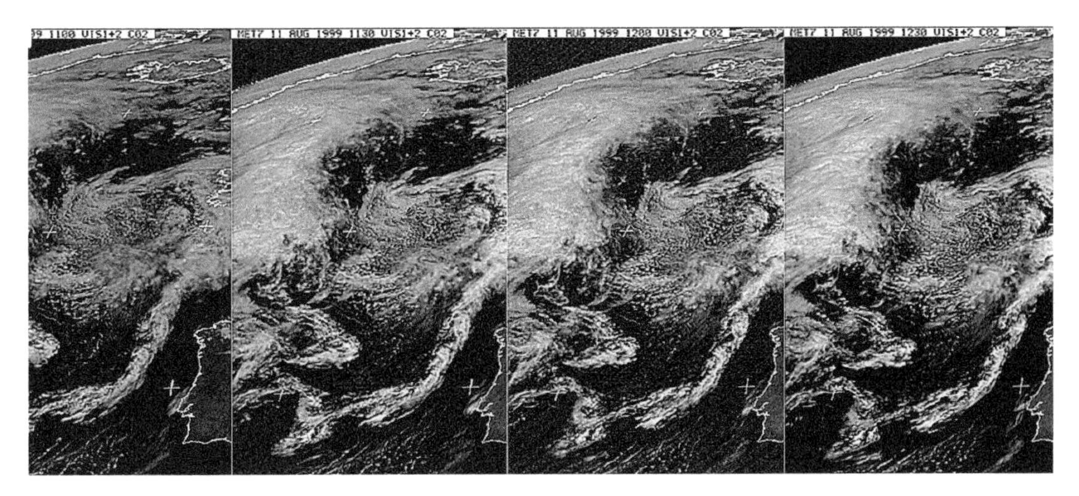

Figure 6.19 Weather satellite METEOSAT 7 every half hour [http://www.seos-project.eu/modules/remotesensing/remotesensing-c03-p05.html].

It includes (i) excessive temporal resolution: < 24 hours–3 days; (ii) medium temporal resolution: 4–16 days; and (iii) low temporal resolution: > 16 days. In this understanding, the earth's surface change rates based on tsunami's effects can be monitored and estimated over different periods.

The MODIS instrument, for instance, monitors the complete surface of the Earth every 1–2 days, whereas ASTER will take 5 years to see the complete surface. Consequently, MODIS has a greater temporal resolution than ASTER. Nevertheless, MODIS will monitor the poles more regularly than it will monitor a given location on the equator. In this regard, its temporal resolution is greater at the poles. Additionally, the repetition rate and the temporal resolution of the earth looking at satellites is 14–16 days (IKONOS: 14 days, LANDSAT 7: 16 days, SPOT: 26 days). On the contrary, meteorological satellites such as METEOSAT 8 with 15 min have extremely shorter repetition rates (Figure 6.19).

Finally, the temporal resolution, which defines the period of time required for a satellite platform to revisit a specific geographic location (also known as the revisit period). Therefore, it is an important factor for detecting changes, and the rate of these changes, occurring on surface of the planet and supports observations of environmental change, natural hazards, urbanization, deforestation, and weather systems. Thus, the actual temporal resolution of a sensor depends on a variety of factors, including the satellite/sensor capabilities, the swath overlap, and latitude.

In other words, the ability to collect imagery of the same area of the Earth's surface at different periods of time is one of the most important elements for applying remote sensing data. In this understanding, spectra characteristics of features may change over time and these changes can be detected by collecting and comparing multi-temporal imagery. However, there are not many satellite images that have been delivered during 2004 tsunami Boxing Day.

Chapter 7

Potential of Optical Remote Sensing Satellite for Monitoring Tsunami

7.1 Introduction

The remote sensing techniques have a tremendous impact on detecting floor changes triggered by catastrophic events. Recognising the capability of acquiring satellite data of massive areas, the satellite remote sensing can provide a synoptic view of the temporal evolution of the 2004 tsunami. Consequently, remote sensing has been broadly used for December 26, 2004, tsunami detection, mapping, monitoring, damage assessment, and disaster management, and would play an important role in devising a warning machine due to its "wide and insightful eyesight" in two observing similar catastrophic disasters.

In this chapter, the surface results of the earthquake/tsunami have been analyzed by the use of optical satellite images, evaluating the acquisitions before and after the seismic event that happened on 26th of December 2004. The investigation has been targeted on the Andaman Islands archipelago and the North of Sumatra. Results from preliminary ground surveys in Sumatra and Andaman Islands confirmed extensive uplift areas and massive amendment of the coastline. In order to visualize the surface effects of the earthquake, RGB coloration composite satellite data have been generated. It is convenient to perceive the risen out lands, prevalent seashores and reefs, as additionally derived from field observations [99].

The small-scale details similar to roads ought to be detected that results in a demand for a specific level of resolution. Landsat and SPOT satellite imaging have an adequate resolution for vegetation mapping. IKONOS provides a lot of helpful information for this sort of demand. This chapter focuses on available data for monitoring 2004 tsunami which are IKONOS, QuickBird, ASTER and MODIS. In fact, three sort of satellite data are presented, high resolution, moderate resolution and low resolution satellite data.

7.2 Tsunami Observation from High Resolution Satellite Images

7.2.1 Spectral Signature Analysis using Optical High-resolution Satellite Imagery

Figure 7.1 depicts the IKONOS satellite of Lhoknga before (Figure 7.1a) and after the earthquake and tsunami (Figure 7.1b). All the trees, vegetation, and buildings in the area were swabbed away. In this view, the low-lying vegetation and wood houses can be damaged and then carried by water into offshore (Figure 7.1b).

Tsunami influences can be determined and monitored using spectral signature. The spectral signature can distinguish between pre and post-tsunami event. In this understanding, spectral reflectance features (shape, position, intensity, width, etc.) of extraordinary objects throughout pre and post-tsunami lead to the various spectral reflectance of the substances from one-of-a-kind surfaces over exceptional periods. The appearance of spectral reflectance irregularity of vegetation is attributed to the various absorption of chlorophyll (0.4–0.6 μm) and excessive reflectance of the cell microstructure (0.7–1.0 μm) [99].

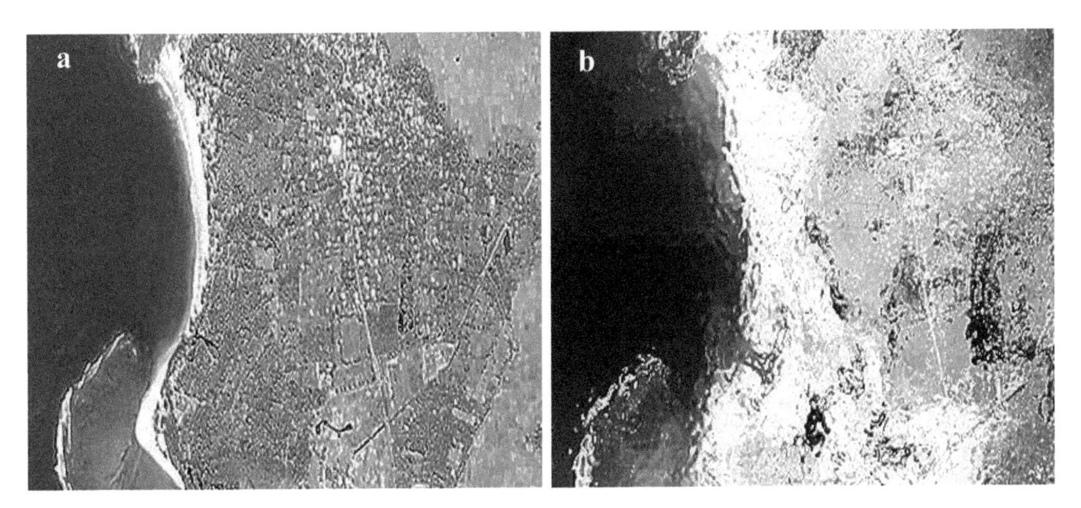

Figure 7.1 IKONOS satellite, shows Lhoknga (a) pre and (b) post 2004 Tsunami.

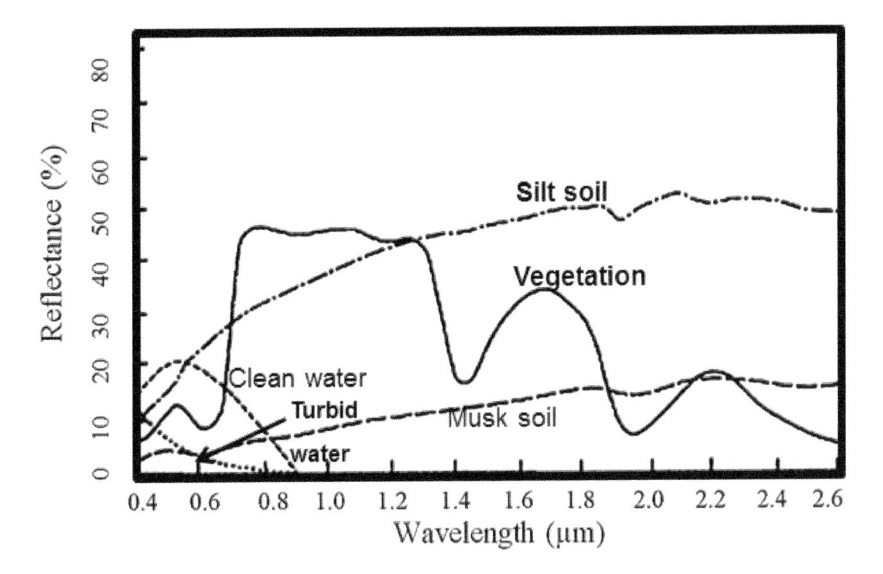

Figure 7.2 Spectral signature for understanding tsunami effects.

Nonetheless, the spectral reflectance of clean water surface prior to tsunami is greater than that of the turbid water surface post-tsunami. This is because the presence of suspended sediment particles in the turbid water post-tsunami led to an improved absorption and dispersion of the spectral radiation than smooth water. This can be clearly identified in Figure 7.2.

Consistent with thermodynamics, the heat capacity of one substance combination due to tsunami is constantly larger than the one of pre-tsunami. Thus, the water body (which is turbid water body) requires greater energy to heat mixture than the one of heating a pure substance to attain the identical temperature. In this case, turbid water has a higher reflectance in the visible region than clear water [101]. In this understanding, the best wavelength for monitoring of turbidity level and total suspended sediment concentrations due to tsunami can be obtained using the hyperspectral sensor between 700 and 900 nm.

Furthermore, the high resolution of IKONOS data of 0.82 m allowed to observe severe damages along the coastal waters of the Coromandel Coast, southeastern India. The IKONOS satellite presents the city of Chennai, a harbor town on the southeastern Indian coast. Post-tsunami, there are rough wave crests that exist along the port (Figure 7.3b). On the contrary, before the tsunami (Figure 7.3a), the houses and infrastructures have been wiped out (Figure 7.3b). In fact, 0.8 m panchromatic of blue spectrum (0.45–0.90 μm) allows the edge detection of arrival waves along the coastal waters [106]. Consequently, this could be based on a wide-field-of-view acquisition of IKONOS [99].

The majority of the radiation incident upon the ocean surface is not reflected but is either absorbed or transmitted. Longer visible wavelengths and near infrared radiation are absorbed more by water than by the visible wavelengths. Thus, water looks blue or blue green due to stronger reflectance at these shorter wavelengths and darker if viewed at red or near infrared wavelengths. The factors that affect the variability in reflectance of a water body are depth of water, materials within water and surface roughness of sea water [100].

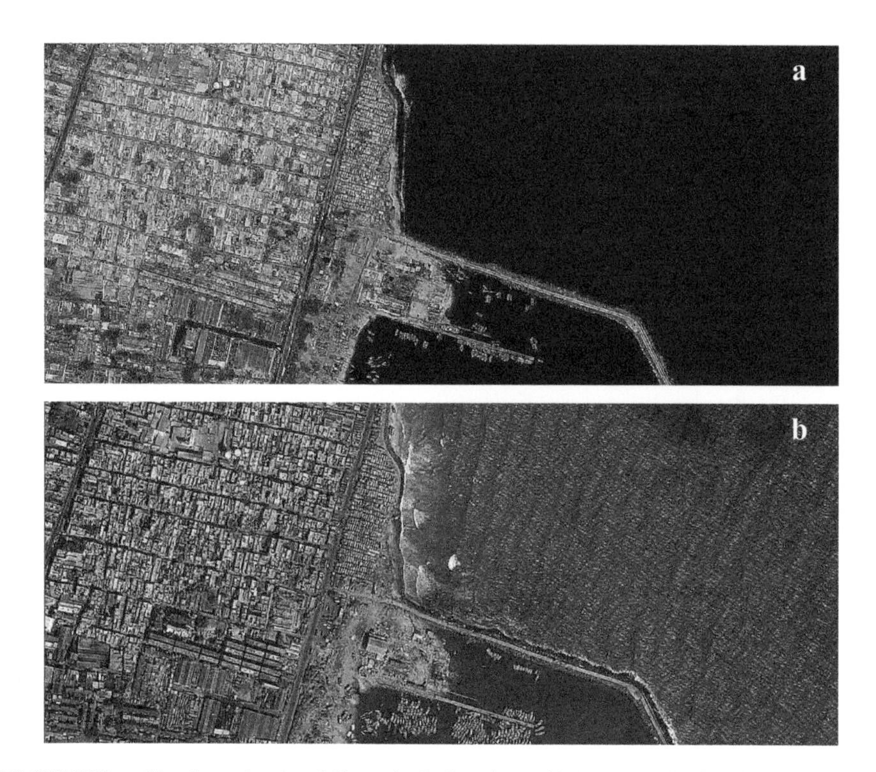

Figure 7.3 IKONOS satellite shows the city of Chennai, a harbor city on (a) August 14, 2002 and (b) December 29, 2004.

7.2.2 NDVI Analysis using Optical High-resolution Satellite Imagery

Contemporary progresses in remote sensing technologies have elevated the competencies of detecting the spatial extent of tsunami-exaggerated zones and structures destruction. The precise spatial resolution of optical data from industrial satellites is up to 0.6–0.7 m (QuickBird owned through DigitalGlobe, Inc.) or 1 m (IKONOS operated by GeoEye). After the 2004 boxing day, these satellites have captured data of tsunami-exaggerated regions, and the data have been exploited for catastrophe management activities, inclusive of disaster response and recovery [100]. To determine the extent of a tsunami inundation zone, NDVI (Normalised Difference Vegetation Index) is the most frequent index acquired from the post-tsunami data, aiming at the vegetation alternate caused by the tsunami run-up on land. The scientific explanation of NDVI can mathematically be written as [99]:

$$NDVI \quad \frac{(NIR \quad R)}{(NIR \quad R)} \tag{7.1}$$

where *R* and *NIR* represent the spectral reflectance or radiance in the red and near-infrared bands, separately. The beaches of Khao Lak, Thailand were struck by a tsunami 2–3 hours after the magnitude 9.0 earthquake of December 26, 2004. As shown in Figure 7.4, the IKONOS data obviously distinguishes the lowest range of NDVI of 0.4 post-tsunami compared to pre-tsunami. The NDVI is used to recognize the vegetation cover where a large change in the land cover and land use between pre-tsunami and post-tsunami is observed. Finally, buildings and vegetation were scoured by the waves, leaving foundations and bare soil. Beach sand was also removed by the tsunamis (Figure 7.4b).

Generally, the internal structure of pre-tsunami healthy trees acts as a diffuse reflector of near infrared wavelengths. Consequently, measuring and monitoring the pre-tsunami and post tsunami near infrared reflectance is one way that scientists determine how did the tsunami actually impact coastal vegetation.

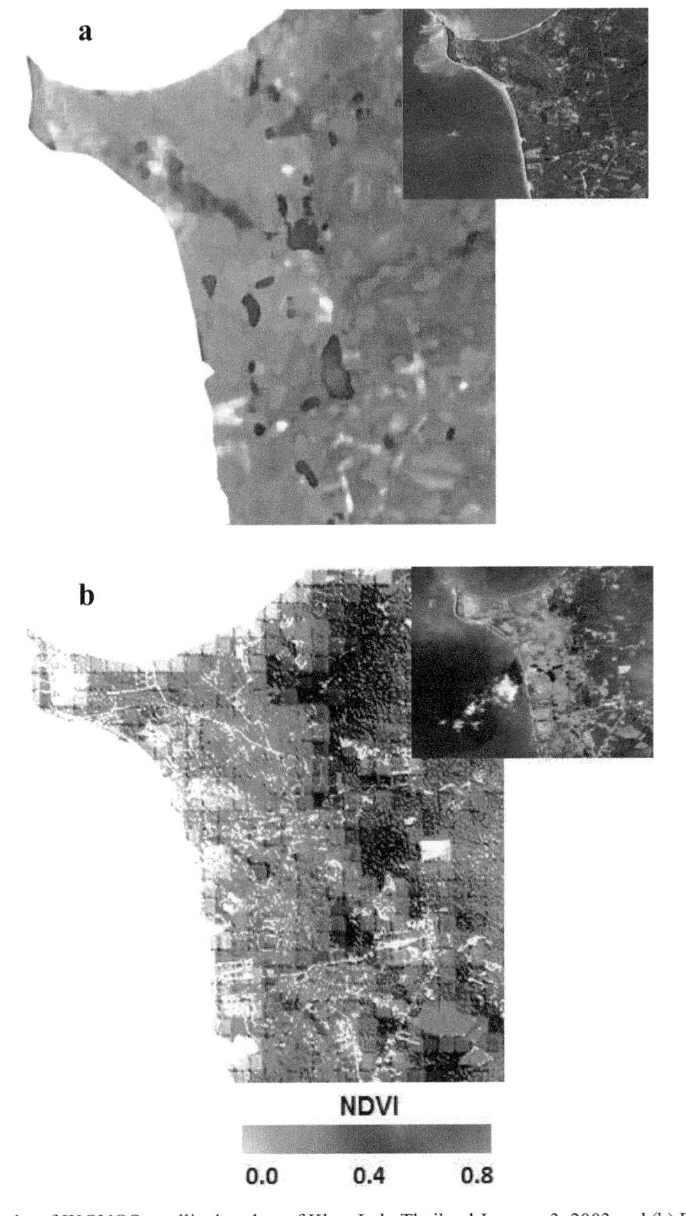

Figure 7.4 NDVI results of IKONOS satellite beaches of Khao Lak, Thailand January 3, 2003 and (b) December 29, 2004.

QuickBird is another high resolution satellite data that shows a great potential for monitoring 2004 boxing day tsunami. In this view, QuickBird image has very high resolution, 0.6 m per pixel (panchromatic-mode) and 2.4 m per pixel (multi-spectral-mode) and has 4 bands: blue, green, red, near-infrared. After the boxing day, QuickBird captures the massive damages along the Gleebruk, a small town located roughly 50 km (31 miles) from Banda Aceh of exaggerated zones on January 2nd, 2005. The NDVI of post-tsunami data of 0.3 is lower than pre-tsunami data. Further, NDVI values along the northwest coast of Sumatra is lower (Figure 7.5b) than one estimated on the coast of Khao Lak, Thailand (Figure 7.4b). In fact, the northwest coast of Sumatra was located 100 km away from the earthquake epicenter and was drowned in waves up to 15 m tall. Figure 7.5b indicates that a tsunami was channelled inland through low-lying

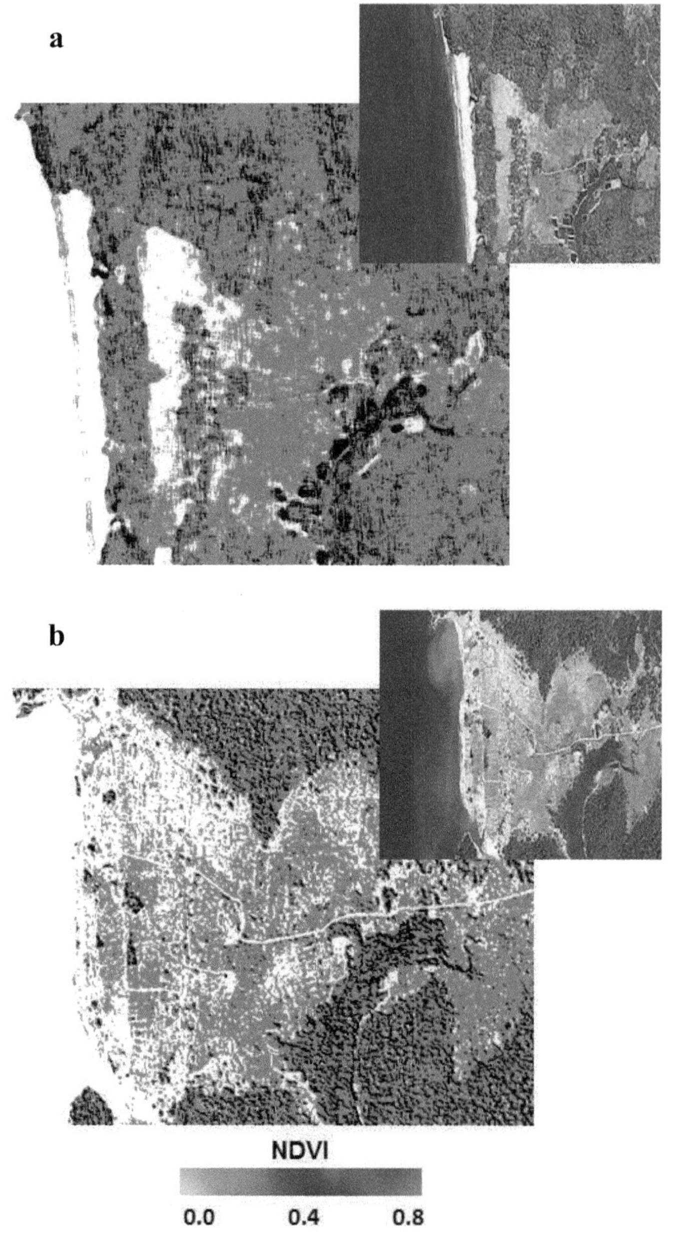

Figure 7.5 NDVI of QuickBird satellite on (a) April 12, 2004 and (b) January 2, 2005 along Aceh.

areas, for instance, stream flood plains. It is clear that Gleebruk exhibited massive destruction. Therefore, buildings, trees, roads, bridges, seashores, and even topsoil, were swept away by the force of the wave.

7.2.3 Damage Index using High Resolution Satellite Data

The advantage of using high-resolution optical satellite data for destruction clarification is the functionality of perceiving structural damage visually. These high resolution data, moreover, allow us to distinguish the spatial extent of massive destructions at the regional scale, where post-tsunami surveys hardly ever penetrate due to restricted survey time and resources. However, be aware that no structural sorts had been identified by using the analysis of the satellite images. Additionally, the characteristic of damage that can be identified from the satellite data is solely the structural destruction or foremost structural failure, which indicates by the changing in a roof shape, particularly "collapsed" and "major or extreme damage" (Table 7.1).

QuickBird satellite image with a 0.6×0.6 m^2 resolution is fine enough for a visual interpretation to differentiate the damage levels of buildings. In this regard, the damage index can be estimated from QuickBird data as compared to IKONOS. In fact, IKONOS satellite data has 1.0×1.0 m^2 resolution which is inadequate to determine the physical building damage features. Let us assign $P_{i,j}$ as the number of pixels of damaged buildings while $N_{i,j}$ being pixels of undamaged buildings, respectively. i and j represent the row and column of QuickBird satellite data. The damage index $D_{i,j}$ is calculated as [255]:

$$D_{i,j} = \frac{P_{i,j}}{N_{i,j} + P_{i,j}} 100 \tag{7.2}$$

Figure 7.6 presents the QuickBird of Meulaboh, Indonesia, collected on January 7, 2005. Meulaboh is located on the coast of Sumatra, roughly 150 kilometers (93 miles) from the epicenter of the magnitude 9.0 earthquake that generated the tsunami. The image shows where the tsunami washed over a narrow peninsula, eroding the beach and destroying many of the town's buildings.

Figure 7.7 depicts the spatial variations of damage index value, post-tsunami. The damage index increased from shoreline and decreased towards inland. The buildings that existed along the coastal waters are completely washed and damaged by the tsunami run-up as indicated by damage index value of 100%. Indeed, the tsunami washed over a narrow peninsula, eroding the beach and destroying many of the town's buildings.

Table 7.1 Example of building damage classification criteria for the 2004 boxing day tsunami.

Damage Levels	Pre	Post	Criteria for Classification
Washed away		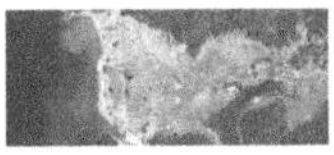	All buildings and structures washed away.
Collapsed		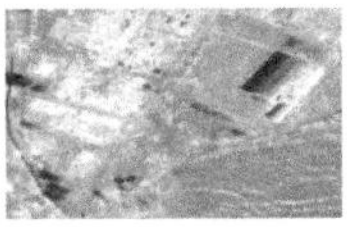	Large scale building collapsing through roofs and walls.
Major			Small scale building collapsing through roofs.
Survived	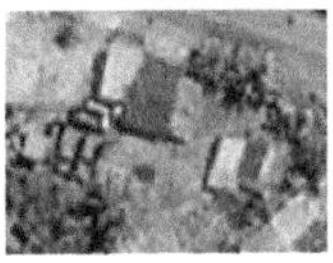	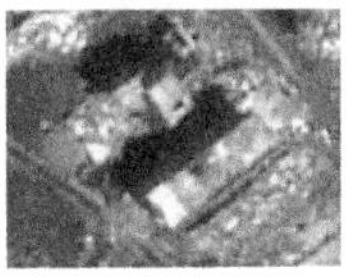	The physical characteristics of the survived building have changed between pre and post-tsunami event.

Figure 7.6 QuickBird data post-tsunami along Meulaboh, Sumatra.

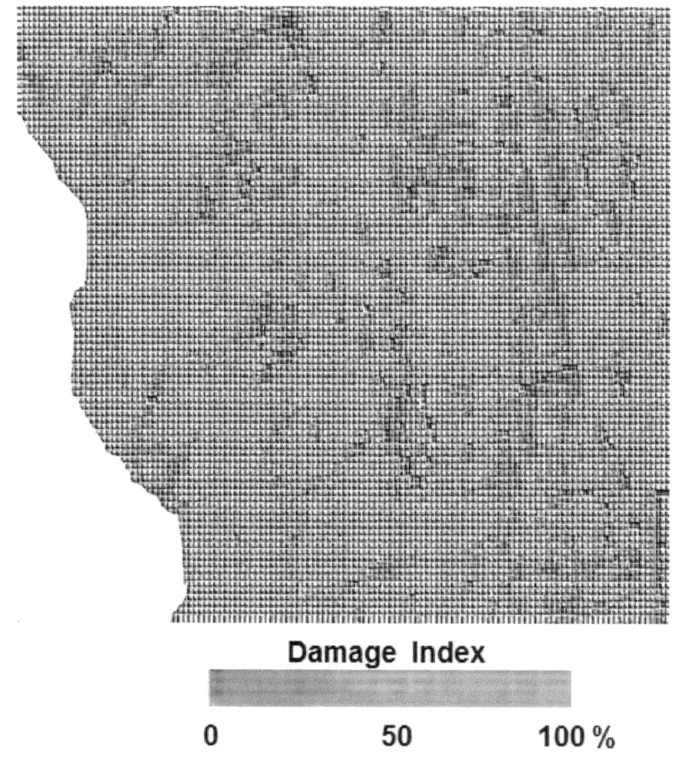

Figure 7.7 Damage index spatial variations on Meulaboh, Sumatra.

7.3 Tsunami Inundation Mapping using Terra-ASTER Images

ASTER (Advanced Spaceborne Thermal Emission and Reflection Radiometer) is one of the sensors on board Terra satellite. ASTER sensor can collect the data of not only visible and near-infrared bands, but also of a short wavelength infrared (SWIR) band. The resolution is 15 m for visible and near-infrared bands and 30 m for SWIR band. Hence, ASTER data with a moderate-resolution is useful for detecting wide tsunami inundation areas quickly because red, near-infrared and short wavelength infrared bands are considered to give the good indicators of land cover characteristics.

The visible and near-infrared radiometer of ASTER produces imagery of ground sampling interval of 15 m, and it has the three nadir-looking bands and one backward-looking band for stereo scoping. Unlike other optical sensors, ASTER does not have a blue color band. Terra-ASTER observed the hard-hit area, including Khao Lak two years before and 5 days after the tsunami, as shown in Figure 7.8 in a false color composite which means the reflectance of the near-infrared (NIR) band is assigned to the red component, that of the red (R) band to the green component, and that of the green (G) band to the blue component, respectively [256].

The comparison shows an interesting pattern of damage along the coast. It is the long, smooth curving beaches that have been devastated by the tsunami, not the land that juts into the ocean. Several factors probably contributed to this pattern. First, the elevation is certainly a factor. The post-tsunami data was dominated by NDVI value of less than 0.66 as compared to pre-tsunami data (Figure 7.9).

However, the affected areas are less affected than the areas which are found along the coastal water of Aceh (Figure 7.6). In fact, the heavy vegetation covers along the coastal water of Thailand acted as natural barriers to reduce the energy of tsunami run-up. Moreover, the headland in the center of the image is probably a high rocky point that would not be easily inundated by a large wave. The wrinkle of inland mountains appears to curve out to the coast between the two damaged beaches. The beaches, on the other hand, probably have a low elevation that gently slopes toward the ocean, allowing any water that comes ashore to sweep further inland [254].

Notwithstanding ASTER determination of 30 m, the 90 m SRTM conveyed exact approximations of run-up statures. Obviously, the areas with regions with soak bluffs, waterfront fields and level lying surge have the most elevated estimation of the run-up. Figure 7.10 shows Khao Lak coastal waters, Thailand, which is constructed using ASTER DEM and SRTM. The inundation zone is delineated by dashed white line. Therefore, Figure 7.11 depicts the simulated run-up heights estimated using DEM. Clearly, regions B, C, and E have the most astounding estimation of the run-up of 15 m when contrasted with areas A and D. Truth be told, areas A and D are spoken to be the headland and high plain which doesn't enable the wave to run-up or spread widely [256].

Despite what might be expected, north Aceh drift, with the ASTER LIB information from Banda Aceh, Lhok Kruet is ruled by little inland slopes (Figure 7.12) and by the greatest run-up tallness of 32 m (Figure 7.13). In any case, the Lhok Kruet have a little inland slope along Lhok Kruet struck by a run-up of 8 m height. In reality, the slopes are amazingly more extreme with higher height, which did not enable the wave to run-up further (Figure 7.12). McAdoo et al. (2007) [257] agrees with these results.

Finally, DEM of ASTER data also can reveal the uplift occurrence due to the 2004 tsunami. Figure 7.14 shows the post-earthquake ASTER images of a small island off the northwest coast of Rutland Island, 38 km east of North Sentinel Island, and the submergence of the coral reef surrounding the island. It is interesting that DEM suggests that the actual uplift is on the order of 1 to 2 m.

Figure 7.8 ASTER satellite data acquired on (a) February 28, 2003 and (b) December 31, 2004 along Khao Lak coastal water, Thailand.

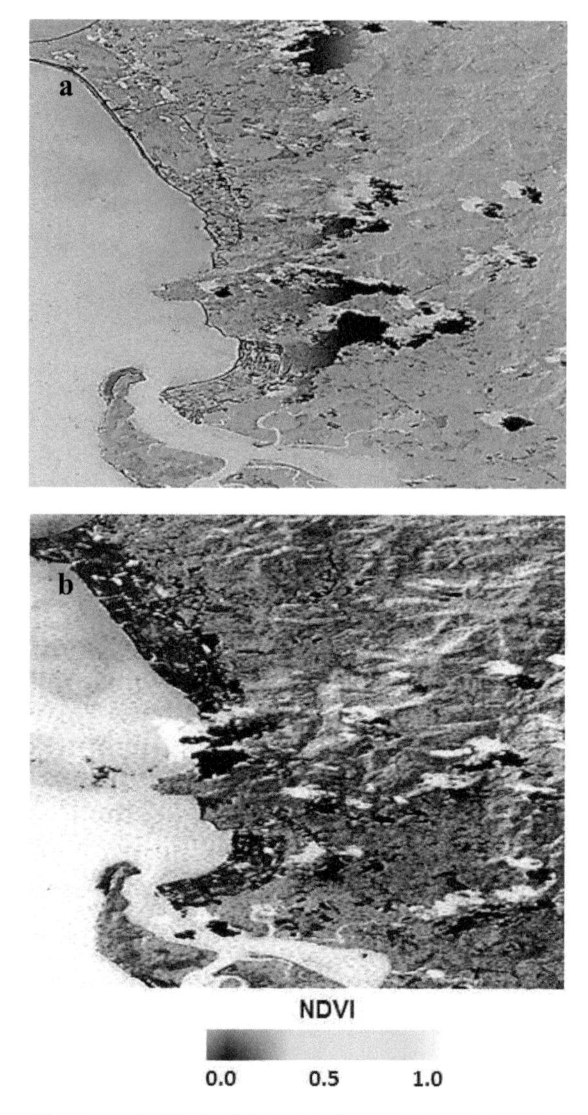

Figure 7.9 NDVI of ASTER data (a) pre and (b) post-tsunami.

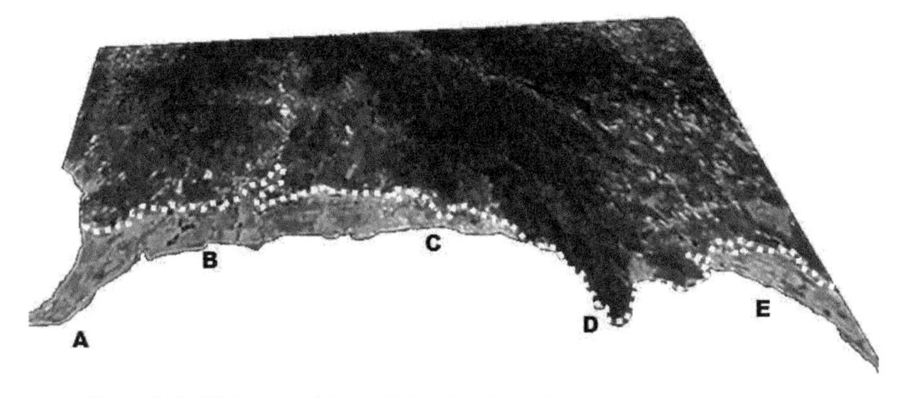

Figure 7.10 DEM extracted from ASTER data along Khao Lak coastal water, Thailand.

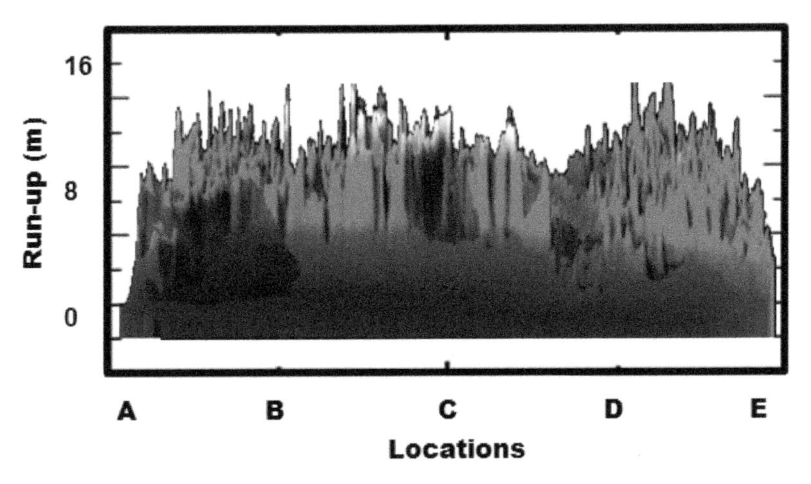

Figure 7.11 Run-up extracted from SRTM along Khao Lak coastal water, Thailand.

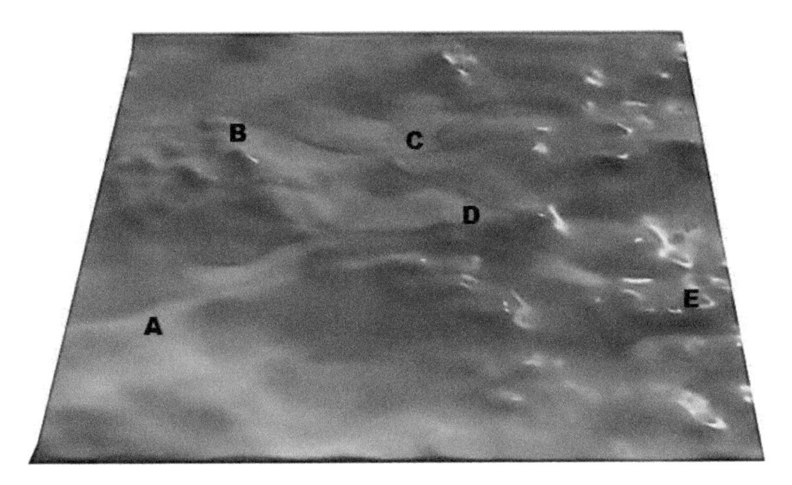

Figure 7.12 DEM extracted from ASTER for Lhok Kruet, Aceh.

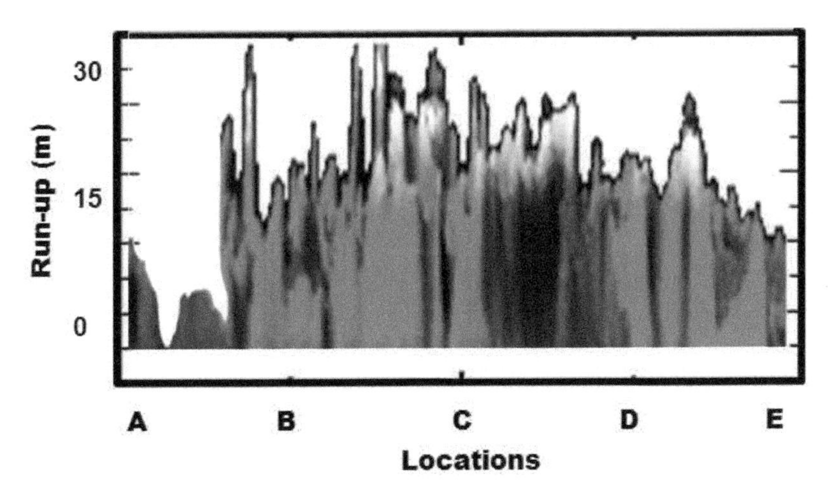

Figure 7.13 Run-up extracted from SRTM for Lhok Kruet, Aceh.

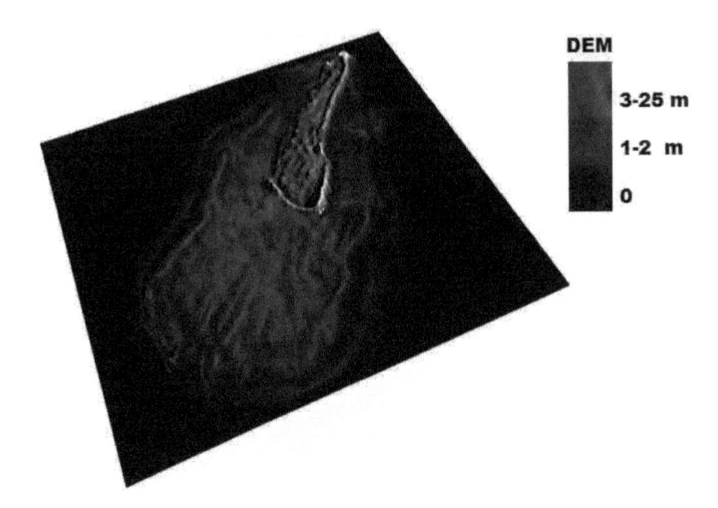

Figure 7.14 DEM extracted from ASTER data on February 4, 2005 along Rutland Island.

Clearly, immersion zone depends on a Digital Elevation Model (DEM) variety. Be that as it may, SRTM inside 90-m cell can not recognize the little landform variety. In any case, utilizing the SRTM information can be actualized to ascertain the geomorphological parameters with poor accuracy [255].

Comparison of Figures 7.9, 7.10 and 7.11 indicates that lower NDVI values are fit with a maximum run-up and lower plain regions. In other words, NDVI decreases in the low plain as vegetation covers are swept away through the highest run-up of 15 m height. The high plain is dominated by high NDVI value of 0.8. Indeed, the tsunami run-up that ranged from 7–15 m cannot spread at more than 20 m DEM [256].

7.4 Tsunami Observation from Low Resolution Satellite Images

A spatial resolution of several hundred meters is the main feature of low resolution satellite data. However, these data are extremely reasonable for ecological checking, farming applications and potential vitality estimation. Among the lower resolution sensor that delivered accurate information about the tsunami was MODIS data. The MODIS satellites operated by NASA started operation in 1999 with the launch of the Terra satellite and was followed by the launch of the Aqua satellite in 2002. The sensors are equipped with 36 bands in different wavelenghts and with different spatial resolutions (250, 500 and 1000 m). The two MODIS satellites delivers one image per day from any location around the globe. This high temporal resolution makes it very attractive for monitoring tsunami. In this view, the MODIS sensors have monitored tsunami affects on marine, landcover and vegetation.

The island of Sumatra experienced both the thundering of the submarine seismic tremor and the waves that were produced on December 26, 2004. Close to the shake, the ocean surged inland causing annihilation of the shorelines of the northern Sumatra. This match of data from the Moderate Resolution Imaging Spectroradiometer (MODIS) on NASA's Terra satellite demonstrates the Aceh area of northern Sumatra, Indonesia, on December 17, 2004 (Figure 7.15a), preceding the tremor, and on December 29, 2004 (Figure 7.15b), three days after the fiasco. In spite of the fact that MODIS was not particularly intended to mention the exceptionally small objective facts that are typically important for mapping coastline changes, the sensor in any case watched clear contrasts in the Sumatran coastline. On December 17 (Figure 7.15a), the green vegetation along the west drift seems to achieve the distance to the ocean, with just an incidental thin extend of white that is most likely sand. After the seismic tremor and waves, the whole western drift is fixed with a perceptible purplish-darker fringe (Figure 7.15b).

On a moderate-resolution image such as Figure 7.15b, the affected area may seem small, nevertheless, each pixel in the full resolution image is 250 by 250 m. In places, the brown strip reaches inland roughly

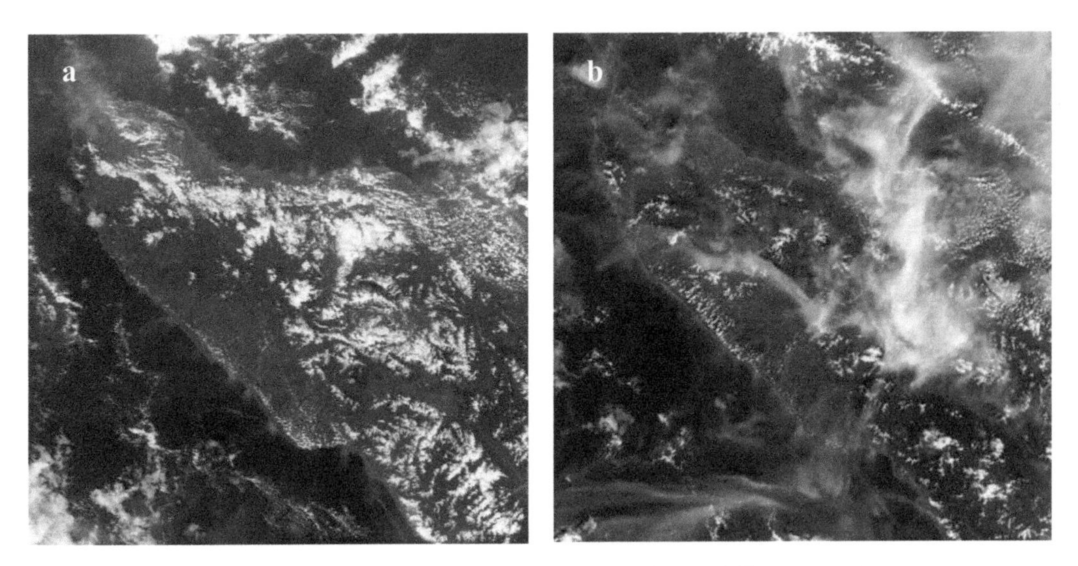

Figure 7.15 MODIS images for Northern Sumatra (a) pre and (b) post tsunami.

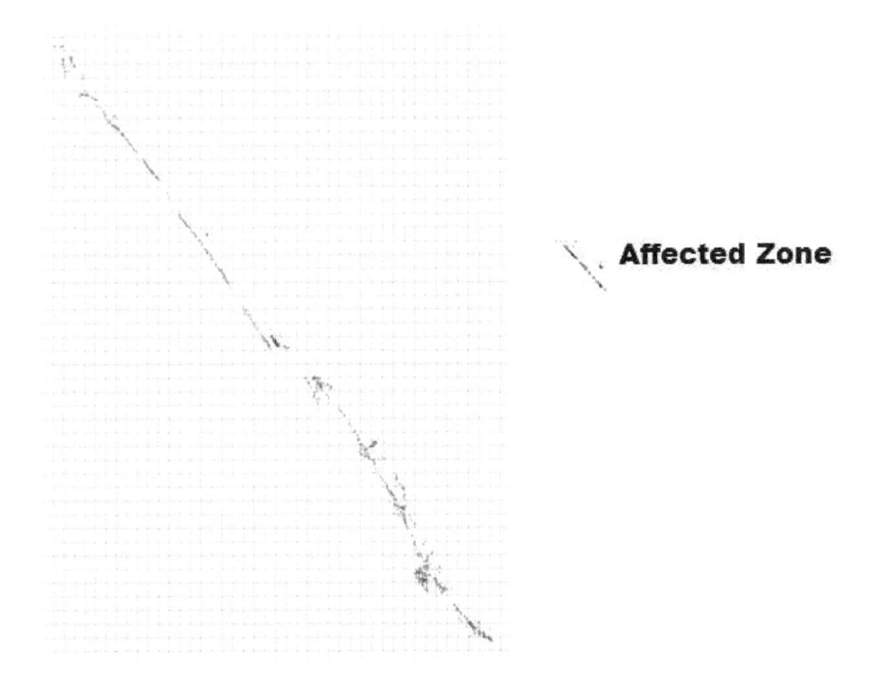

Figure 7.16 Dusky strip of affected zones from MODIS data post-tsunami.

13 pixels, equal to a distance of 3.25 km, or about 2 miles. Moreover, the low resolution MODIS satellite data reveals the influenced zones as a tiny strip of dusky along the coast (Figure 7.16).

The Andaman Islands which are located 850 km north of the epicenter of the quake were the most affected zones by 2004 tsunami. Therefore, those islands were not just among the primary land masses to be cleared under the wave, they have also been shaken by a progression of delayed repercussions. Combined MODIS data demonstrates a bit of the Andaman Islands on January 23, 2003 (Figure 7.17a),

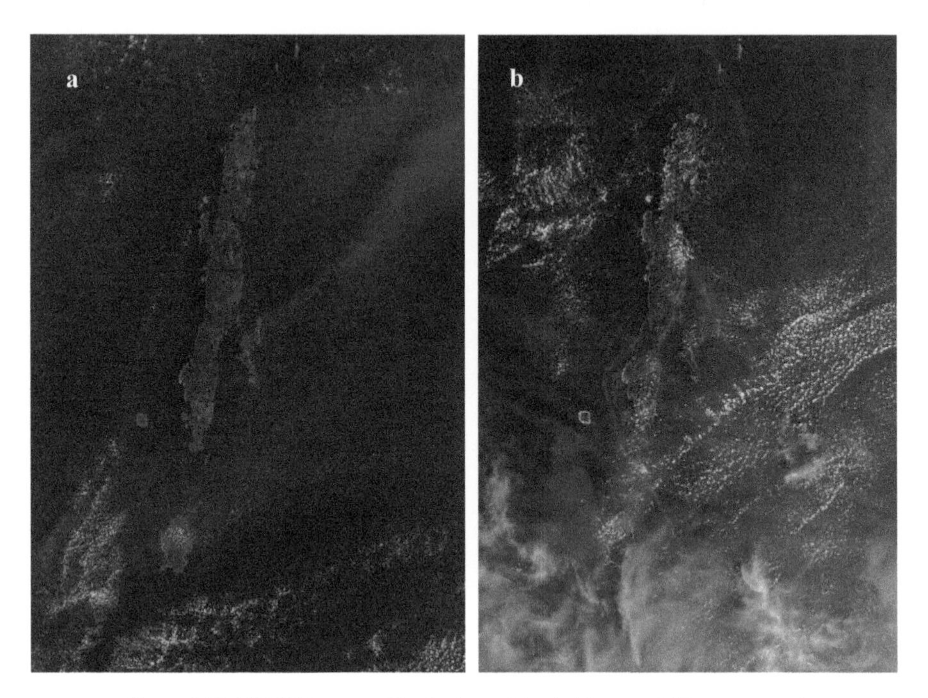

Figure 7.17 MODIS images of the Andaman Islands (a) pre and (b) post tsunami.

and January 3, 2005 (Figure 7.17b), respectively. In the latest image (Figure 7.17b), the shorelines along the west side of the islands have been exposed, leaving a portion of splendid tan land along the drift. The change is most prominent on North Sentinel Island, home of the Sentinels aboriginals, and on Interview Island. In this view, the vegetation covers along the coastline have been supplanted with the massive sand depositions post the tsunami disaster. North Sentinel is bordered by a coral reef, which shows up turquoise in the pre-wave MODIS image which, however, may have been covered in sand and dregs, post-tsunami.

Chapter 8

Modelling Shoreline Change Rates Due to the Tsunami Impact

8.1 Shoreline Definition Regarding Tsunami

The shoreline is a sign for tsunami influences on coastal zones. In fact, tsunami can cause the rapid changes of shorelines in less than an hour. It is the first boundary which is receiving the massive energy destructive of tsunami.

A novel definition of shoreline is the reality it accords to the bodily boundary of land and water (Figure 8.1). Regardless of its apparent easiness, it is barely to impose the boundary of shoreline. The changes of shoreline occur rapidly because of the quick hydrodynamic nature of water levels at the coastal zones. The coastal hydrodynamic parameters, for instance, are waves, tides, storm surge, run-up, etc., which can alter the shoreline inside much less than a few hours. For instance, a swash area is changeable per second or few minutes. In fact, a swash zone (Figure 8.1) is a feature of the wind, swell and infragravity wave durations, which range from seconds to a few minutes.

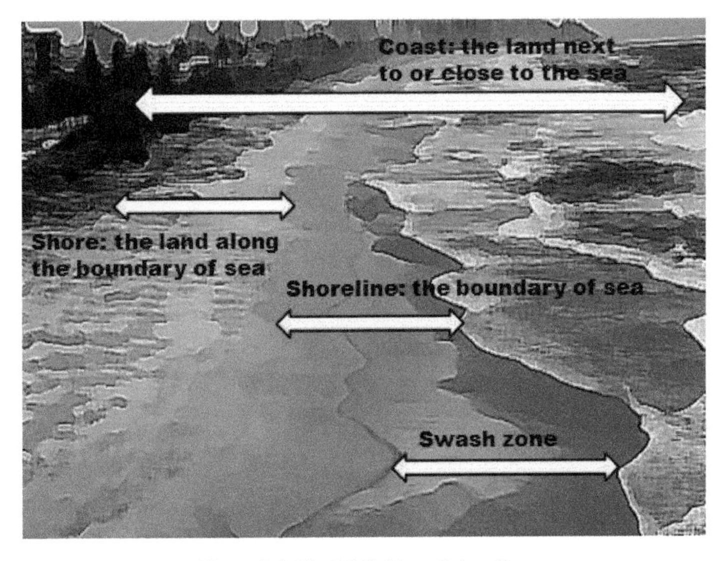

Figure 8.1 Ideal definition of shoreline.

The appropriate definition of the shoreline in such tropical zone wherein the boxing day happened is the end of vegetation borders. Under this circumstance, the shoreline is defined as the end of vegetation covers (Figure 8.2). This definition absolutely inaugurates the vegetation line as the favored location for the shoreline and the particles line as a secondary alternative. Therefore, tsunami can cause the massive destructive of coastal zones. This event happens over short periods. The acceleration of shoreline modifications can occur in a few hours because the massive amount of tsunami's energy strikes the coastline.

The deep sea earthquake creates the tsunami as a result of the sudden vertical rise of the ocean bottom. In this regard, the sea level rises by several meters because of the displacement of the massive volume of water. Under this circumstance, the rapid propagation and run-up of a tsunami can cause the demolishment of the coastal ecosystems (Figure 8.3). As a rule-of-thumb, a sandy shoreline retreats about 100 meters for every meter rise in sea level. It means a certain amount of sea level rise, as a result of the earthquake that triggered the tsunami waves, will change the shoreline from the original position [128].

As a matter of fact, it is very important to investigate the coastline deformations, especially after the tsunami waves defeated the coastal regions. The most deformed shorelines are along the Banda Aceh, Indonesia and Kalatura, Sri Lanka due to the tsunami run-up of 30 (Figure 8.4). This deadliest disaster

Figure 8.2 Shoreline at the end of vegetation cover.

Figure 8.3 Aceh coastal zone damages due 2004 tsunami.

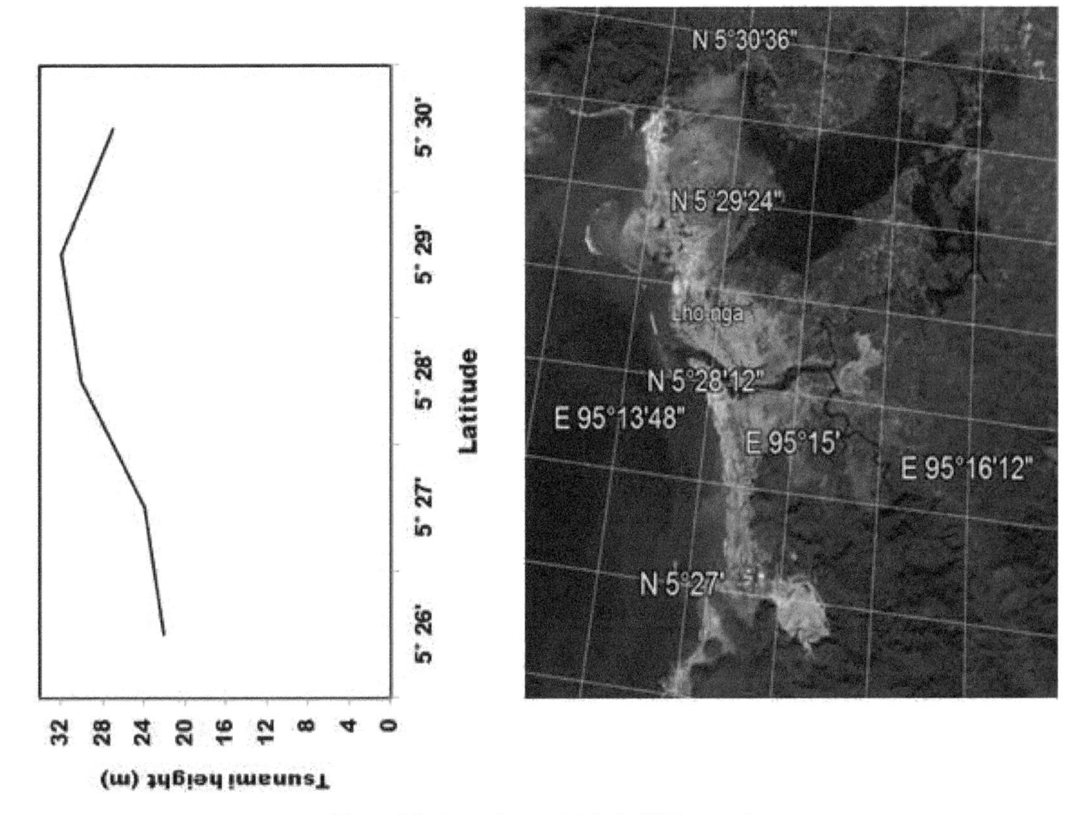

Figure 8.4 Tsunami wave height had hit Indonesia.

has caused massive devastation to the environment and the world ecosystems despite killing millions of people all over the world [121].

8.1.1 Optical Remote Sensing for Shoreline Extraction

For many decades, automated or semi-automatic shoreline detection and mapping are established using optical remote sensing data. For instance, band 5 of LANDSAT Thematic Mapper (TM) is used to delineate shoreline by the aids of threshold level-slicing and image classification techniques. Shoreline extraction is mainly based on visible and near IR spectra. In this context, visible and near Infrared (NIR) bands are used commonly due to their wavelength that is capable of producing a contrasting image between water bodies and land area.

Recently, precise detection and mapping of land covers are possible to be delivered by high-resolution remote sensing data. In this view, the number of bands is increased in such Worldview-2 to 8 bands. This can deliver reliable information on land covers. Further, the pixel dimensions are reduced to less than 1.0 m in such the Quickbird satellite data. On the contrary, shoreline extraction from high-resolution data is immobile barley mission. On this understanding, an extraordinary investigation is required to distinguish between different objects in high-resolution data because of their small resolution, for instance, soil, and seawater.

Therefore, conventional image processing tools involved unsupervised and supervised classification, segmentation, NDVI (Normalized Difference Vegetation Index) and NDWI (Normalized Difference Water Index) which are examined for shoreline extraction from high-resolution data. However, these procedures can enclose mixed pixels that belong to one or more classes [258]. Under this circumstance, Chien (2007) and Giannini and Parente (2015) developed novel procedures for image processing which are based on object-based approach. In this approach, image segmentation was compiled with rule-based or supervised classification. In this regard, the fuzzy logic is considered the precise algorithm to avoid the uncertainty of mixed pixels. For instance, the membership degree of [0, 1] is deliberated as a certainty map to separate water and vegetation classes in high-resolution data. In this understanding, positive NDVI presents soils, and dense and healthy vegetation has the maximum NDVI values. On the contrary, the negative NDVI values which close to zero presents water (McAdoo et al. 2007).

Guariglia et al. (2006) developed step of computer program code which involved visible and near IR spectra bands for automatic detection of the shoreline geographical location. This code is stated as: If Near IR/(Green−Red) < 1, then 255 else 0. In fact, the land-water interface is clearly defined as the absorbed near IR radiation energy by water and, thereby, contributed nearly no energy returns in case of near IR.

8.1.2 Hypotheses and Objective

Concerning above perspective, we address the question of tsunami rule on shoreline deformation. This is demonstrated with high-resolution satellite data, i.e., the Quickbird. Three hypotheses are examined:

(i) Segmentation based on distance matrix algorithm can be used to separate land and water boundary in the Quickbird satellite data;
(ii) Shoreline can be extracted automatically by using edge algorithms, i.e., Sobel and Canny;
(iii) Coastal vegetation diversities have role to reduce the damage along the coastal studies.

This chapter hypothesizes that the tsunami deformed the shoreline rapidly. In this regard, the main objective is to model the changing rates of shoreline along the most affected coastal areas such as Aceh and Sri Lanka. The advanced technology of remote sensing pre and post tsunami can deliver accurate information on the shoreline deformation.

8.2 Study Areas and Data Acquisitions

Different locations of Gleebruk, Banda Aceh, Indonesia and Kalutara, Sri-Lanka are examined for tsunami effects shoreline deformation. Gleebruk, Banda Aceh, Indonesia is located in the west coast of Banda Aceh between 5°15’ 53.14” N to 5°17’ 57.42” N and 95°13’ 05.84” E to 95°15’ 01.21” E, facing the Indian Ocean (Figure 8.5). Additionally, this area is located on the ring of fire which makes it very vulnerable to earthquake disasters. Furthermore, it is a developed area with suburban and wooden houses. The Gleebruk coastal area is mostly full of agricultural and residential development.

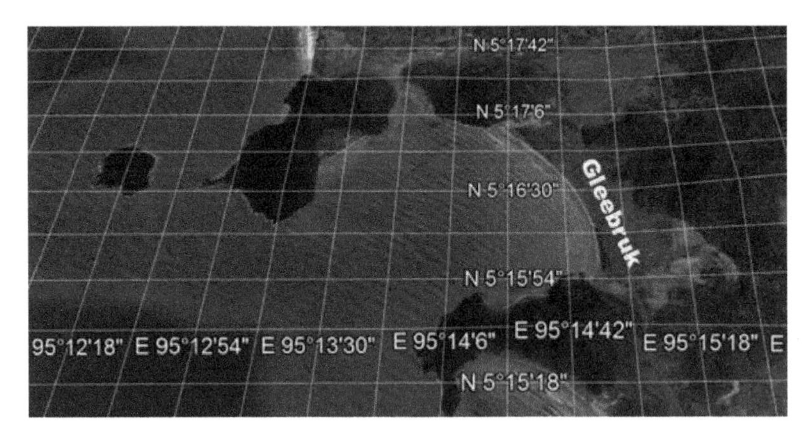

Figure 8.5 Geographical location of Gleebruk.

Kalutara coastline is located in Sri-Lanka between 6°34' 21.03" N to 6°34' 57.28" N and 79°57' 13.63" E to 79°58' 04.87" E (Figure 8.6). Moreover, Sri Lanka is dominated by two monsoon periods. Indeed, southwest monsoon brings rain mainly from May to July to the western, southern and central regions of the Sri Lankan Island. On the contrary, the northeast monsoon rains occur in the northern and eastern regions in December and January which are frequently influenced by the coastal zone of Sri Lanka.

During this investigation, the high-resolution Quickbird satellite data were acquired from Digital Global archive data. The images of Quickbird were acquired pre and post tsunami disaster of Gleebruk coastline, Indonesia (Figure 8.7) and Kalutara coastline, Sri Lanka (Figure 8.8).

One of most commercial high resolution earth observation data was a Quickbird. It launched in 2001 and ended in 2015. It collected panchromatic data at a resolution of 0.61 m. Therefore, it collected

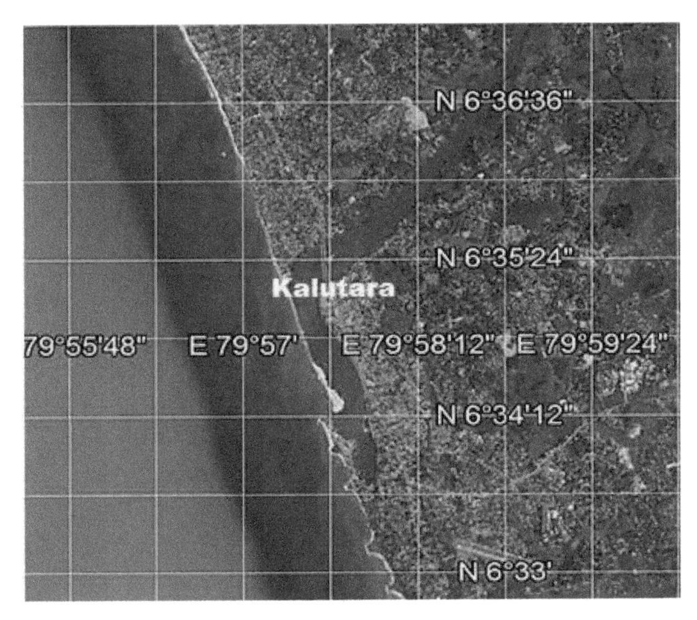

Figure 8.6 Geographical location of Kalutara, Sri-Lanka.

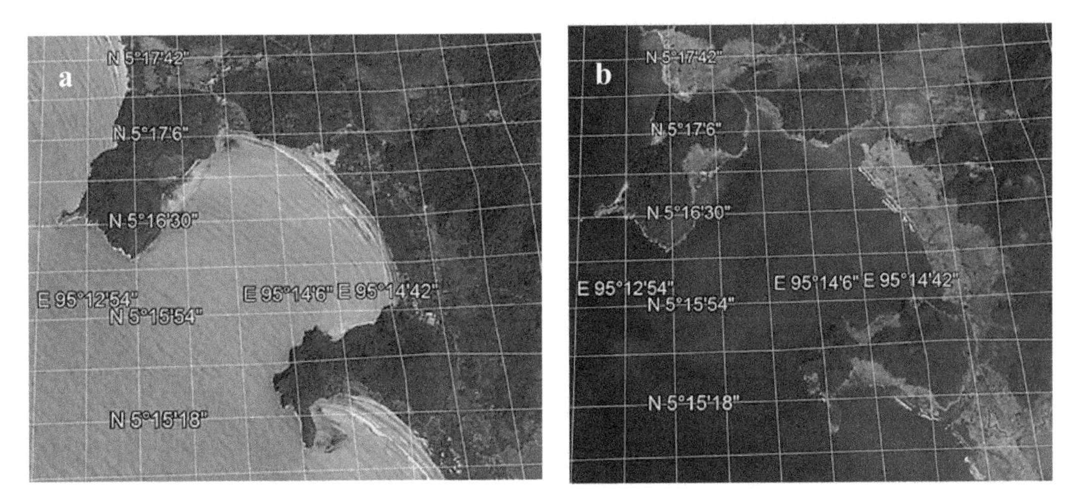

Figure 8.7 Quickbird satellite data pre and post tsunami along Gleebruk in (a) January 3, 2003, and (b) January 1, 2005.

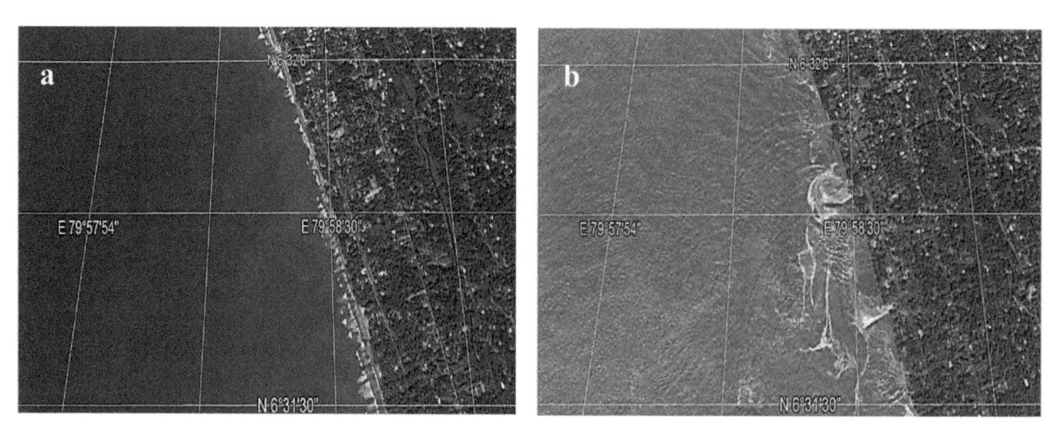

Figure 8.8 Quickbird satellite data pre and post tsunami along Kalutara in (a) January 1, 2004, and (b) December 26, 2004.

Table 8.1 Characteristics of Quickbird images.

Band	Wavelength Region (μm)	Pixel Resolution (m)
1	0.45–0.52 (blue)	2.44–2.88
2	0.52–0.60 (green)	2.44–2.88
3	0.63–0.69 (red)	2.44–2.88

multispectral data with different resolutions as a function of its latitude variations. At an altitude of 300 km, the image resolution is 1.61 m while the resolution is 2.44 m at 450 km. In this regard, the off-nadir viewing angle, which is ranged between 0 to 25°, ruled the pixel resolution. The advantages of Quickbird sensor is the coverage of 16.5 km in the cross-track direction. Consequently, the Quickbird sensor has a high revisits frequency of 1 to 3.5 days and delivers an excellent stereo geometry data. At high resolution, it can provide precise, detailed information for shoreline or vegetation line mapping and detection of changes.

8.3 Automatic Detection of Shoreline Extraction

Rate change of shoreline is appraised by using the displacement vector technique. In doing so, edge detection algorithms of Soble and Canny are used to extract historical vector layers from Quickbird data. The Sobel edge detector is used to identify the edge pixel location of the vegetation lines on the image. This operation performs a 2-D spatial gradient measurement on the image and emphasizes regions of high spatial gradient that corresponds to edges. Consistent with Di et al. (2003) [122], Sobel algorithm enhances edges and outlines the features along boundaries. The operator consists of a pair of 3 × 3 convolution, horizontal and vertical masks.

8.3.1 Image Segmentation

Image segmentation is mainly defined as an image partitioning technique which mobilizes the data into a grid of linked arrangements of pixels. It requires prior edge detection techniques such as Sobel and Canny. Consequently, the Quickbird data are segmented using Hierarchical clustering algorithm. In this view, there is no preferred number of clustering that is required. Under this circumstance, the Quickbird data is segmented into two clusters which involves water and land. Then the distance matrix was used to separate the water from the land and also to split the coastal zone into eroded cluster due to tsunami effects. The distance matrix can mathematically be written as:

$$D_{cl}\left(W_i,S_j\right)=\max_{x,y}\left\{d(x,y)\middle|x\in W_i,y\in S_j\right\} \tag{8.1}$$

where D_{cl} is the maximum distance between the dissimilar clusters of water and sea, W_i and S_j, respectively. On the contrary, the D_{sl} is the minimum distance between the two similar effected E_i and E_j shoreline areas, respectively, due to tsunami run-up. This is expressed as:

$$D_{sl}\left(E_i,E_j\right)=\min_{x,y}\left\{d(x,y)\middle|x\in E_i,y\in E_j\right\} \tag{8.2}$$

Figures 8.9 and 8.10 indicate an excellent clustering procedure for separating water and land boundary. In this regard, Hierarchical clustering algorithm determined the dissimilarity criteria between the land and water boundaries. Consequently, the Hierarchical clustering based minimum distance is able to identify the massively eroded shoreline due to tsunami effects (Figures 8.9b and 8.10b).

Additionally, the water features are then eliminated from the image to separate water from the land boundary. Certainly, the land and water obstacles are created within the image. The coastal pixels (sand region) are segmented to separate from the vegetation line. Indeed, the hierarchical clustering algorithm incorporates both minimum and maximum distance. Furthermore, the spatial variations of space and time aid the visible overall performance of hierarchical clustering to break up water and land boundary. In this view, the damaged coastlines are perfectly identified by way of boundary disconnected of the coastal pixels (Figure 8.10b).

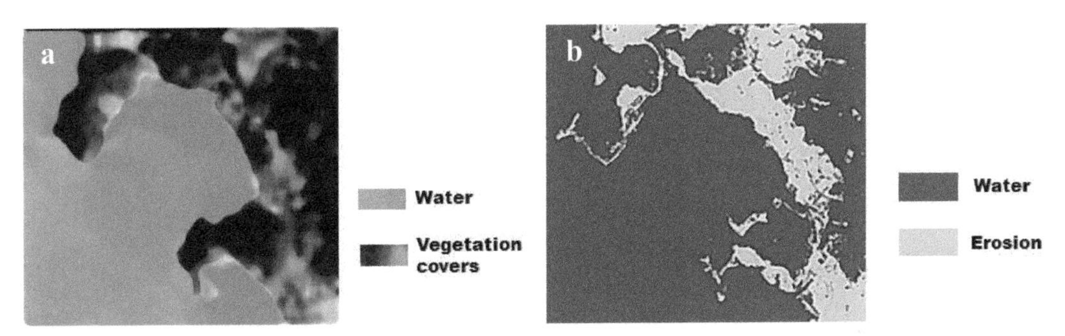

Figure 8.9 Image segmentation (a) pre and (b) post tsunami for Gleebruk coastal zone.

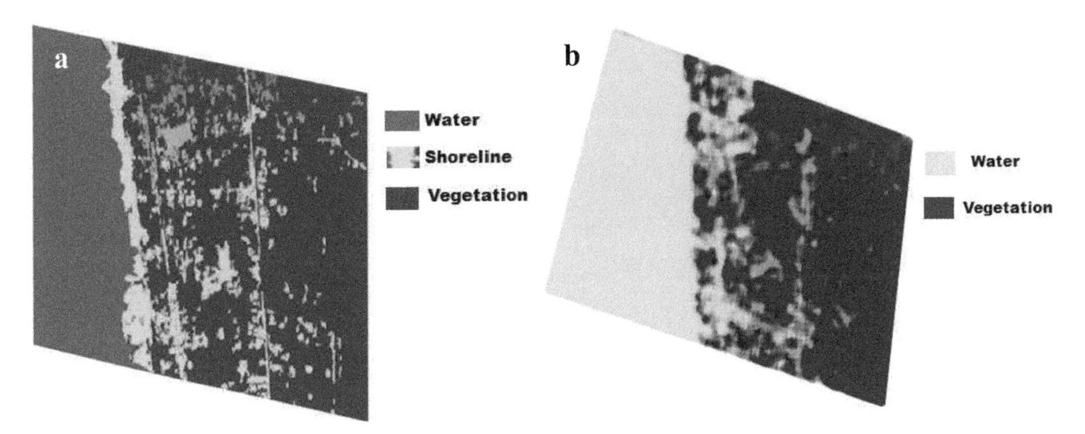

Figure 8.10 Image segmentation (a) pre and (b) post-tsunami for Kalutara coastal zone.

8.3.2 Theory of Edge Detection

Edge detection is an expression in image processing and computer vision, principally in the fields of feature recognition and feature mining, to denote the algorithms which propose for detecting points in a digital image at which the image brightness varies abruptly or, has incoherence. The mining of edges or shapes from a two dimensional array of pixels (a gray-scale image) is a precarious step in many image processing techniques. A variance of computation is available which defines the degree of contrast changes and their orientation.

There are various techniques to attain edge detection. Nevertheless, the main edge detection of distinctive techniques can be categorized into (i) Laplacian; and (ii) Gradient. In gradient technique, the discriminated edge is spotted by estimating the maximum and minimum of pixel intensities in the accurate result of the satellite data. Nonetheless, zero crossing is a keystone of the Laplacian algorithm within the second derivative of the image to automatic edge detection. Consequently, an edge presents the one-dimensional slope of the object and estimating the spinoff of the image can emphasize its region [125].

Let us assume that the boundary of coastline and water pixels have a swift gradient path which may be casted through the perspective of the gradient vector that is located at boundary pixels. At an ideal frame pixel, there are rapid intensity modifications from 0 to 255. This is shown on the gradient path (Figure 8.11). The magnitude of the gradient exposes the energy of the edge. If we calculate the gradient at uniform regions, we end with a zero vector which means that there is no edge pixel [121–126].

It is obvious that the discontinuity boundary is absent in derivate images or along the uniform pixels (Figure 8.11). Then, to detect the pixel edges, the gradient magnitude must be estimated. Therefore, the most naive edge processing is a function of a threshold. In fact, the corresponding pixels are concerned as an edge pixel when the magnitude of the gradient intensity is larger than compared to the threshold. In this view, it can be described as (i) edge strength, which is equivalent to the magnitude of the gradient; and (ii) edge direction, which is equivalent to the angle of the gradient. Basically, the gradient is not defined at all for a discrete function. In this regard, the edge gradient can be explained by an ideal continuous pixels and can be estimated using some operators such as the Gaussian and Laplacian operators [124–126].

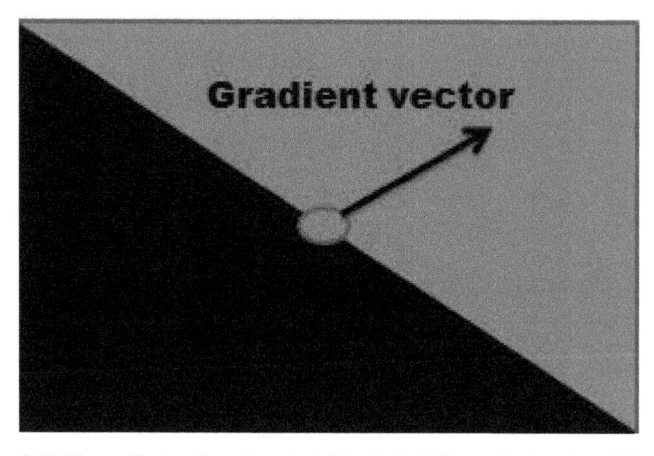

Figure 8.11 The gradient and an edge pixel. The circle indicates the location of the pixel.

8.3.3 Sobel Algorithm

Consistent with the Sobel attributes, the entire factor of the algorithm corresponds to zero, which, in certain circumstances, can reduce the run-off awkwardness. The Sobel operator instigates a 2-D spatial gradient magnitude on an image and so accentuates regions of high spatial frequency that resembles edges.

Usually, it is exploited to acquire the approximate absolute gradient magnitude at every pixel in an input grayscale Quickbird image [125].

The Sobel operator relies on central differences which is viewed as a guess of the first Gaussian derivative. This is equivalent to the first derivate of Gaussian blurring image ($\frac{\partial}{\partial x}G$) acquired by implementing a 3 × 3 mask to the image. Convolution is both commutative and associative, and is mathematically be written as:

$$\frac{\partial}{\partial x}=(I*G)=I*\frac{\partial}{\partial x}G \tag{8.3}$$

where I is the input Quickbird image, and G is the Gaussian blurs image. A 3 × 3 digital approximation of the Sobel operator is given as

$$\nabla I=\left|(z_7+2z_8+z_9)-(z_1+2z_2+z_3)\right|+\left|(z_3+2z_6+z_9)-(z_1+2z_4+z_7)\right| \tag{8.4}$$

These masks determine the vertical and horizontal edges through each pixel, respectively. The absolute magnitude of the gradient at each pixel and its orientation can be detected through the gradient component of ∇I. Then, the pseudo-convolution operator is used to compute the two components of the gradient and also combine in a single pass over the input image (Figure 8.12).

Then the masks are executed as follows:

$$M_x=\begin{bmatrix}-1 & -2 & -1\\ 0 & 0 & 0\\ 1 & 2 & 1\end{bmatrix} \text{ and } M_y=\begin{bmatrix}-1 & 0 & 1\\ -2 & 0 & 2\\ -1 & 0 & 1\end{bmatrix} \tag{8.5}$$

where M is masking in x and y, respectively. On other words, the $\boldsymbol{M_x}$ mask emphasizes the edges in the horizontal direction while the $\boldsymbol{M_y}$ masking underlines the edges in the vertical direction. Then the edge is generated by computing the resulting of masks in x and y directions.

2-D spatial mapping of gradient intensity at each pixel is generated by the Sobel operator. The Sobel mask assigns the most significance on the core pixel (Figure 8.13).

The mask is fallen over a window of the input image, changes that pixel's value and then shifts one pixel to the right and continues to the right until it reaches the end of a row. It then starts at the beginning of the next row. Figure 8.12 demonstrations the mask being felt over the top left portion of the input image. The formula shows how a particular pixel in the output image would be calculated. The center of the mask is placed over the manipulated pixel in the Quickbird image. In this understanding, pixel (a_{22})

z_1	z_2	z_3
z_4	z_5	z_6
z_7	z_8	z_9

Figure 8.12 Pseudo-convolution masks used to quickly compute approximate gradient magnitude.

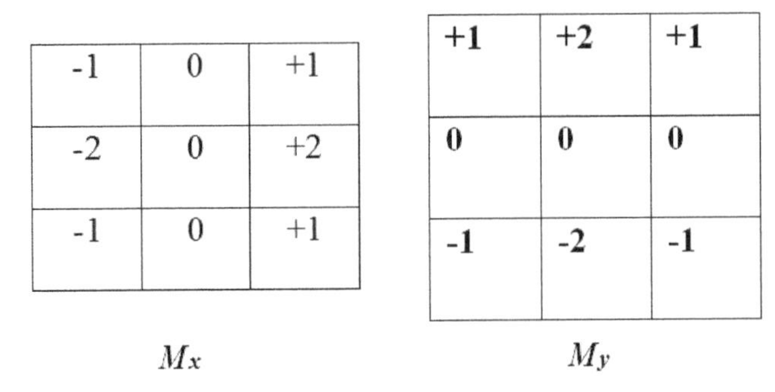

Figure 8.13 A 3 × 3 Sobel mask.

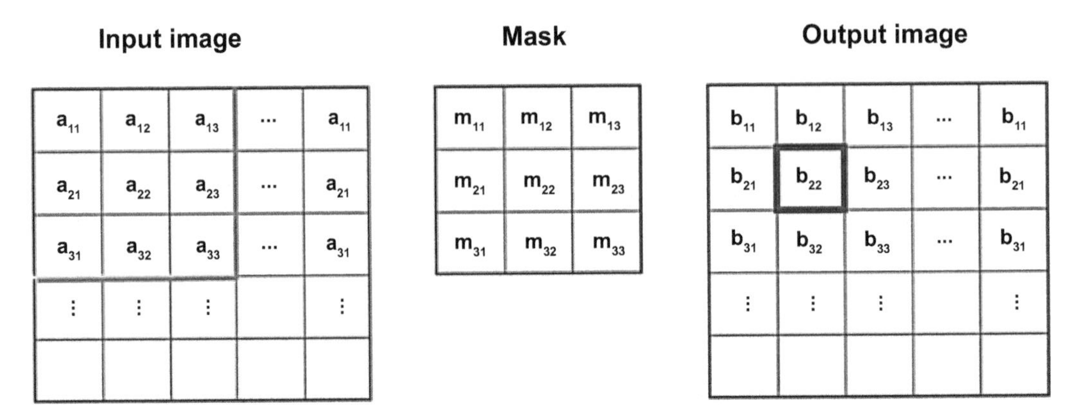

Figure 8.14 Example of Soble algorithm implementation.

can multiply by the resultant mask value (m_{22}) (Figure 8.14). It is imperative to observe that pixels in the first and last rows, along with the first and last columns cannot be manipulated by a 3 × 3 mask. Indeed, the mask cannot match the image edges as it is located over a pixel in the first row [121].

8.3.3.1 Sobel Algorithm Output

Shoreline automatic detection, pre and post-tsunami, are performed by the Sobel algorithm (Figures 8.15 and 8.16). The Sobel algorithm detects the different pattern of shoreline, pre and post-tsunami. During the pre-tsunami, the shorelines are presented in the form of zigzag pattern. During the post-tsunami, the shorelines tended to modify in the form of zigzag pattern because of the widespread of the tsunami run-up. However, the shoreline edge detection is presented in discontinuous lines during post-tsunami. The missing pixels along the shoreline boundaries are due to tsunami run-up impact which caused both sedimentation and erosion (Figure 8.16). In this view, the Sobel algorithm produces image with the areas of a high gradient (the likely edges), visible as thick dark lines.

The automatic detection of shoreline along Kalutara, Sri Lanka presents perfect boundaries than the one detected of Gleebruk (Figures 8.17 and 8.18). It is interesting to notice that post tsunami, Gleebruk's shoreline shifted more widely (Figure 8.16) than pre-tsunami. This is attributed to a massive destruction that occurred along Aceh coastline.

The Sobel operator smooths the input image to a great extent where natural edges are formed into lines in the output image with several pixels wide and less sensitive to noise. Using a Sobel filter, the edges of vegetation lines are detected to form a vector line representing shoreline, pre and post tsunami.

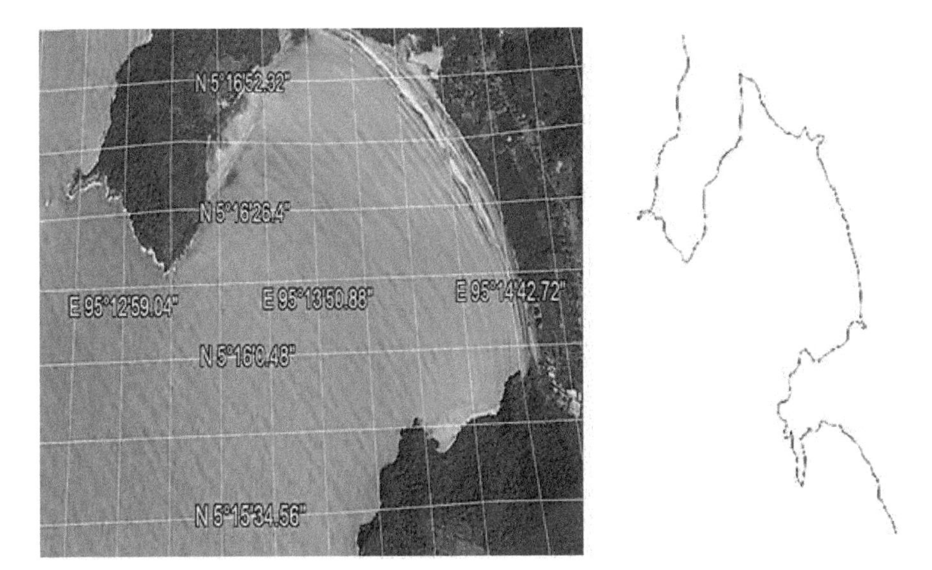

Figure 8.15 Shoreline of Gleebruk coastal zone detected by Sobel algorithm, pre tsunami.

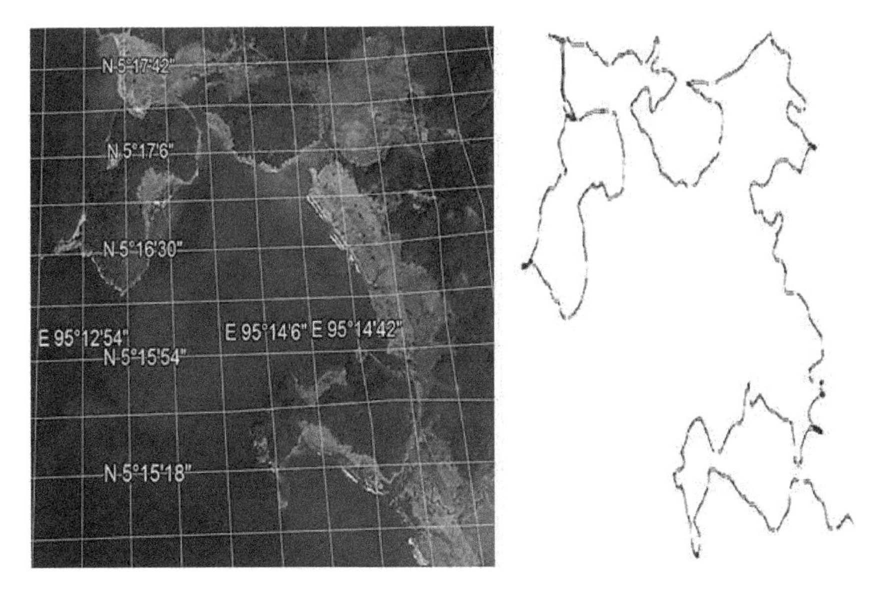

Figure 8.16 Shoreline of Gleebruk coastal zone detected by Sobel algorithm, post tsunami.

The basic statistical analysis of standard deviation is used to evaluate the Sobel algorithm performance (Table 8.2). It is noteworthy to realize that the standard deviation of shoreline edge detection earlier than tsunami disaster is smaller than the post-tsunami catastrophe. Pre-tsunami, the standard deviation ranges between 33.5 and 55.93 while the standard deviation post-tsunami ranges between 52.85 and 76.98. Certainly, the magnitude of edge detection is elevated because of the gradient in the pixel bright values along edge boundary. This could be noticed in the pre-tsunami images. The image of Gleebruk shows lower standard deviation as compared to Kalutara image because of the nonexistence of brighter pixels in shoreline border because of mud flow as a result of receding the water. It is far more visible in Figure 8.16.

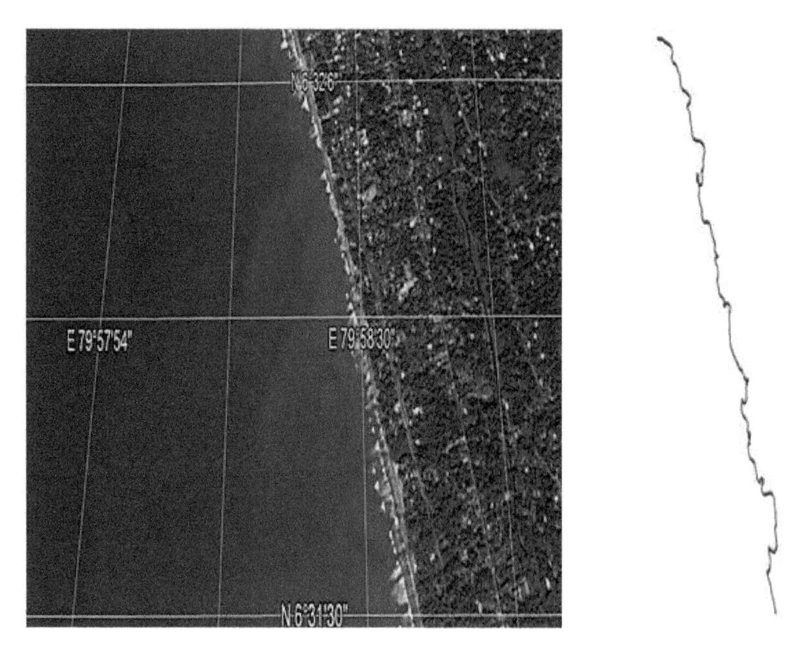

Figure 8.17 Shoreline of Kalutara, Sri Lanka detected by Sobel algorithm, pre-tsunami.

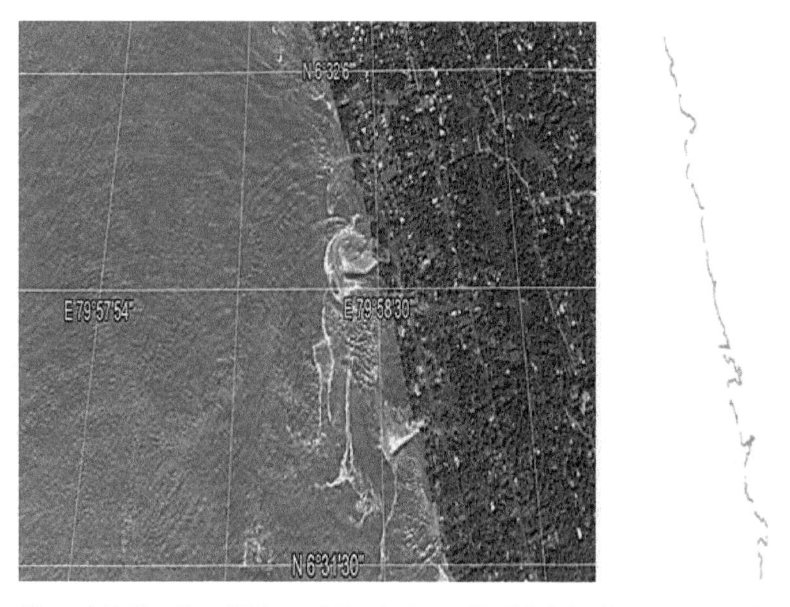

Figure 8.18 Shoreline of Kalutara, Sri Lanka detected by Sobel algorithm, post-tsunami.

Table 8.2 Standard deviation for Sobel edge detection algorithm pre and post-tsunami in Kalutara and Gleebruk.

Data	**Standard Deviation**	
	Pre-tsunami	**Post-tsunami**
Kalutara, Sri Lanka	55.93	76.98
Gleebruk, Indonesia	33.89	52.85

8.3.4 Canny Algorithm

Canny algorithm detection is a method to extract beneficial structural data from distinct vision items and dramatically lessen the quantity of facts to be processed. It has been extensively implemented in numerous computer vision systems. Canny has determined that the necessities for the utility of edge detection on diverse vision systems are particularly comparable. Consequently, an edge detection technique to deal with these necessities may be carried out in a wide range of conditions. The overall criteria for edge detection consists of: (i) detection of edge with low mistakes rate: this means that the detection ought to appropriately capture as many edges proven within the image as feasible; (ii) the brink point detected from the operator ought to appropriately localize at the center of the edge; and (iii) a given edge within the image have to be marked as soon as an edge gradient is determined, and wherein feasible, image noise should not create fake edges [123].

Primarily based on those criteria, the canny edge detector first smooths the Quickbird data to cast off any noise. It then reveals the image gradient to spotlight pixels with excessive spatial derivatives. The set of rules then tracks alongside those pixels and suppresses any pixel that is not always at the maximum suppression. The gradient array is now further decreased by hysteresis. Hysteresis is used to detect alongside the closing pixels which can no longer be suppressed. Hysteresis makes use of thresholds and if the magnitude is lower than the primary threshold, it is set to 0 (made a non-edge). If the magnitude is above the excessive threshold, it is made an edge. And if the magnitude is between the two thresholds, then it is set to 0 except there may be a route from this pixel to a pixel with a gradient above the second edge [125].

To fulfill those necessities, Canny used the calculus of variations—a method which reveals the feature which optimizes a given feature. The most reliable feature in Canny detector is defined through the sum of four exponential terms; however, it could be approximated by means of the primary spinoff of a Gaussian operator [124].

The Canny detector of edge detection splits into five distinctive steps:

1. Implement a Gaussian algorithm to smooth the image in order to dispose of the noise [127].
2. Locate the intensity gradients of the Quickbird data.
3. Implement non-maximum suppression to remove spurious reaction to edge detection.
4. Determine a double threshold to identify perfect edges [124].
5. Tune pixels by means of hysteresis. Finalize the detection of edges by way of suppressing all of the different edges which are susceptible and no longer linked to robust edges [125–127].

8.3.4.1 Apply Gaussian Filter to Smooth the Image in Order to Remove the Noise

A Gaussian low pass filter G is applied within the Canny algorithm to smooth the Quickbird images. In this view, the width of the algorithm and hence the amount of smoothing are determined by the standard deviation σ. Let $I(x, y)$ denote the input of QuickBird images. Then the convoluted image output is an array of smooth data [124],

$$S(x, y) = G(x, y, \sigma) * I(x, y) \tag{8.6}$$

where $S(x, y)$ is the spread of the Gaussian low pass filter and controls the degree of smoothing.

8.3.4.2 Finding the Intensity Gradient of the Image

Assume the window kernel size of 3×3 pixels and lines along the row x and column y. Estimating the gradient vector of smooth image $S(x, y)$ is a particularly induced edge enhancement. The mathematical description of this step as a function of Equation 8.6 is given as [125]:

$$g_x(x, y) \approx S(x + 1, y) - S(x - 1, y) \text{ and } g_y(x, y) \approx S(x, y + 1) - S(x, y - 1) \tag{8.7}$$

where the kernel window sizes of 3 × 3 pixels and lines are [127]

$$h_x = \begin{bmatrix} -1 & 0 & 1 \\ -1 & 0 & 1 \\ -1 & 0 & 1 \end{bmatrix} \text{ and } h_y = \begin{bmatrix} -1 & -1 & -1 \\ 0 & 0 & 0 \\ 1 & 1 & 1 \end{bmatrix} \tag{8.8}$$

Efficient implementation of the Canny detector combines the smoothing and enhancement steps with the aid of convolving the Quickbird data with the spinoff of the Gaussian kernel (G_{ij}). Considering all boundary detection, consequences are effortlessly tormented by the Quickbird data noise. It is important to remove the noise to prevent false alarm edge pixel detection as a result of noise. To smooth the image, a Gaussian algorithm is carried out to convoy with the Quickbird data. This step can slightly smooth the data to reduce the consequences of obvious noise on the threshold detector. The equation for a Gaussian algorithm kernel of size (2k + 1) × (2k + 1) is given by means of:

$$G_{ij} = \frac{1}{2\Pi\sigma^2}\exp\left(-\frac{(i-(k+1))^2 + (j-(k+1))^2}{\sigma^2}\right); 1 \le i, j \le (2k+1) \tag{8.9}$$

A sequence of accurate steps are required to implement effective Canny edge detector algorithm. First, the authentic image is smoothed by the Gaussian algorithm to locate and decide any edges. Due to the reality that the Gaussian algorithm can be computed by using a simple mask, it is used absolutely within the Canny algorithm. As soon as an appropriate mask is computed, the Gaussian mask can be performed by fashionable convolution techniques. A convolution mask is commonly smaller than the original image. As a result, the mask is slid over the image, manipulating a rectangle of pixels at a time. In this understanding, the larger the width of the Gaussian mask decreases the image noise. The localization error inside the detected edges, moreover, increases slightly because the Gaussian width is expanded.

Once the edge direction is known, the next step is to relate the edge direction to a direction that can be traced in an image. Therefore, there are only four directions which describes the surrounding pixel of the shoreline boundaries. The horizontal direction is assigned with 0° while the vertical direction is ordinated with 90°. Thus, other directions involve the positive diagonal within 45° and along the negative diagonal within 135°, respectively. In this view, the shoreline edge orientation must resolve into one of these directions. In other words, this orientation is a function of the close direction. For instance, the orientation angle of 3° must resolve into 0°. The magnitude and orientation of the gradient, therefore, are computed using:

$$Magnitude(x, y) = |g| = \sqrt{g_x^2 + g_y^2} \tag{8.10}$$

The edge direction is calculated using:

$$\theta = \tan^{-1}\left(\frac{g_y}{g_x}\right) \tag{8.11}$$

However the gradient in the *x* and *y* directions can involve an error when $\sum x = 0$. Under this circumstance, the edge direction must be equal to 90° or 0°, which depends on what the gradient's value in the *y*-direction is equal to. If g_y has a value of zero, the edge direction equals 0°. Else, the edge direction equals 90° [127].

8.3.4.3 Non-maximum Suppression

The non-maximum suppression applies to estimated edge direction. In this regard, non-maximum suppression traces the edge direction and suppresses any pixel value not belonging to a known edge. In this context, a thin line of known edge is generated in the output Quickbird images [126].

8.3.4.4 Double Threshold

Double threshold is applied to accurate edge which is produced by non-maximum suppression. This procedure aims at reducing the noise along the shoreline detected edge. Further, double threshold removes the weak gradient value and preserves the edge with the high gradient value.

As a result, threshold values are set to elucidate the special varieties of edge pixels: one is known as high threshold rate and the other is known as the low threshold value. For instance, if the threshold pixel's gradient rate is higher than the immoderate threshold rate, they are marked as robust facet pixels. If the threshold pixel's gradient value is smaller than the excessive threshold rate and larger than the low threshold rate, they may be marked as weak edge pixels. If the pixel rate is smaller than the low threshold rate, they may be suppressed [127].

8.3.4.5 Edge Tracking by Hysteresis

Lastly, hysteresis is cast off as a method of excluding streaking. In fact, Streaking is the splitting of an edge contour as a result of the operator output fluctuating above and under the threshold. If a single threshold, T_1 is implemented to Quickbird image, and an edge has a median strength identical to T_1, then because of noise, there may be instances wherein the threshold dips under the threshold. Similarly, it is going to additionally increase above the threshold, making an edge look like a dashed line. To avoid this, hysteresis makes use of a high and a low threshold. Any pixel within the shoreline that has a value greater than T_1 is presumed to be an edge pixel and marks shoreline boundary. Then, any pixel which might be connected to this edge pixel and which has a value more than T_2 is also decided on as edge pixels of shoreline boundary. In this view, the gradient must be under T_1 to detect the shoreline edge pixels.

8.3.4.6 Canny Algorithm Output

The Quickbird pre and post-tsunami images are examined by Canny algorithm. Canny algorithm is able to detect very thin edges and there are no false alarm edge points along the shorelines because integrity is not sensitive to noise (Figures 8.19 to 8.22). It is shown that the edges of shoreline detected are very smooth and fine. However, there are a lot of lines generated from other edges of buildings and roads as the Canny algorithm is very sensitive to edges (Figures 8.21 and 8.22). The inner lines represent the infrastructures while the outer line represents the shoreline (Figures 8.21 and 8.22).

Indeed, Canny algorithm is an optimal step for edges corrupted by white noise. The maximum of the detector is related to three criteria (i) the detection criterion that expresses the fact that important edges should not be missed; (ii) the localization criterion considers that the distance between the actual and local position of the edge should be minimal; and (iii) the one response criterion minimizes multiple responses to a single edge. This is partly covered by the first criterion, since when there are two responses to a single edge, one of the them should be considered as false which solves the problem of an edge corrupted by noise and works against non-smooth operators. Figures 8.21 and 8.22 show a feature synthesis approach. This explains that all significant edges from the operator with the smallest scale are marked first, and the edges of a hypothetical operator with the larger standard deviation are synthesized from them. Further, additional edges are marked only if they have a significantly stronger response than that predicted from synthetic output.

Table 8.3 shows the statistical analysis of Canny algorithm. It is interesting to find that the standard deviation of shoreline edge detection before tsunami disaster is smaller than the one observed after the

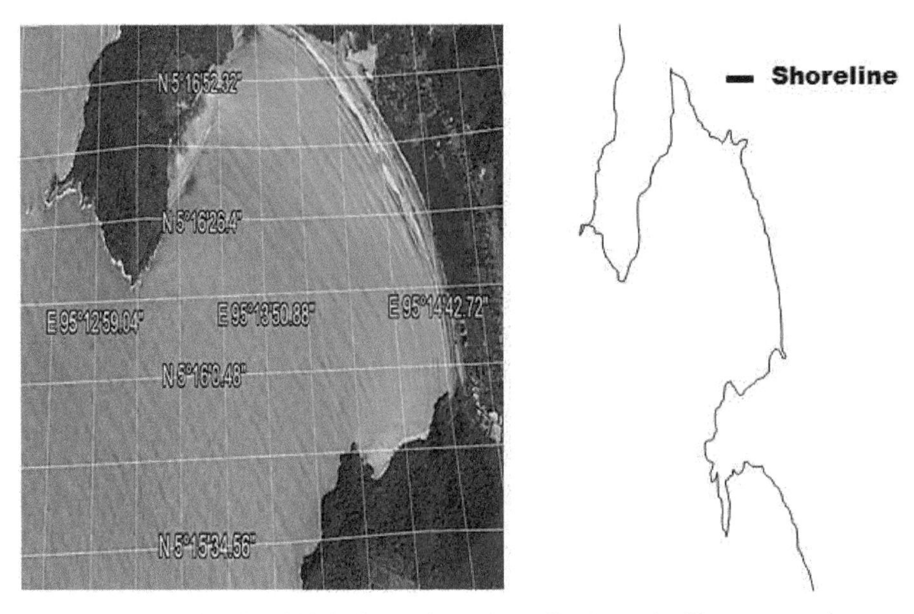

Figure 8.19 Shoreline of Gleebruk coastal zone detected by Canny algorithm, pre-tsunami.

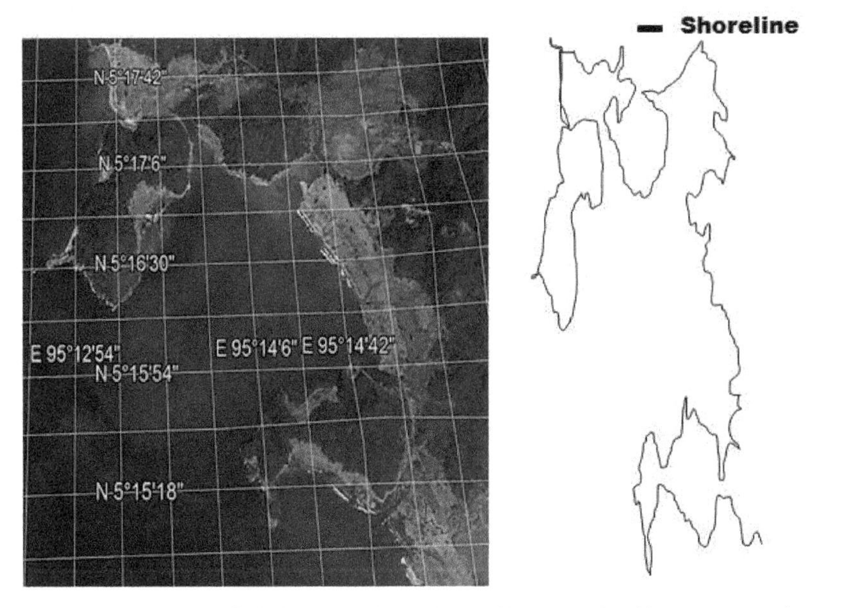

Figure 8.20 Shoreline of Gleebruk coastal zone detected by Canny algorithm, post-tsunami.

tsunami disaster. This finding is similar to Sobel algorithm. Nevertheless, Canny algorithm shoreline edge detection prior and after the tsunami disaster shows lower standard deviation values as compared to a Sobel algorithm (Table 8.3). This proves that the Canny algorithm detects very high details of shoreline edges (Figures 8.19 to 8.22) and automatically generates continuous vector lines.

Generally, Canny algorithm has a more advantageous performance as compared to Sobel algorithm because it implements probability for error reduction. Further, the Canny algorithms also improve signal to noise ratio in addition to noise condition detection.

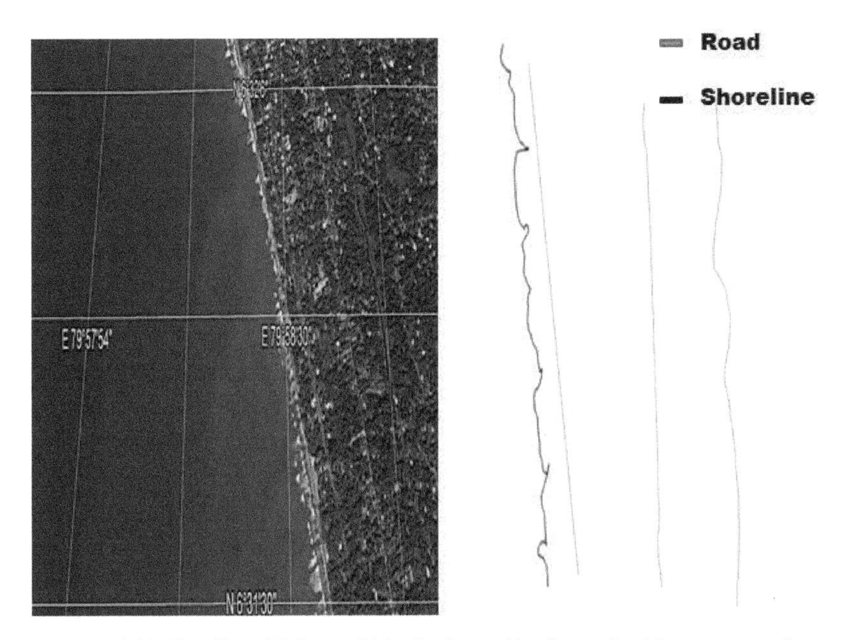

Figure 8.21 Shoreline of Kalutara, Sri Lanka detected by Canny algorithm, pre-tsunami.

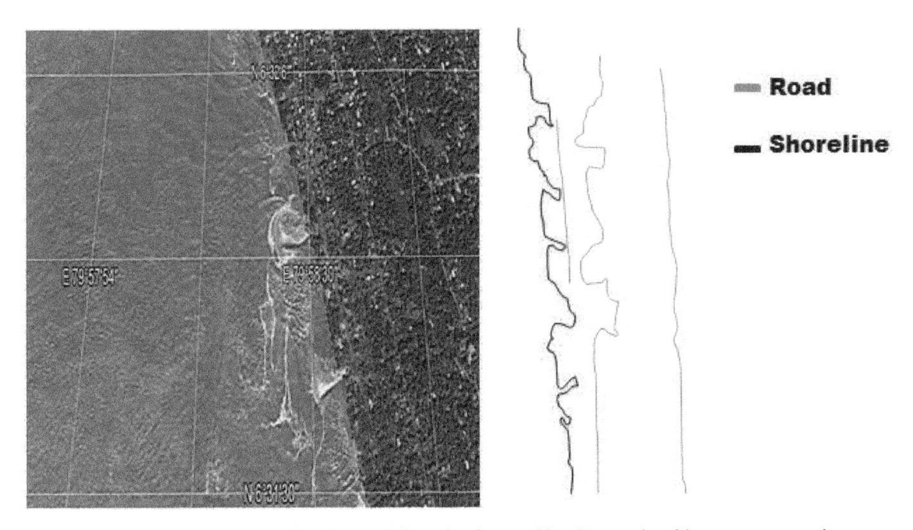

Figure 8.22 Shoreline of Kalutara, Sri Lanka detected by Canny algorithm, post-tsunami.

Table 8.3 Standard deviation obtained by Canny edge detection algorithm for before and after tsunami image in Kalutara and Gleebruk.

Image	**Standard Deviation**	
	Pre-tsunami	**Post-tsunami**
Kalutara, Sri Lanka	35.20	50.71
Gleebruk, Indonesia	26.99	23.67

8.4 Tsunami Impacts on Shoreline Deformation

The plants' line within the Quickbird data, pre and post-tsunami, are digitized into the vector layers which might be overlaid with the one detected by using the Sobel and Canny algorithms. On each vector, the road junctions are selected then the perpendicular line from the center of road junctions to shoreline is drawn.

Then, the distance from the middle of the intersection of the vector where the road and coastline intersect becomes decided. With the intention to compute the shoreline modification rate, 30 transects are plotted upright to the coastline to divert the overlapped historical shoreline vectors. The distance from every interception to the baseline is restrained. Following Chien (2007) [257], the pointwise technique is applied to estimate coastline modification charge. Consequently, the shoreline has a single cell-sized ΔX and ΔY and the angle Φ is used to determine its orientation (Figure 8.23). The deserter distance ΔR in x-coordinate is cast as:

$$\Delta R = \Delta R_x \cos^{-1} \Phi \tag{8.12}$$

The mathematical model for dissimilar retreat distances r_1 and r_2 for points Y1 and Y2 is formulated as:

$$r_1 + r_2 = 2 (\Delta R_x \cos^{-1} \Phi) \tag{8.13}$$

Equations 8.12 and 8.13 can be given as pseudocode to guess the shoreline change rates:

```
-----------------------------------------------------------------
Input dissimilar period vector layers of shoreline
 For every time step do
Estimate deformation rate ΔR for j = 1,2,......,N_y,
Estimate a different location of shoreline r[i] = 0.5 (ΔR_x,i −r_i−1),
End
-----------------------------------------------------------------
```

Statistical evaluation of two-way ANOVA and T-test are used to decide the accuracy of these techniques. In doing so, the envisioned shoreline rate changes through the use of automated edge detection algorithms and manual digitizing are compared.

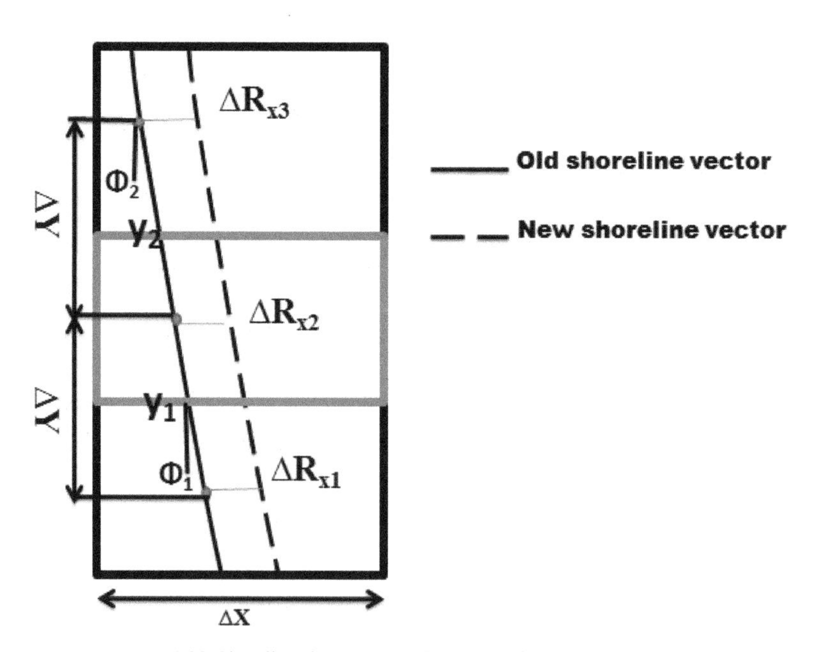

Figure 8.23 Shoreline change rate estimations using a pointwise model.

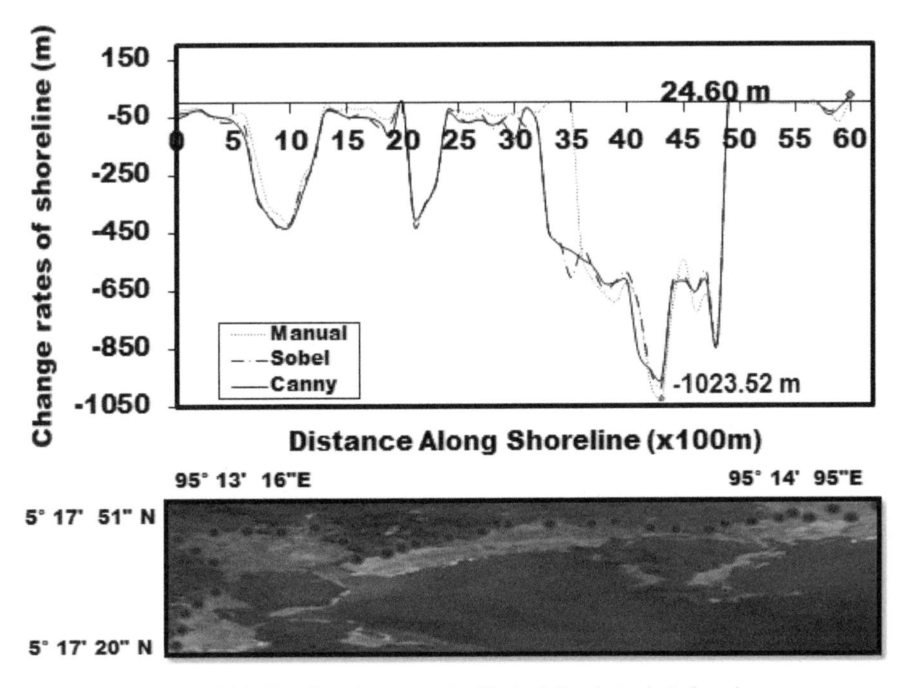

Figure 8.24 Shoreline change rate for Gleebruk Banda Aceh, Indonesia.

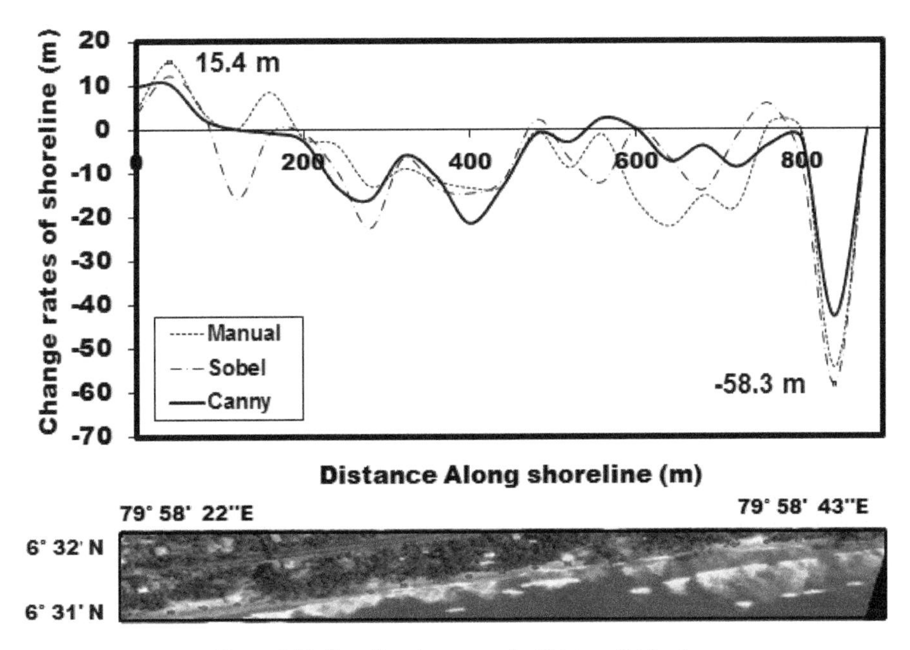

Figure 8.25 Shoreline change rate for Kalutara, Sri Lanka.

Figure 8.24 reveals that the Gleebruk coastline has the highest change rate of –1023.52 m than Kalutara. On the contrary, Kalutara is dominated by the maximum rate of accretion of 1 m (Figure 8.25). Consequently, the accretion is absent along the Gleebruk's shoreline. In reality, Gleebruk coastal zone is located close to the epicenter of the 2004 earthquake which created the tsunami. In this regard, the Banda

Table 8.4 ANOVA test for automatic detection algorithm and manual digitizing.

Data	Vector Layers	F_{stat}	F_{sig}	P-value	Significant
Kalutara	Manual *vs.* Sobel	0.0003	4.01	0.28	Sig
	Manual *vs.* Canny	0.41	4.06	0.03	Sig
Gleebruk	Manual *vs.* Sobel	0.34	3.92	0.44	Sig
	Manual *vs.* Canny	0.007	3.92	0.01	Sig

Table 8.5 The T-test shows significant difference existing between Sobel edge detection and canny edge detection algorithms.

Image	Analysis	df	t_{stat}	t_{crit}	P value	Output
Kalutara	Canny *vs.* Sobel	22	0.3	2.08	0.4	$t_{stat} < t_{crit}$ $P < 0.5$ Sig
Gleebruk	Canny *vs.* Sobel	60	–3.0	2.001	0.002	$t_{stat} < t_{crit}$ $P < 0.5$ Sig

Table 8.6 RMSE obtained for comparison of automatic edge detection algorithm with manual digitization.

Image	Analysis	R^2	RMSE (±m)	Std Error
Kalutara	Manual *vs.* Sobel	0.68	7.90	1.65
	Manual *vs.* Canny	0.72	6.82	1.42
Gleebruk	Manual *vs.* Sobel	0.84	18.94	2.42
	Manual *vs.* Canny	0.85	16.35	1.09

Aceh was the primary region struck by the tsunami wave energy of 5 megatons which triggered massive destruction run-up of 34 m which moved as far as more than 10 km inland [116–120]. In short, the tsunami wave washed up all of the wooden houses and infrastructures which are placed along the coastline and inland, also, inflicting the dramatic destruction of coastal geomorphology [256].

It is noteworthy to understand that there are significant differences between manual digitizing and automatic detection algorithms (Table 8.4). Indeed, the F_{sig} is larger than F_{stat} where p value is less than 0.5. Table 8.4 designates that the Canny algorithm performs better than Soble algorithm. In fact, t_{crit} is larger than t_{stat}. For instance, t_{crit} is 2.08 whereas t_{stat} is 0.3 in case of QuickBird data of Kalutara.

The shoreline rate changes of automatic edge detection algorithm are assessed with manual digitization to determine the accuracy of the outputs. The Root Mean Square Error (RMSE) is calculated to determine the accuracy of shoreline rate changes obtained using automatic detection with manual digitization. The ANOVA analysis is used to determine the significant relationship existing between the automatic edge detection algorithms and manual digitization. The T-test is applied to determine the significant difference existing between Sobel edge detection and canny edge detection algorithms. Further, Table 8.6 confirms that the Canny algorithm provides an accurate estimate of shoreline change rate as its RMSE equals to ± 6.82 m, which is smaller than Sobel algorithm. In fact, the Canny algorithm produces vector lines close to manual digitizing. This is because canny algorithm generates shorelines very close to true edges when compared to manual digitizing vectors. Canny is able to detect a sequence of pixels along the curve of the images and generate vector lines. It is able to link pairs of polylines and forms a smooth continuous line. Lastly, the smallest standard deviation for edge detection algorithm gives the precise edge detection.

8.5 The Role of Vegetation Covers on Tsunami Wave Energy Reduction

The high densities of coastal vegetation covers have performed a wonderful role to reduce the massive tsunami demolitions. Therefore, the strength of tsunami destruction is based not on their run-up effects and offshore traits but also on landscape appearances, for example, topography, land cover, and coastal geomorphology. The dense plants' covers have assisted humans to escape from tsunami catastrophe waves and saved human lives. Nonetheless, the growth of urban in the coastal zone by means of wooden houses, for instance alongside the Aceh coastal water, have elevated casualties. In contrast, the dense vegetation covers in Thailand, Sri Lanka, and India have created momentous obstacles between the coastline and higher ground, obstructing human beings' functionality to avoid natural disaster such as a tsunami. In this regard coastal inland infrastructures can be protected from a tsunami by planting heavy cover of agroforests.

This would favor the protection of heavy, dense vegetation covers along the coastal zone and not allow the reduction of naturally scattered trees and mangrove. Coastal settlements should be built away from the shoreline at high elevated areas. In fact, coastal vegetation covers have demonstrated drop of casualties because of dense agroforest covers in the front of settlements [256]. Under these circumstances, the successive tsunami wave energy has been trapped and reduced due to the dense vegetation covers as seen in Kalutara. Distribution of extra dense agroforests (i.e., cacao, rubber and multi-layered home gardens) in between the sea and districts should be an essential spatial planning measure. Bayas et al. (2011) [129] agrees with the above comments.

The Kalutara shoreline, however, is dominated by a low rate of coastline destruction because of the existence of the massive intensity of coconut trees alongside the coastline. These bushes, consequently, act as natural wave breakers and reduce the intensity of tsunami run-up into inland. In this view, this observation confirms the necessity of shielding the vegetation covers with a height of 30 m alongside any coastal region to prevent from the effect of natural disasters. Consequently, those flora covers can act as a defensive wall of a sort from natural disasters, including a tsunami. It is also advised that all coastal cities and their infrastructures should be located some distance away from the coastline,approximately at the distance of 20 km to avoid massive damages. It can be said that computerized edge detection techniques can be used as an automatic geomatic solution for quick feedback on feature modifications because of natural disasters.

Chapter 9

Modelling of Tsunami Impacts on Physical Properties of Water using MODIS Data

A Study Case of Aceh, Indonesia

9.1 Introduction

It is needless to mention that remote sensing is highly useful for modelling, monitoring and mapping the tsunami drifts on the earth's surface [17–22]. It is a great potential for appraising, monitoring and mapping various parameters relating physical properties of sea surface. Recent advancement in remote sensing towards data acquisition and integration of spatial and temporal physical properties of sea water models provided a renewed prospect for managing and evaluating the sea surface water quality problems due to tsunami appearances. This chapter proposes some novel mathematical algorithms to retrieve numerous sea water quality parameters using Moderate Resolution Imaging Spectrometer (MODIS). The developed model is implemented to assess and map these water quality parameters distributions in the context of Aceh coastal waters.

The advances made in water tsunami modeling using remote sensing data and information systems coupled with decision support systems in the management process are increasingly being recognized. This chapter combines the optical remote sensing data with mathematical modeling to develop and support system concerning tsunami impacts on the surface water quality in Aceh coastal waters. It also explores the development of these tools to solve particular water quality problems. The main scope of the integration is to evidently comprehend the effects of tsunami on the sea surface water quality of different types. This integration is expected to deliver a precise evaluation of the water quality problems due to tsunami drifts and to develop remedial management actions for marine environmental protection in the future.

In order to study the consequences of the 2004 tsunami on sea surface salinity (SSS), sea surface temperature (SST), suspended sediment (SS) and ocean phytoplankton, a remote sensing analysis has been carried out using the MODIS instrument on-board the TERRA and AQUA satellites. The 2004 tsunami was of unprecedented strength and affected the largest portion of the Indian Ocean in recorded history. For instance, sea surface temperature, suspended sediment, and chlorophyll growths were observed using satellite remote sensing data over a large area, extending from the shallow areas along the coast to the deep water hundreds of kilometers away. By contrast, the water column temperature decreases dramatically from onshore to offshore during the tsunami and post tsunami periods.

Endlessly, the catastrophe of 2004 Boxing studies has augmented certainly [1]. Scientists recognize the mechanisms of the Indian Ocean tsunami of 2004. Beside, the Japanese tsunami is considered as a massive destructive disaster because of the M_w 9.0 earthquake that struck off the coast of Honshu, Japan on March 11 2011 [4].

Initial reports were similar to those on December 26, 2004, when a massive underwater earthquake off the coast of Indonesia's Sumatra Island rattled the Earth in its orbit. The 2004 quake, with a magnitude of M_w 9.1, was the largest one since 1964. But as in Japan, the most powerful and destructive aftermath of this massive earthquake was the tsunami that it caused. The death toll reached higher than 220,000 [1–3, 10].

Our objective is to address the following questions: (i) how the 2004 tsunami impacted coastal water properties? (ii) can remote sensing technology based on the MODIS data monitor the coastal water properties prior to and post the tsunami event? and (iii) do the statistically predicted algorithms obtain accurate answers on the tsunami impacts on the coastal water properties?

Since the interaction between the tsunami mechanism and coastal water body can be complex, we start with a simplified problem, and then progress to a more realistic setting. Our focus is on the impacts of the tsunami in sea surface salinity, suspended sediment, chlorophyll concentration, and sea surface salinity.

9.2 Coastal Water of Aceh

The examined area is placed along the western coastal zone of Aceh with boundaries of latitudes 3° 30' N to 6° 30' N and longitudes of 93° 30' E to 99° 30' E (Figure 9.1). The Sunda trench is strolling north-south alongside the coastal waters of Aceh closer to the Andaman Sea. Strolling in a roughly north-south line on the seabed of the Andaman Sea is the boundary between tectonic plates, the Burma plate and the Sunda Plate. These plates (or microplates) are believed to have formerly been a part of the bigger Eurasian Plate; however, they have been shaped whilst transforming fault activity intensified as the Indian Plate started out its substantial collision with the Eurasian continent [10, 13]. As a result, a returned-arc basin center was created, which commenced shaping the marginal-basin, which might end in the Andaman Sea, the modern-day levels of which started about 3–4 million years in the past. On December 26, 2004, a massive part of the boundary between the Burma Plate and the Indo-Australian Plate slipped, inflicting the 2004 Indian Ocean earthquake. This megathrust earthquake had a value of 9.3. Between 1300 to 1600 km of the boundary underwent thrust faulting and shifted by approximately 20 meters, with the ocean floor being uplifted by numerous meters. The spreading uplift of the seafloor generated a massive tsunami with an anticipated peak of 28 meters (30 ft) [14].

Consistent with Abbott (2008) [11], the Indian-Australian plate moves obliquely toward western Indonesia at 5.3 to 5.9 cm/yr (2 to 2.3 in/yr). The enormous, ongoing collision results in subduction-caused (Figure 9.2) earthquakes that are frequent and huge. Many of these earthquakes send off a tsunami. On 26 December 2004, 1.200 km (740 mi) long fault rupture began as 100 km (62 mi) long portion of the plate tectonic boundary, ruptured and slipped during a minute. The rupture then moved northward at 3 km/s (6,700 mph) for four minutes. It then slowed to 2.5 km/s (5,6000) during the next six minutes. At the northern end of the rupture, the fault movement slowed drastically and only traveled tens of meters during the next hour. It appears that a bend or scissors like tear in the subduction plate may have delayed the full rupture in December 2004.

The shallow part is located at the northern and eastern regions of the Andaman Sea with less than 180 m. Therefore, the average water depth is approximately 1000 m. However, the deepest water is located in west with approximately 3000 m. This part is dominated by deep submarine valleys which is placed at the east of the Andaman-Nicobar Ridge with 4000 m. Further, the ocean floor consists of pebbles, gravel and sand [14].

The monsoons of the Southeast Asia govern the climate and sea surface's physical properties. For instance, the average sea surface temperature (SST) ranges from 26–28°C in February and increases to 29°C in May. On the contrary, the deep water temperature layer of 1600 m is stable with 4.8°C. Consequently, the summer sea surface salinity ranges between 31.5 to 32.5‰. On the contrary, the winter sea surface salinity ranges between 30.0–33.0‰ along the southern part. In the northern part, however, the sea surface salinity reduces to the range of 20–25‰ because of the inflow of fresh water from the Irrawaddy River [23].

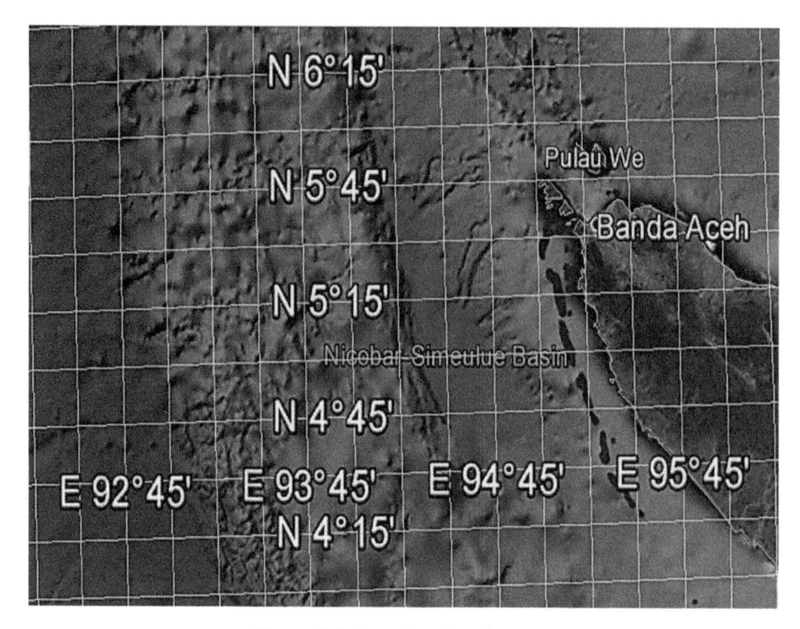

Figure 9.1 Location of study area.

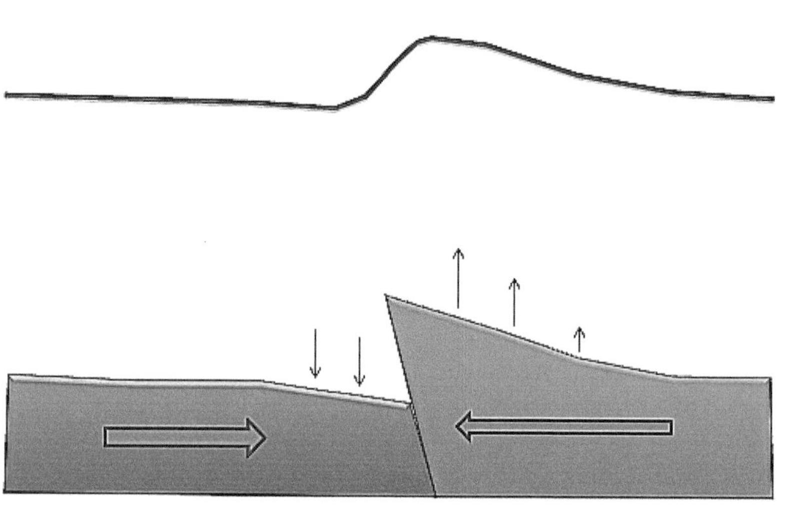

Figure 9.2 Deformation of sea floor and sea surface.

9.3 MODIS Satellite Data

Moderate Resolution Imaging Spectrometer (MODIS) is the sensor on-board platform of Terra and Aqua. This sensor has 36 spectral bands at moderate resolution (0.25–1 km). The spatial is 250 m for channel 1 and channel 2, 500 m for channel 3 to 7 and 1000 m for channel 8 to 36. These various bandwidths of each spectral bands are optimized for imaging specific surface and atmospheric features. The description for each spectral band is as shown in the Table 9.1. MODIS has the capability to view the surface of the Earth every 1–2 days making observation over land, ocean surface and clouds. MODIS is also able to measure various parameters regarding the surface of the Earth such as surface temperature (land and ocean) and fire detection, ocean color (sediment, phytoplankton), global vegetation maps and change detection,

Table 9.1 NOAA AVHRR spectral bands.

Channel	Width of Band (μm)	Region of Electromagnetic Spectrum	Use in General Remote Sensing
1	0.58–0.68	Visible	Visible/albedo/cloud
2	0.72–1.1	Near infrared	Vegetation/water/cloud
3	3.55–3.93	Middle infrared	Fire/warm surfaces
4	10.3–11.3	Far infrared	Sea surface temperature
5	11.5–12.5	Far infrared	Sea surface temperature

cloud characteristics, aerosol concentrations and properties, temperature and moisture soundings, physical characteristics of snow cover and ocean currents [99].

MODIS operates continuously during the day and night portions of each orbit. Data from all bands are collected during the day portion of an orbit, whereas only the thermal infrared data are collected during the night portion of an orbit. The instrument is calibrated periodically using the three internal targets-solar diffuser, black body, and spectroradiometric calibration assembly. The MODIS instrument consists of a cross-track scan mirror, collecting optics and individual detector elements. MODIS has the swath of 2330 km (across track) by 10 km (along track at nadir). Along track is the optical set up as well as the scanning mechanism of MODIS. This sensor makes observation within one scan ten lines of 1 km spatial resolution (40 lines of 250 m resolution and 20 lines of 500 m resolution, respectively). All MODIS data are recorded in 12 bits of radiometric resolution. The radiometric resolution indicates the grey levels of sensor in terms of the capability of the sensor to detect any small differences of the measured parameter from the earth's surface [100].

Satellite remote sensing observations from the NASA MODerate Resolution Imaging Spectroradiometer (MODIS) instrument can be used to derive chlorophyll. MODIS is a moderate resolution multi-spectral sensor currently flying on two NASA satellites, AQUA and TERRA. MODIS uses mid and thermal-IR for measuring the emissivity of the surface. MODIS chlorophyll products are corrected for atmospheric disturbances. The AQUA satellite was designed to observe the earth's water cycle. Its sun-synchronous polar orbit (south to north) passes over the Equator in the afternoon (13:30 p.m.), and it acquires data for nearly the entire earth, each day. The TERRA satellite was designed to collect data relating to the earth's biogeochemical and energy systems. TERRA is also in a sun-synchronous orbit and crosses the Equator at approximately 10:30 a.m. local time [99].

This study is based on standard AQUA and TERRA MODIS Level 3, 4.63 km gridded 3-day and 8-day composite products, generated through surface emissions. The gridded data are generated by binning and averaging the nominal 1 km swath observations, yielding a $\approx$ 4 km gridded global data. The data are distributed in HDF-EOS format. The use of MODIS data is limited to cloud-free conditions. Since MODIS is installed on both the TERRA and AQUA satellites, four data points are available per day. MODIS data can be freely obtained through direct broadcast, which requires an X-band antenna and its control equipment, or from the NASA MODIS website. The data are distributed in the HDF5-EOS format [99–101].

9.3.1 Comparison between MODIS and other Optical Satellite Sensors

The advanced remote sensing technology, nowadays, has been recognized as a powerful tool for retrieving physical ocean parameters such as SST, SSS and SSC. Several sensors are implemented accurately for this task, for instance, MODIS, NOAA AVHRR, and SeaWiFS. Nevertheless, MODIS data are acquired to retrieve Sea Surface Temperature (SST), Chlorophyll-a concentration and salinity due its superior characteristics as compared to NOAA AVHRR and SeaWiFS. In fact, MODIS data have an excellent spectral resolution than NOAA AVHRR and SeaWiFS. In this context, both NOAA and SeaWiFS have the spatial resolution of approximately 1 km at the nadir point. However, their global covaerges are about 4 km (Tables 9.1 and 9.2). On the contrary, MODIS data have different resolution for the 36 different spectral

Table 9.2 SeaWiFS spectral bands.

Bands	Wavelength Units (nm)	Wavelength Colors	Primary Uses
1	412	Violet	Dissolved organic matter
2	443	Blue	Chlorophyll absorption
3	490	Blue-green	Pigment absorption (case 2)
4	510	Blue-green	Chlorophyll absorption
5	555	Green	Pigments, optical properties, sediments
6	670	Red	Atmospheric correction and sediments (CZCZ) heritage
7	765	Near-IR	Atmospheric correction, aerosol radiance
8	865	Near-IR	Atmospheric correction, aerosol radiance

bands at moderate resolution of 0.25 km to 1 km. In this regard, bands 1 and 2 have 250 m spatial resolution, while bands 3 to 7 have a spatial resolution of 500 m. However, the spatial resolution increases to 1000 m through bands 8 to 36. Unlike MODIS and SeaWiFS, the NOAA AVHRR has a temporal resolution every two hours per day. Therefore, MODIS and SeaWiFS revisit the same place every 1–2 days [99].

MODIS data have similar ocean color bands as SeaWiFS bands but they are narrow (Table 9.3). The narrow bands improve the atmospheric correction. Under this circumstance, the MODIS ocean band at 748 nm [31] is about half the width of the equivalent SeaWiFS band and consequently, it avoids the nearby atmospheric oxygen absorption (Table 9.3).

Extra spectral bands in MODIS can permit supplementary research in an extensive range of use. Both MODIS and NOAA AVHRR can quantify SST even as SeaWiFS can't. Nonetheless, MODIS SST is superior to AVHRR SST because of the precise sensitivity and lower signal-to-noise attributes of the MODIS instrument (Table 9.3). The mid-IR channels are particularly beneficial within the excessive water vapor, low-latitude areas as compared with preceding radiometers. Ocean color tracking may be performed by the use of both sensors, NOAA AVHRR and MODIS [101]. As a consequence, the MODIS sensor is anticipated to provide precise information because of the narrower bandwidths, distinctive for ocean color and phytoplankton measurement as compared to the corresponding spectral bands in NOAA AVHRR [98–101].

9.4 Impact of Tsunami on Coastal Physical Properties

9.4.1 Retrieving Sea Surface Salinity and Suspended Sediment

The statistical learning theory provides a framework for salvaging the coastal water parameters of MODIS data prior to and post the tsunami event. In short, it accurately retrieves the linear correlation of MODIS bands and real coastal parameters, for instance sea surface salinity (SSS) and suspended sediment (SS).

Let $S \in I$, where $S = \{(SSS, SS)\} \in I$. In this view, SSS and SS are sea surface salinity and suspended sediment thats belong to the MODIS radiance data I. Let S have a linear relationship with the radiance data I. This can be expressed mathematically as:

$$S = b_0 + b_1 I_1 + b_2 I_2 + b_3 I_3 + \dots\dots\dots\dots + b_k I_k + \varepsilon \quad (9.1)$$

Therefore, Equation 9.1 can be written in terms of the observations as

$$S = b_0 + \sum_{i=1}^{k} b_i I_i + \varepsilon \quad (9.2)$$

Table 9.3 MODIS spectral bands.

Primary Use	Band	Bandwidth
Land/Cloud/Aerosols Boundaries	1	620–670 nm
	2	841–876 nm
Land/Cloud/Aerosols Properties	3	459–479 nm
	4	545–565 nm
	5	1230–1250 nm
	6	1628–1652 nm
	7	2105–2155 nm
Ocean Color Phytoplankton Biogeochemistry	8	405–420 nm
	9	438–448 nm
	10	483–493 nm
	11	526–536 nm
	12	546–556 nm
	13	662–672 nm
	14	673–683 nm
	15	743–753 nm
	16	862–877 nm
Atmospheric Water Vapour	17	890–920 nm
	18	931–941 nm
	19	915–965 nm
Surface/Cloud Temperature	20	3.660–3.840 μm
	21	3.929–3.989 μm
	22	3.929–3.989 μm
	23	4.020–4.080 μm
Atmospheric Temperature	24	4.433–4.498 μm
	25	4.482–4.549 μm
Cirrus Clouds Water Vapor	26	1.360–1.390 μm
	27	6.535–6.895 μm
	28	7.175–7.475 μm
Cloud Properties	29	8.400–8.700 μm
Ozone	30	9.580–9.880 μm
Surface/Cloud Temperature	31	10.780–11.280 μm
	32	11.770–12.270 μm
Cloud Top Attitude	33	13.185–13.485 μm
	34	13.485–13.785 μm
	35	13.785–14.085 μm
	36	14.085–14.385 μm

where S is the calculated sea surface salinity and suspended sediment concentration, k is a number of MODIS radiance bands which equals to 7 bands, b_0, b_i are constant coefficient of linear relationship between MODIS radiance data (I) and S. ε is residual error of S which are estimated from the selected MODIS bands. The unknown parameters in Equation 9.2, that are b_0 and b_i, may be estimated by a general least square iterative algorithm. This procedure requires certain assumptions about the model error component ε. Stated simply, we assume that the errors are uncorrelated and their variance is σ_ε^2. In general, if ε_i and ε_j are two un-correlated errors, then their covariance is zero, where we define the covariance as

$$Cov\,(\varepsilon_i, \varepsilon_j) = E\,(\varepsilon_i, \varepsilon_j) \tag{9.3}$$

The least-square estimator E of the b_i minimizes the sum of squares of the errors, say

$$E = \sum_{j=1}^{n} (S_j - b_0 - \sum_{i=1}^{k} b_i I_i)^2 = \sum_{j=1}^{n} \varepsilon_j^{\,2} \tag{9.4}$$

where S_j is the value of S measured at I_i, n is the total number of data points and $n \geq k$. It is necessary that the least squares estimators satisfy the equations given by the k first partial derivatives $\frac{\partial E}{\partial b_i} = 0$, $i=1,2,3,.....,k$ and $j=1,2,3,.....,n$. Therefore, differentiating Equation 9.4 with respect to b_i and equating the result to zero, we obtain

$$n\hat{b}_1 + (\sum_{j=1}^{n} I_{2j})\hat{b}_2 + (\sum_{j=1}^{n} I_{3j})\hat{b}_3 + ... + (\sum_{j=1}^{n} I_{kj})\hat{b}_k = S_j$$

$$(\sum_{j=1}^{n} I_{2j})\hat{b}_1 + (\sum_{j=1}^{n} I^2_{2j})\hat{b}_2 + ... + (\sum_{j=1}^{n} I_{2j} I_{kj})\hat{b}_k = \sum_{j=1}^{n} I_{2j} S_j$$

....

$$(\sum_{j=1}^{n} I_{kj})\hat{b}_1 + (\sum_{j=1}^{n} I_{kj} I_{2j})\hat{b}_2 + ... + (\sum_{j=1}^{n} I^2_{kj})\hat{b}_k = \sum_{j=1}^{n} I_{kj} S_j \tag{9.5}$$

The Equations (9.5) are called the least-squares normal equations. The $\hat{b}_k$ found by solving the normal Equations (9.5) are the least-squares estimators of the parameters b_i. The only convenient way to express the solution to the normal equations is in matrix notation. Note that the normal Equations (9.5) are just a k x k set of simultaneous linear equations in k unknowns (the $\{b_k\}$). They may be written in matrix notation as

$$H\,\hat{b} = h \tag{9.6}$$

where

$$H = \begin{bmatrix} n & \sum I_{2j} & ..... & \sum I_{kj} \\ \sum I_{2j} & \sum I_{2j}^2 & ..... & \sum I_{2j} I_{kj} \\ \sum I_{3j} & \sum I_{3j} I_{2j} & ..... & \sum I_{3j} I_{kj} \\ .... & .... & ...... & .... \\ \sum I_{kj} & \sum I_{kj} I_{2j} & .... & \sum I_{kj}^2 \end{bmatrix}$$

$$\hat{b} = \begin{bmatrix} \hat{b}_1 \\ \hat{b}_2 \\ \vdots \\ \hat{b}_k \end{bmatrix} \text{ and } h = \begin{bmatrix} \sum S_j \\ \sum I_{2j} S_j \\ \vdots \\ \sum I_{kj} S_j \end{bmatrix}$$

Thus, H is a k x k estimated matrix of MODIS radiance bands that are used to estimate sea surface salinity, and b and h are both $k \times 1$ column vectors. The solution to the least-squares normal equation is

$$\hat{b} = H^{-1} h \tag{9.7}$$

where H^{-1} denotes the inverse of the matrix H. Given a solution of least squares normal equations, the retrieval $(S)_{MODIS}$ is estimated using the fitted multiple regression model of Equation 9.2 as is given by

$$S_{MODIS} = \hat{b}_1 + \sum_{i=1}^{k} \hat{b}_i I_i \tag{9.8}$$

Following Marghany [17, 18], ε errors that represent the difference between retrieved and *in situ* SSS are computed within 10 km grid point interval and then averaged over all grid points having the same range of distance to coast, where the bias ε on the retrieved *SSS* or SS_{MODIS} is given by:

$$\varepsilon = \frac{\sum_{i=1}^{N} (S_{MODIS} - S_{in-situ})}{N} \tag{9.9}$$

where S_{MODIS} is the retrieved *SSS* and *SS* of MODIS satellite data, $S_{in\text{-}situ}$ is the reference of ground data of *SSS* and *SS*, respectively. Thus, the empirical formula of SSS_{MODIS} (psu) which is based on Equations 9.8 and 9.9 is

$$SSS_{MODIS}(\text{psu}) = 27.40 + 2.0\, I_1 - 3.4\, I_2 + 2.0\, I_3 + 2.2\, I_4 + 1.8\, I_5 + 0.3\, I_6 + -1.8\, I_7 \pm 1.1 \tag{9.10}$$

Further dissolved salts and suspended substances have a major impact on the electromagnetic radiation attenuation outside the visible spectrum range [21]. In this context, the electromagnetic wavelength larger than 700 nm is increasingly absorbed, whereas the wavelength less than 300 nm is scattered by non-absorbing particles such as zooplankton, suspended sediments and dissolved salts [17, 19].

Therefore, SS (mg/l) can be derived from the first seven bands of MODIS data as following [21],

$$SS_{MODIS}(\text{mg/l}) = -4.281 + 23.628\, I_1 - 15.675\, I_2 - 14.653\, I_3 + 79.251\, I_4 + 21.303\, I_5 + 9.709\, I_6 + 10.963\, I_7 \tag{9.11}$$

This equation is different than the equation that was obtained by Wong et al. (2007) [21] in terms of constant coefficient of linear models and involving the retrieved S_{MODIS} bias value. Finally, root mean square of bias (RMS) is used to determine the level of algorithm accuracy by comparing with *in situ* sea surface salinity. Further, linear regression model is used to investigate the level of linearity of sea surface salinity estimation from MODIS data. The root mean square of bias equals [17],

$$RMS = [N^{-1} \sum_{i=1}^{N} (S_{MODIS} - S_{situ})^2]^{0.5} \tag{9.12}$$

9.4.1.1 Tsunami Impact on Sea Surface Salinity and Suspended Sediment Variations

The tsunami impact on sea surface salinity has been examined on three MODIS satellite data along Aceh coastal waters. These data are acquired on 23rd, 26th and 27th, 2004, respectively (Figure 9.3). Consistent with Marghany (2009) [16], Moderate Resolution Imaging Spectroradiometer (MODIS) has 1 km spatial resolution and have 36 bands which range from 0.405 to 14.285 μm [24]. Further, the MODIS satellite takes 1 to 2 days to capture all the scenes in the entire world, acquiring data in 36 spectral bands over a 2330 km swath.

It is worth mentioning the existence of heavy cloud covers over Aceh on December 23th, 2004 (Figure 9.3a) as compared to after the tsunami on December 27th 2004 (Figure 9.3c). Truthfully, Aceh is situated in a tropical zone where the main atmospheric feature is the dense cloud covers. Figure 9.4 exposes the spatial variation of the salinity distribution along Aceh, which are derived using the linear least square algorithm. On December 23th, 2004, the sea surface salinity ranged between 28 psu to 31 psu. Nevertheless, during the tsunami on 26th December, 2004, the sea surface salinity dramatically increased and ranged between 34 psu to 36 psu (Figure 9.4b). The sea surface salinity continued to increase post tsunami and ranged between 37 psu to 38 psu.

The isohaline contours of sea surface salinity were derived from *in situ* data. These data are acquired from http://aquarius.nasa.gov/. Figure 9.5 demonstrates that the *in situ* sea surface salinity increased dramatically. Prior to the tsunami, the isohaline contours ranged between 28.5 to 29.0 psu (Figure 9.5a). The isohaline contours, however, increased to 36.7 psu during the tsunami (Figure 9.5b) and continued to increase to 37 psu through the post-tsunami (Figure 9.5c).

Both MODIS satellite data and *in situ* data confirm the homogenous sea surface salinity, pre-tsunami and post-tsunami, with the radical increment of sea surface salinity from 28.5 psu to 38.0 psu in coastal waters of Aceh. Figure 9.6 shows the comparison between *in situ* sea surface salinity measurements and

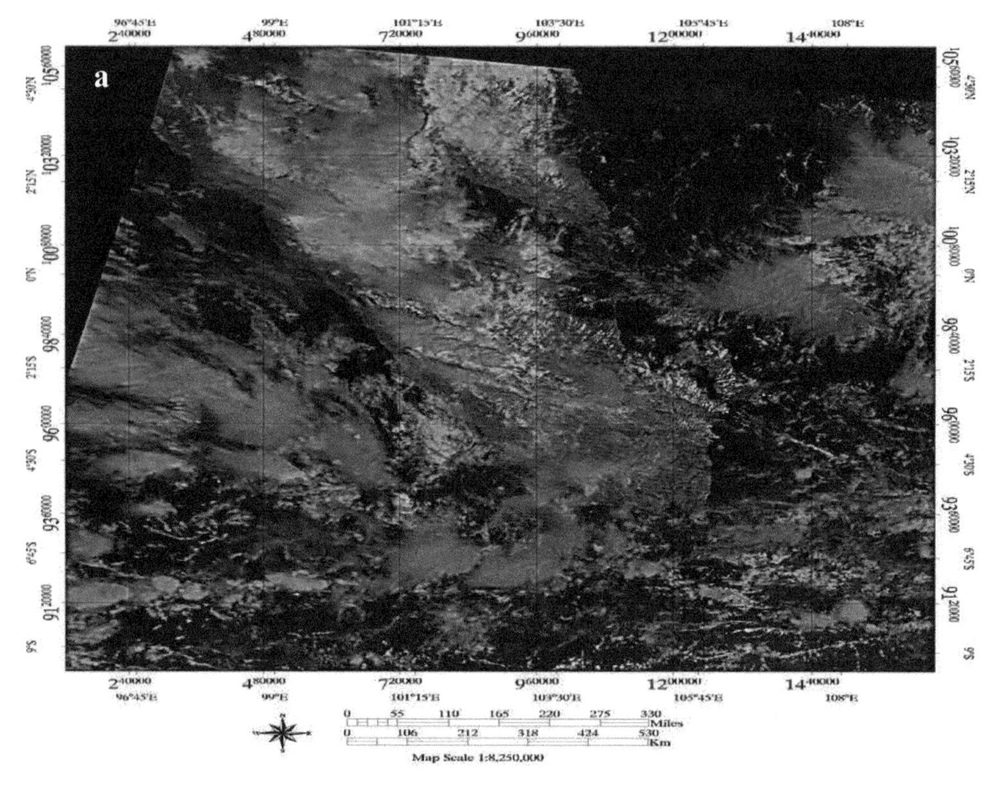

Figure 9.3 contd. ...

...Figure 9.3 contd.

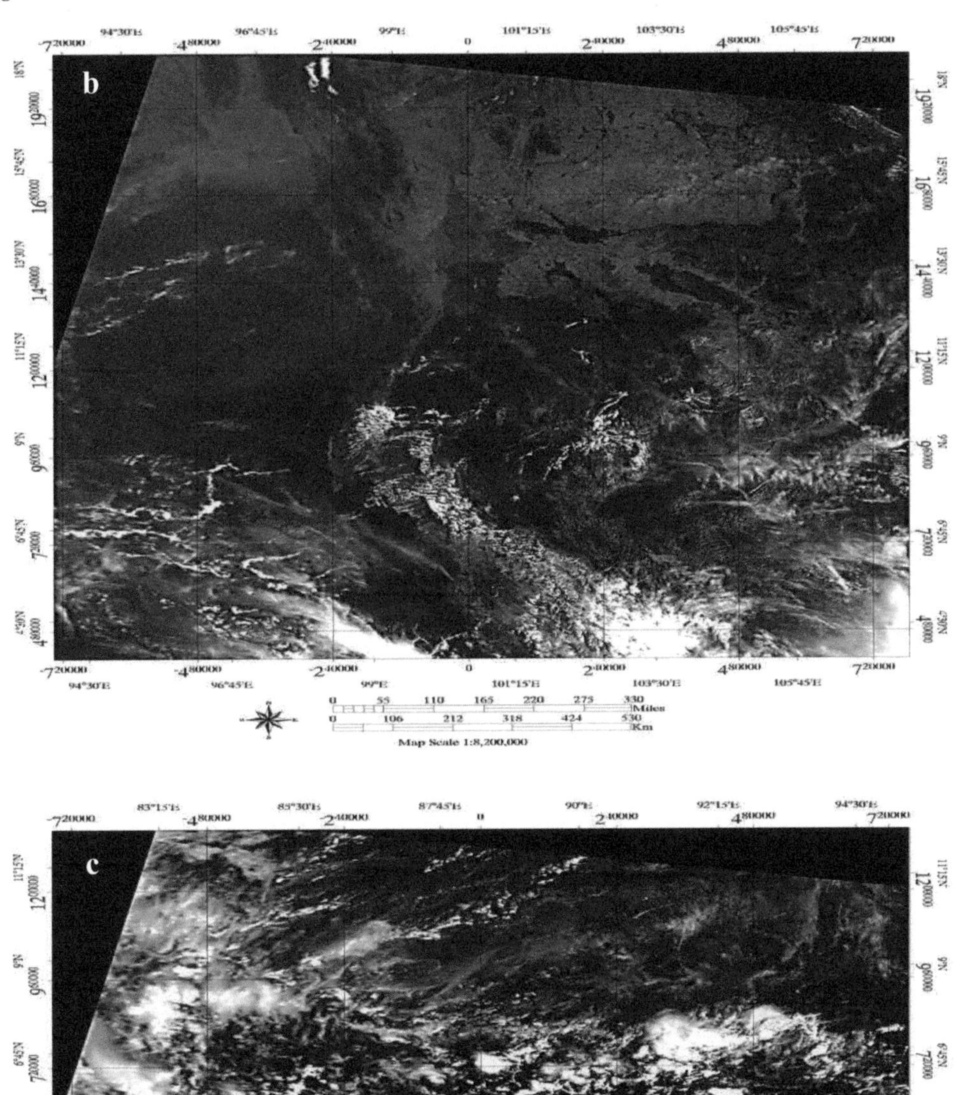

Figure 9.3 MODIS satellite data (a) pre tsunami, (b) during the tsunami and (c) post tsunami.

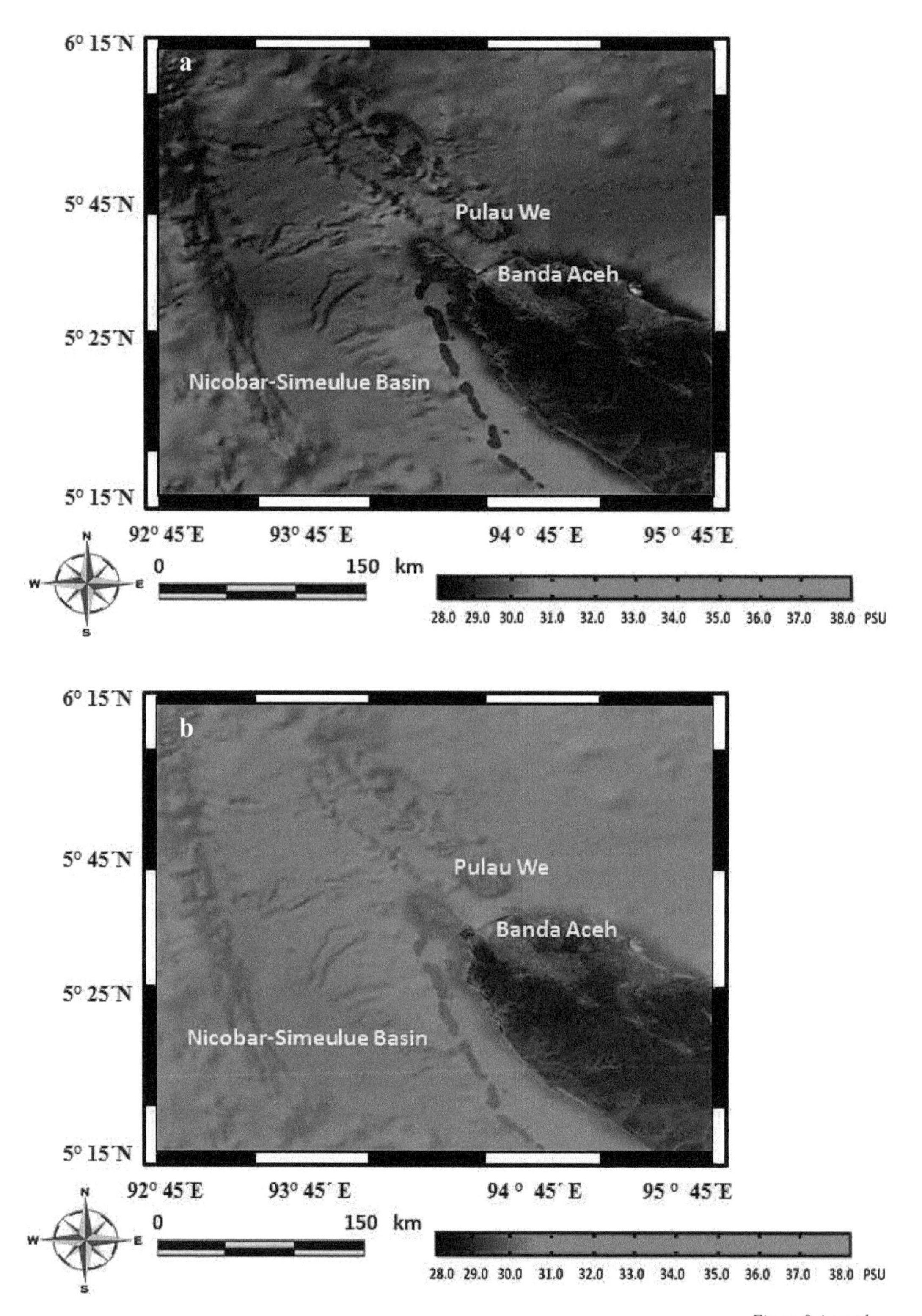

Figure 9.4 contd. ...

...Figure 9.4 contd.

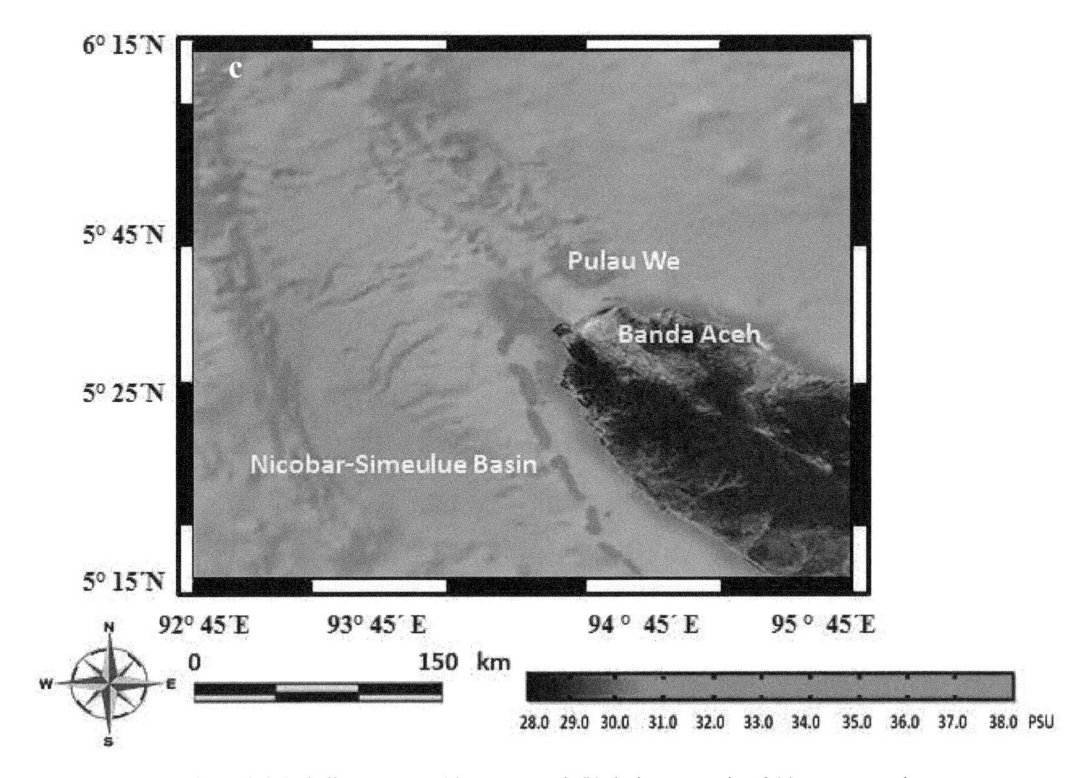

Figure 9.4 Isohaline contours (a) pre-tsunami, (b) during tsunami and (c) post-tsunami.

SSS modeled from MODIS data. Regression model shows that the SSS modeled by using a linear least square method is in good agreement with *in situ* data measurements. The degree of correlation is a function of r^2, probability (p) and root mean square of bias (RMSE). The relationship between estimated SSS from MODIS data and *in situ* data shows positive correlation as r^2 value is 0.96 with $p < 0.00007$ and RMS value of ± 1.1 psu. Further, accurate results of sea surface salinity in recent study can be explained as: using multiple MODIS bands, i.e., 1 to 7 bands, is a useful extension of linear regression model in the case where SSS is a linear function of 7 independent bands. Such a practice is particularly useful when modeling SSS of MODIS data. This statement is agreed upon by Qing et al. (2013) [24].

In addition, the least square algorithm minimizes the error between the retrieved SSS from MODIS data and *in situ* measurement [16]. This means that the use of a brand new approach based totally on the least squares algorithm decreases the sum of the residual errors for the estimating SSS from MODIS data. Besides, this investigation designates the possibilities of direct retrieving of the SSS from visual bands of MODIS data without using such parameter of color dissolved natural matter, a_{CDOM} [15]. This endorses the studies executed by means of Marghany (2010) [17], Marghany et al. (2010) [18] and Marghany and Mazlan (2011) [19].

The MODIS satellite data also have a great potential for retrieving suspended sediment. Figure 9.7 depicts the suspended sediment spatial concentrations, pre-tsunami (Figure 9.7a) and post-tsunami (Figure 9.7c). Prior to the tsunami, the SS ranged from 3 to 5 mg/l (Figure 9.7a). Nevertheless, during the tsunami, on 26th December, 2004, the SS dramatically increased and ranged from 15 to 22 mg/L (Figure 9.7b). Thus, the SS continued to increase 32 mg/l on December 27th, 2004.

The mechanism of the December 2004 tsunami deposits involves four phases. First, the horizontal shearing drift resulted from spilling the front of the wave which changed into resuspending the deposits of

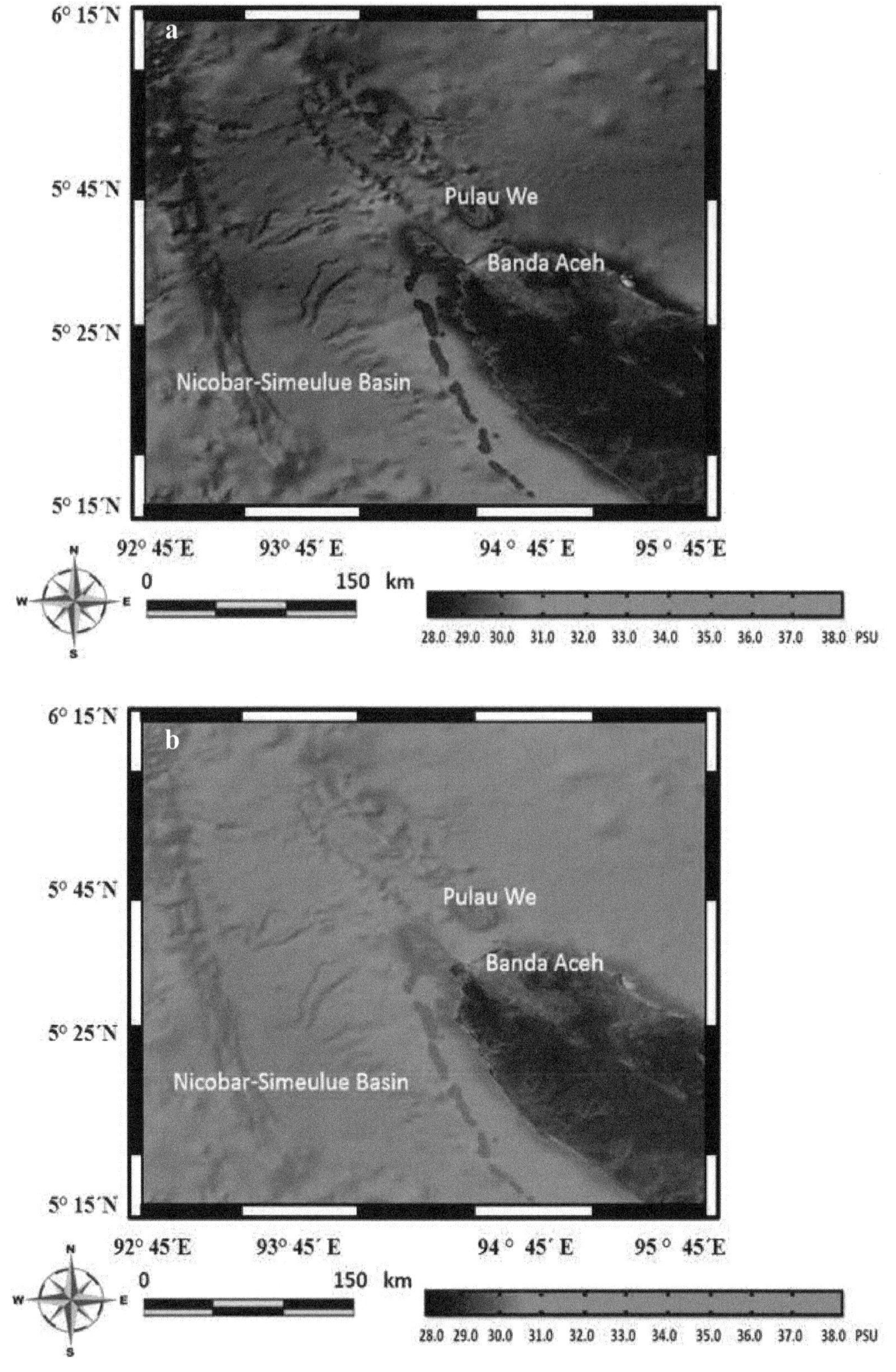

Figure 9.5 contd. ...

...Figure 9.5 contd.

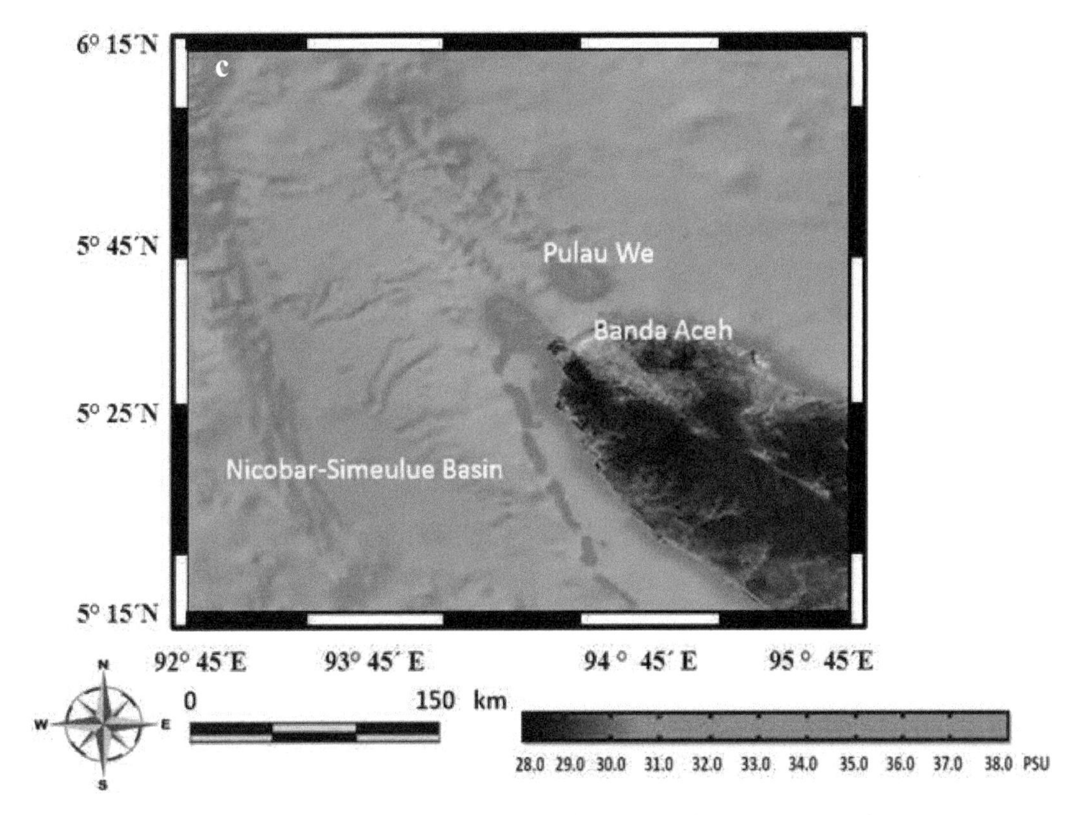

Figure 9.5 *In situ* SSS (a) pre-tsunami, (b) during tsunami and (c) post-tsunami.

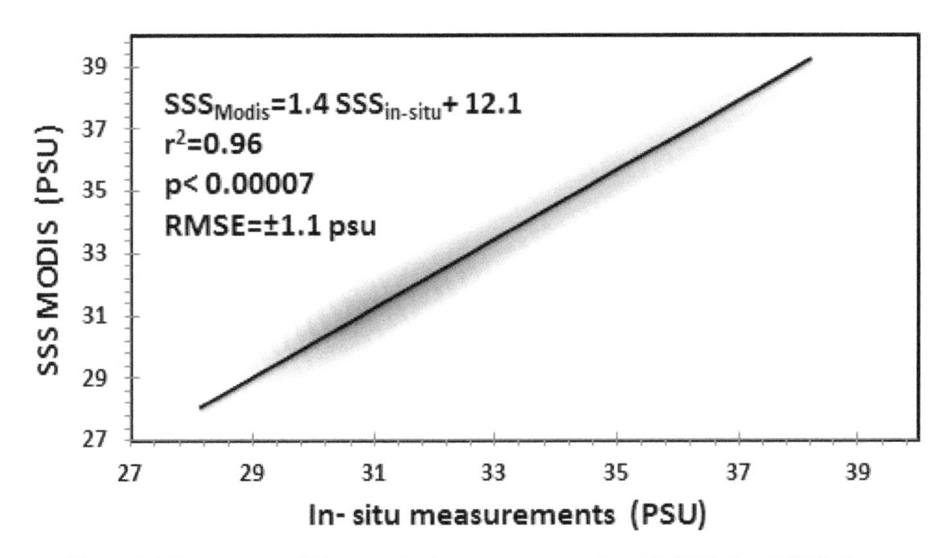

Figure 9.6 Regression model between *in-situ* measurement and modeled SSS from MODIS data.

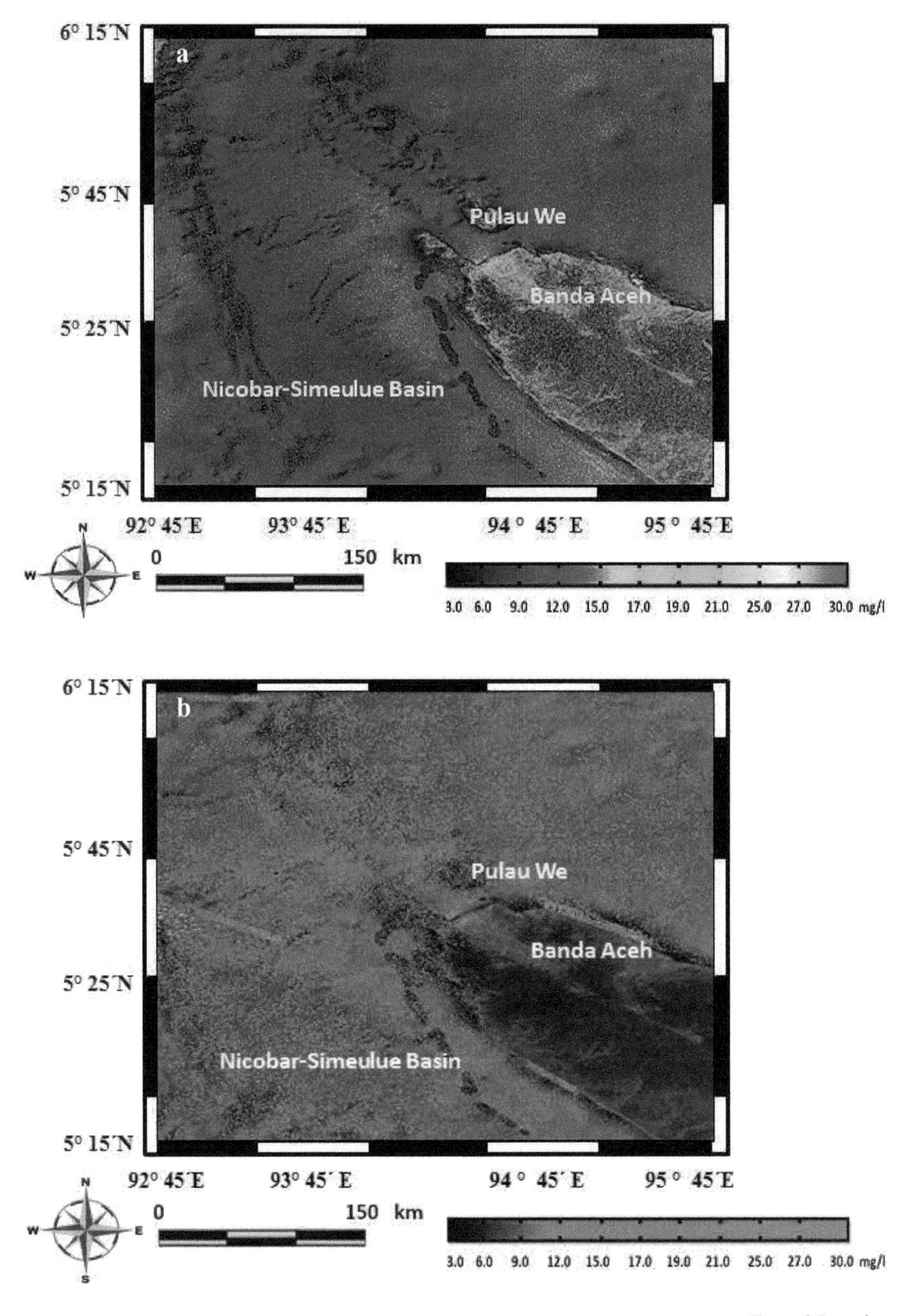

Figure 9.7 contd. ...

...*Figure 9.7 contd.*

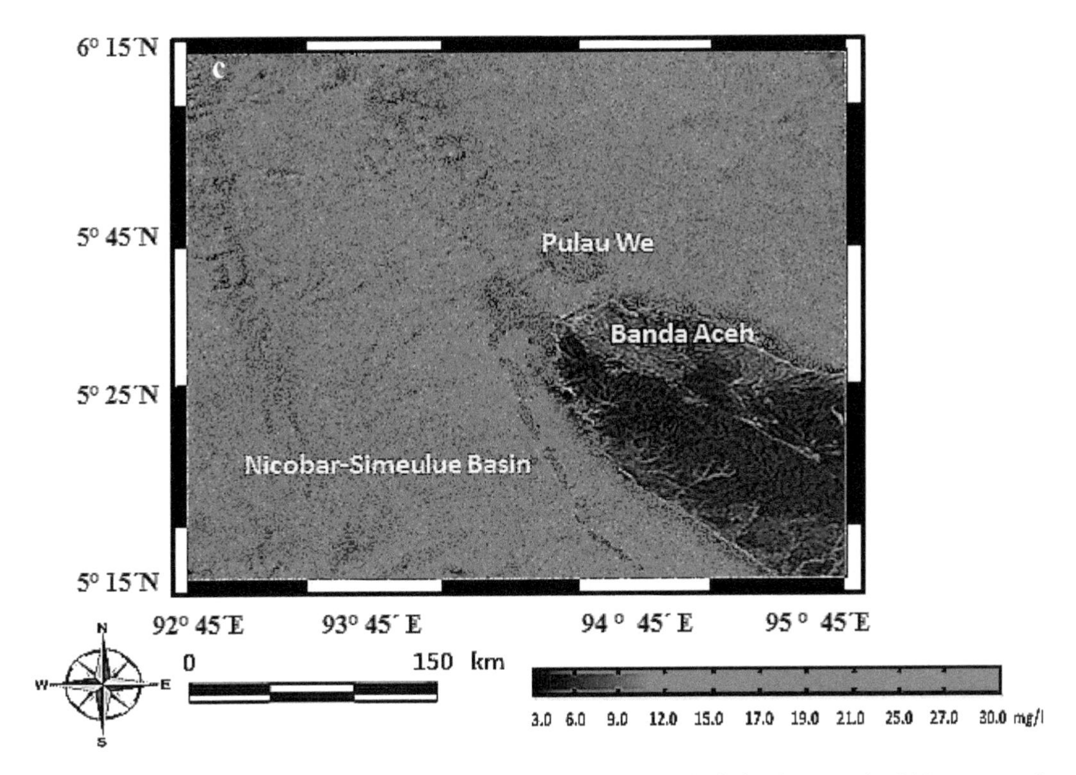

Figure 9.7 Suspended sediment retrieved from MODIS data (a) pre-tsunami, (b) during the tsunami and (c) post-tsunami.

a previous wave. Then the turbulent flow triggered massive sand plumes because of extraordinary erosion of the soil which can cause damage to coral reefs. Under these circumstances, the massive volumes of sediment transports have existed (Figure 9.6). Subsequently, the second phase of deposition has occurred after the spilling wave has surpassed. In this regard, the different sediment particles inversely sorted in the top parcel of the couplets or triplets of deposit layers. Further, as the shearing velocity and turbulent flow reduced, the coarser sediments descended prior to the finer sediments.

The ultimate layering of triplets is shaped in the third phase of deposition. Prior the backwash and the next wave, the last run-up sedimentation occurred. The third phase of sedimentation is characterized by thickness deposits. Consequently, the last phase of deposits is characterized by a gravity-driven drift, which is a function of quantity sediment concentration and topography. Moreover, the primary dominant characteristic is the decreasing of coarser sorting degree along the seaward due to the backwash. Under this circumstance, the coastal water of Aceh becomes more erosive and revises the upper portion of the run-up deposit [261].

Finally, Figure 9.8 proves the linear correlation between salinity increment and high suspended sediment concentration. This can be seen by the highest r^2 of 0.96 and a low probability (p) less than 0.00005. The maximum salinity of 38 psu ensued because of high suspended sediment concentration of 32 mg/l which added to the coastal water of Aceh.

According to Ilayaraja (2010) [143], the studied tsunamigenic sediments, deposited by the 26th December 2004 tsunami in the Andaman group of islands consist of poorly sorted, coarse sand to medium sands, and are similar to depositional effects of previously reported earthquake-generated tsunami waves. The tsunami genic sediment consists of a coarse sand layer with abundant reworked shell and other carbonate fragments. The tsunami sediments were mainly composed of boulders of corals and sand which determined

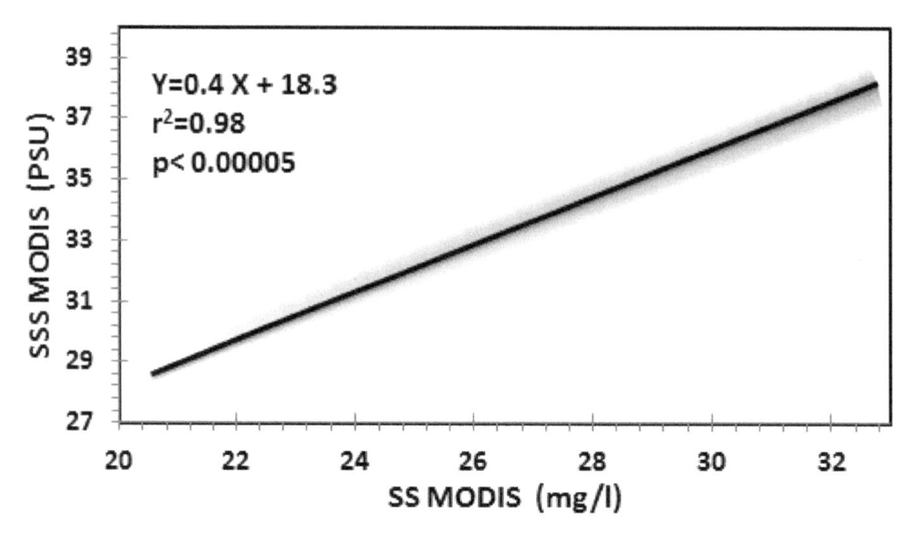

Figure 9.8 Correlation between SSS and SS are retrieved from MODIS data.

the high-energy environment throughout the study area. The variation in mean size, therefore, reveals the differential energy conditions that lead to the deposition of these kind of sediments in different locations.

9.4.1.2 Sediment Impacts on Sea Surface Salinity

The Sea Surface Salinity along Banda Aceh coastal waters changed spectacularly, pre and post-tsunami event of 2004. During earlier run-up phases, the flows were extremely strong to erode sediments deposited especially when the coastal topography and bathymetry canalized the tsunami flow. In this regard, the backwash flows were potentially more erosive and powerful than run-up flows because of hyper-concentrated flow routing by coastal morphology. Therefore, tsunami resulted in massive deposits of sand, silt and fine gravel containing isolated boulders [25]. Consistent with Font et al. (2010) [22], the areas close to the shoreline are eroded by the passing wave front and deposition occurred when the wave front passed by. Just 6 min later, the direction of the flow close to the shoreline began to change in the seaward direction before the wave front reached the inundation limit. A portion of the sediment was entrained by the wave front. In other words, tsunami of 2004 caused brief coastal flooding with high overland flow velocities and strong abrasion and reworking of the nearshore materials (Figure 9.9). Under this circumstance, the distinctive sediment compositions of sand, mud, and gravel have been deposited in most nearshore areas. In this regard, the salinity values are highly increased due to tremendous genetic sediment differences carried by a wave from inland. According to Moore et al. (2006) [23], geochemical proxies have provided evidence for saltwater inundation, associated coral and/or shell material, high-energy flows (heavy minerals, if present), and possible contamination associated with tsunamis.

The Quickbird satellite data acquired on the 28th December, 2004 confirms the heavy deposits along the coastal waters of Aceh. Figure 9.10 demonstrates extremely eroded shoreline, completely destroyed buildings and infrastructures which are caused to intensive deposits. Moreover, the 2004 tsunami caused spatial variation of suspended sediments which increased the coastal water salinities. In this regard, sediments need to be assessed for salinity before any crops are planted in the affected zones.

In general, different grain size of sediments deposited in the coastal waters of Banda Aceh increased the salinity in the sea surface. Indeed, these sediments contain different levels of salts and mineral concentration. Post-tsunami, SSS has increased extremely because the sea coast was rough and turbid with suspended sediments (Figures 9.11 and 9.12). This is excellent evidence of addition of the extremely high amount of salts and minerals due to high levels of sediment deposits in the coastal waters of Banda Aceh [26, 31].

Figure 9.9 High deposits flow due to tsunami 2004 along Aceh coastal waters.

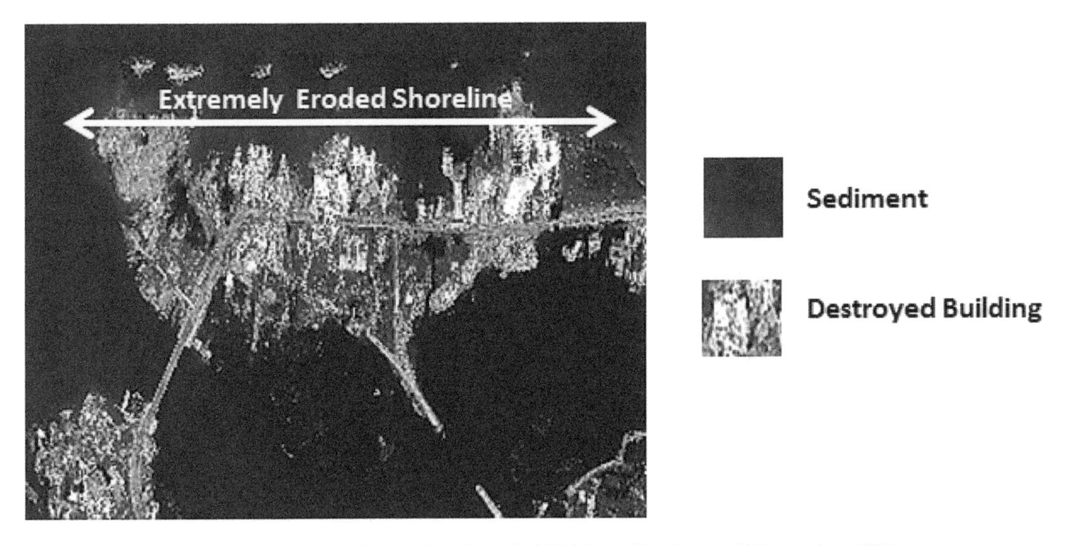

Figure 9.10 High sedimentations from Quickbird satellite data on 28 December, 2004.

9.4.2 Tsunami Impact on Chlorophyll-a

The retrieval of ocean color parameters such as phytoplankton pigment (chlorophyll-a) in oceanic waters involves two major steps: atmospheric correction of visible channels to obtain normalized water, leaving radiances in shorter wavelengths and a second application of the bio-optical algorithm for retrieval of phytoplankton pigment concentration. Chlorophyll products are generated by analyzing spectral measurements in the blue and green parts of the spectrum, roughly corresponding to the phytoplankton absorption of minimum and maximum of blue and green of EM spectrum, respectively. Most algorithms have been developed for deriving concentrations in deep water where the reflection from the bottom can be neglected [131–144]. The MODIS algorithms used in this study attempt to compensate the usually higher values of chlorophyll concentrations in coastal regions [145].

Datt (1998) [146] showed how chlorophyll a absorbs energy primarily in the blue-violet (400–475 nm) and orange-red (590–650 nm) parts of the electromagnetic (EM) spectrum, and reflects energy primarily in the green part of the EM spectrum. In contrast, chlorophyll b absorbs energy primarily in the green (510 nm) part of the EM spectrum. Chlorophyll b complements chlorophyll a by increasing the absorption of energy in the green part of the EM spectrum.

Two major steps are required to derive chlorophyll pigment using reflected radiation. First, atmospheric correction must be carried out to remove artifacts introduced by chemicals and aerosols [147]. Then, an algorithm is applied to generate the chlorophyll concentration product [148]. A major limitation of ocean color remote sensing is the sensitivity to meteorological conditions, particularly clouds.

9.4.2.1 Chlorophyll Algorithm

A number of bio-optical algorithms for retrieval of chlorophyll have been developed to relate measurements of ocean radiance to the *in situ* concentrations of phytoplankton pigments. Recently, O'Reilly and Maritorena (1998) have proposed an empirical algorithm (also known as Ocean Chlorophyll 2 or OC2) to be operated on NASA Sea viewing wide field of view sensor (SeaWiFS) ocean color data. This algorithm captures the inherent sigmoid relationship between Rrs490/Rrs555 band ratio and chlorophyll concentration C where Rrs is the remote sensing reflectance. Prior to Chl-a modelling from MODIS satellite data, radiometric correction, land and cloud masking are performed (Figure 9.11). Then selected band widths of 443 μm, 448 μm, 531 μm and 551 μm are used to estimate Chl-a a concentrations (Table 9.4).

Table 9.4 Empirical algorithm of Chl-a.

Algorithm	Type	Equations	Band Ratio (R) and Coefficient (a)
Aiken's	Hyperbolic	Chl = exp[a+b*ln (R)]	R = (L_w 448/L_w 551) a = 0.8933
Clark	Hyperbolic	Ln (chl) = a + b (R)	R = ln[(L_w 443+L_w 531)/(L_w 551)] a = 3.7656
Gordon	Hyperbolic + Power	Chl = a (L_w 551/L_w 443)^b	a = 1.9333
NDCI	Hyperbolic	Ln (chl) = a (R) – b	R = [(L_w 443/L_w 448)–(L_w551/L_w488)] a = –3.1773

In practice, chlorophyll-a concentration can be estimated by Gordon algorithm, Clark-3-bands algorithm, Normalized Difference Chlorophyll Index (NDCI) and Aiken's algorithm. The algorithms which were used in a single or multiple bands of MODIS data are based on hyperbolic and power function forms. This involves four algorithms which are evaluated in this work (Table 9.4). These algorithms are Aiken's algorithm and NDCI algorithm [149]. Both algorithms are based on the concept of band ratio and having hyperbolic and power function form. The Aiken's hyperbolic model estimates chlorophyll by a hyperbolic function using two band ratio of band 9 and band 10. The Normalized Difference Chlorophyll Index (NDCI) algorithm, however, was implemented by three bands as compared to the Aiken's algorithm. The existing coefficients in both algorithms are derived from the regression model of MODIS data and *in situ* measurements.

For each pixel and a wavelength (λ), remote sensing reflectances (R_{rs}) were derived from normalized water, leaving radiance [145–151] which is given by

$$R_{rs}(\lambda) = \left[\frac{nL_W}{F_0(\lambda)}\right] \tag{9.13}$$

where F_0 is the extraterrestrial solar irradiance and is L_w downwelling radiance. Validation of satellite derived chlorophyll-a concentration was carried out with *in situ* samples of chlorophyll-a fraction. In practice, chlorophyll-a concentration can be estimated by Gordon algorithm, Clark-3-bands algorithm, Normalized Difference Chlorophyll Index (NDCI) and Aiken's algorithm. The algorithm used for measuring chlorophyll-a concentration [152] can be calculated as

$$C_a = 10^{0.283 - 2.753R + 1.457R^2 + 0.659R^4} \tag{9.14}$$

$$\text{where R= log 10} \left(\frac{R_{rs443} > R_{rs448}}{R_{rs551}} \right) \tag{9.14.1}$$

where C_a is Chlorophyll-*a* concentration (mg.m^{-3}), R_{rs} is the remote sensing reflectance and R is blue-green band ration [dimensionless].

9.4.2.2 Tsunami Impact on Chlorophyll-a Variations

The synoptic maps of chlorophyll concentrations pre, during and post tsunami periods are shown in Figure 9.11. During the pre tsunami period on 23rd December 2004, the chlorophyll concentration pattern seems to have a homogenous variation along the coastal waters of Aceh. The maximum chlorophyll-a concentration value of 0.2 mg/m^3, however, is found on 23rd December, 2004 (Figure 9.11a).

In contrast, the period during the tsunami on the 26th December, 2004 suggests the highest chlorophyll concentration value of 0.5 mg/m^3 as compared to the pre-tsunami period (Figure 9.11b). It is interesting to find that the chlorophyll concentration continues to increase and spread gradually during the post tsunami period of 27th December 2004 (Figure 9.11c) as it moved away from the coastline. This might explain the possibilities of upwelling and occurrences along the coastline, especially during the tsunami and post-tsunami.

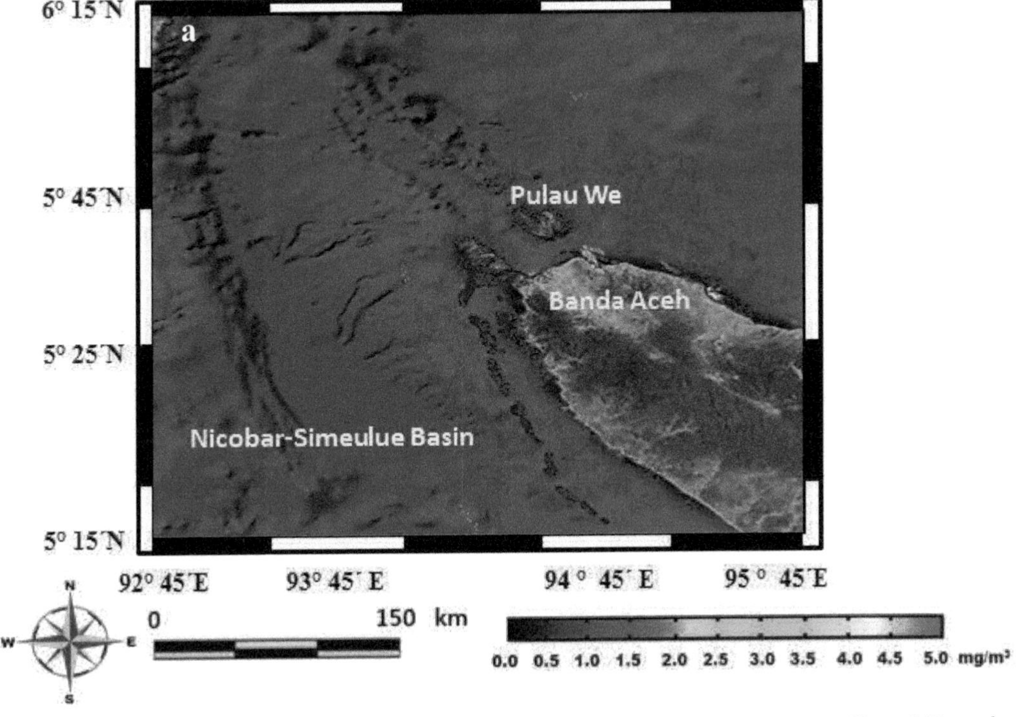

Figure 9.11 contd. ...

...Figure 9.11 contd.

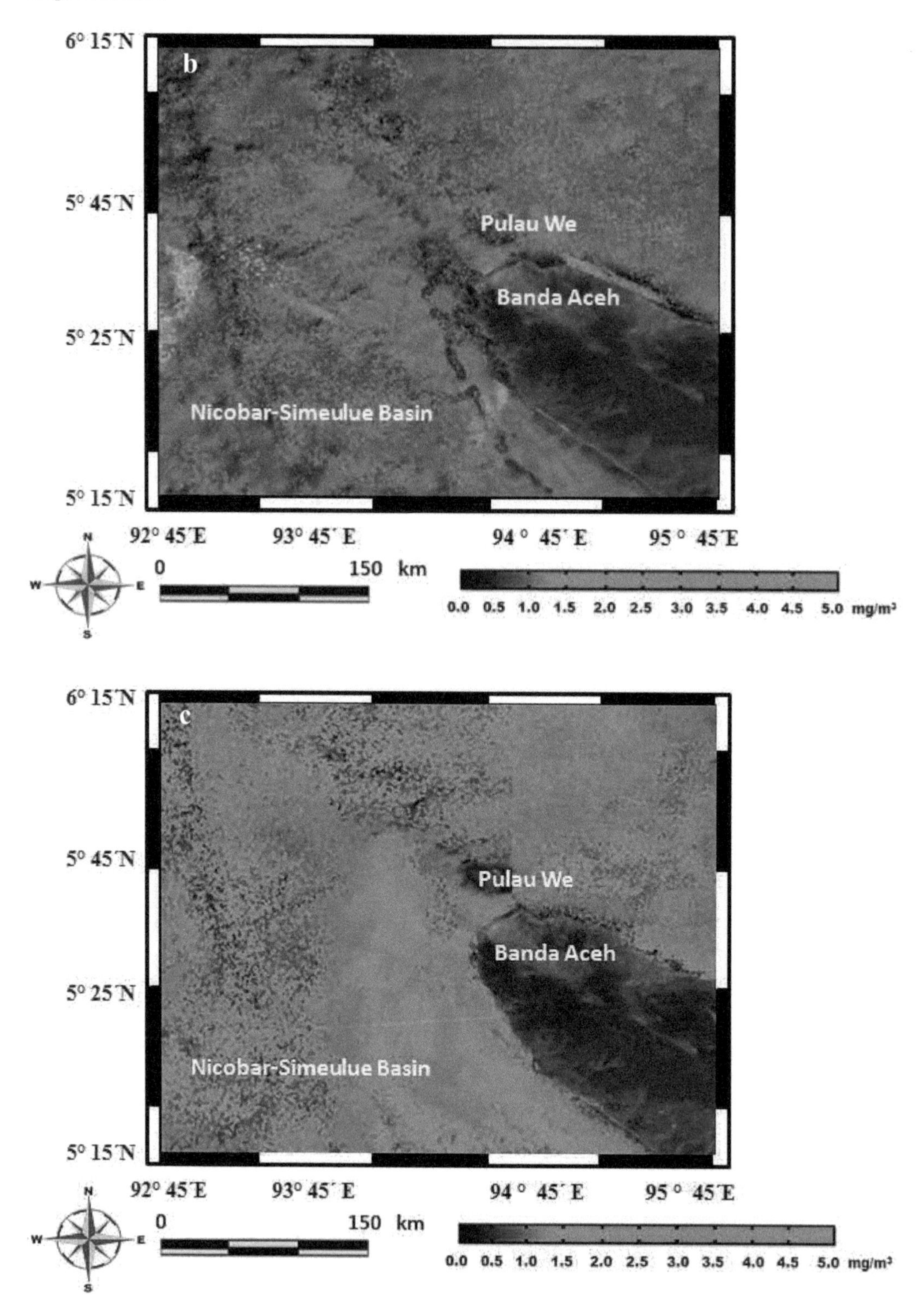

Figure 9.11 Chlorophyll concentration retrieved from MODIS data (a) pre-tsunami, (b) during the tsunami and (c) post-tsunami.

Consistent with ICMAM (2005) [131], Sarangi et al. (2011) [132], Anil et al. (2006) [133], and Tan et al. (2007) [153], Chl-a and normalized water, leaving radiance nlw 551, showed a sudden increase of Chl-a and low nlw 551 at the northwestern Sumatra and middle of the Malacca Straits after the tsunami. These results are agreed upon by Sarangi et al. (2011) [132]. Therefore, Sarangi et al. [132] stated that the high chlorophyll concentration (1.0–5.0 mg/m^3) has been observed and spread. This high chlorophyll concentration has continued uptil January 1, 2005. The spreading of chlorophyll has been reduced on the January 3 chlorophyll image in the Andaman Sea. Further, they reported an evidence of an increase in the chlorophyll concentration in the tsunami-affected regions in the Bay of Bengal, the Andaman Sea, and around the Sri Lankan coastal waters. The chlorophyll concentration was observed to be around 1.0–5.0 mg/m^3 in the tsunami-affected areas off Cuddalore, Nagapattinam, around the Sri Lankan east coast, and Andaman Sea water, which was < 1.0 mg/m^3 prior to the tsunami.

The increment of Chl-a could be due to the nutrient enrichments from the land which was delivered by the ebbing waves. In this context, the upsurge in coastal nutrients because of the tsunami might have accelerated phytoplankton growing along the coastal water. The bloom was observed to the north of Aceh Province. A large offshore phytoplankton bloom was observed because of the rapid propagation of the tsunami wave and its receding. A comprehensive investigation is required to comprehend the procedures accompanying the chlorophyll blooming because a tsunami created such upwelling flow which increased the chlorophyll concentrations along the Sumatra, Indian, and the Malacca Straits coastal waters [154].

9.4.3 Tsunami Impact on the Sea Surface Temperature

Satellite technology has improved our ability to measure SST by allowing frequent and global coverage. In this view, the Moderate Resolution Imaging Spectroradiometer (MODIS) on board of the Terra satellite is the sensor can measure SST accurately. MODIS has 36 bands over the electromagnetic spectrum from 0.41–14.39 μm and ranging from 0.25–1.0 km in spatial resolution. The MODIS design and Terra orbit result in multiple image acquisitions over the oceans every day, both day and night [157].

9.4.3.1 Sea Surface Temperature Retrieving from MODIS Data

MODIS data are selected to determine the SST in the study since it has very high temporal resolution where it provides two images (day and night) in a day. The data are of good quality with 36 bands. For instance, the first two bands can be used to detect cloud-covered area and the bands 8–16 are suitable for ocean applications such as ocean color and phytoplankton mapping through appropriate bands of 20, 21, 22, 23, 31, and 32 for retrieving sea surface temperature [155].

The ocean depth has a major impact on the SST value because the temperature gradients are below the ocean's surface. Measurements made at only a depth of one or two molecules below the ocean's surface are considered the "interface SST" and cannot be realistically measured. However, at a depth of roughly 10 μm, it is known as the "skin SST". The reduced length of thermal infrared radiation corresponds to this depth of the ocean. The "sub-skin SST" is at a depth of ~ 1 mm and corresponds to the attenuation length of microwave radiation. Beyond this depth is what is commonly referred to as the "bulk SST", "near-surface SST", or "SST_{depth}" [155–157].

The thermal band data received were in thermal radiance values. The sea surface extraction from MODIS data required values of brightness temperature to be applied in the SST algorithm. Therefore, the conversion of radiance to brightness temperature was carried using thermal band radiance (band 31 and band 32) from MODIS image. This could be done by using inversion of Planck's equation. The equation to convert radiances to brightness temperature is given by [157]:

$$BT = C_2/(\lambda \ln (C_1/\lambda^5 \text{radiance}) + 1) \tag{9.15}$$

where BT is the brightness temperature for band 31 and band 32, respectively, C_1 is 1191.0659 mWm^{-2}srm^{-4}, C_2 is 0.01438833 deg K and λ is wavelength in micrometre (μm). The SST extraction was carried out using the derived brightness temperature data. For MODIS, the bands used to measure the SST are in the atmospheric windows at a wavelength of 3.5 μm to 4.2 μm and 10 to 12 μm whereas there are three mid-infrared bands (20, 22 and 23) and two thermal infrared bands (31 and 32). These bands have noise equivalent around 0.05 K and a spatial resolution of 1 Km^2 at the nadir. The bands located in wavelength of 4 μm (20, 22 and 23) exhibit high sensitivity and are placed where the influence of column water vapor is minimal on the sensed radiances. Although the mid infrared window is cleaner than those in the thermal infrared, measurement of the shorter wavelengths are susceptible to contamination by solar radiation reflected at the surface [156].

On the other hand, bands in the far-infrared between 10 and 12 μm for bands 31 and 32 are located near the maximum emission for 300 K blackbody and placed where there is significant difference in integrated water vapor absorption for the two bands. These bands have larger bandwidth, but suffer from large water absorption in the tropical narrow bandwidth. The integrated atmospheric transmissivity over each of the MODIS infrared bands (20, 22, 23, 31 and 32) differs. Consequently, algorithms can be constructed which depend on the differences in measured temperature among these bands. Reviews on several SST algorithms have been made and the best SST algorithm is selected. The non linear sea surface temperature (NLSST) algorithm can be applied for daytime. The mathematical algorithm used to retrieve the sea surface temperature is calculated as [156].

$$\text{NLSST (day)} = A_1 \text{ x } T_{11} + A_2 \text{ x } T_{sfc} (T_{11} - T_{12}) + A_3 \text{X} (T_{11} - T_{12}) \text{ x } (\sec\theta - 1) + A_4 \quad (9.16)$$

where T_{11}, T_{12} are equivalent brightness temperature in Kelvin for wavelengths of 11 and 12 μm, respectively. T_{sfc} is 29°C, A_1, A_2, A_3, and A_4 are 1.228552, 0.9576555, 0.1182196, and 1.774631, respectively. Finally, θ is the satellite zenith angle from nadir which is 29.67° [157].

9.4.3.2 Sea Surface Temperature Variations

The synoptic variations of SST, which was retrieved from MODIS data pre-tsunami, during the tsunami and post tsunami are shown in Figure 9.12. On 23rd December 2004, the SST varied between 29.0°C and

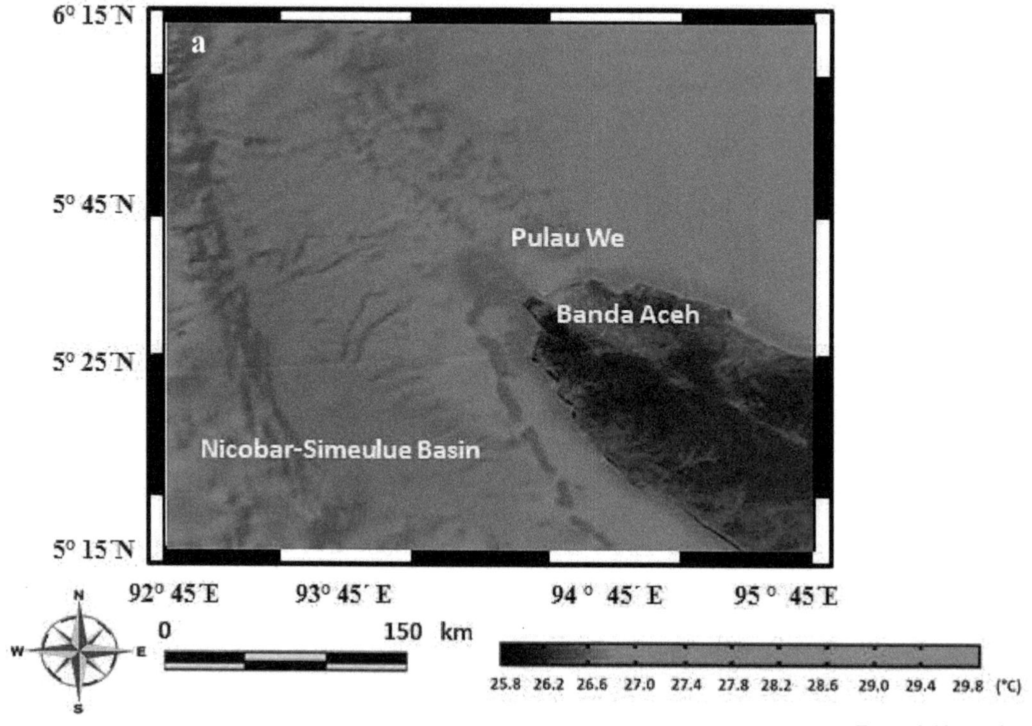

Figure 9.12 contd. ...

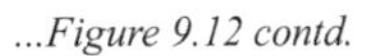

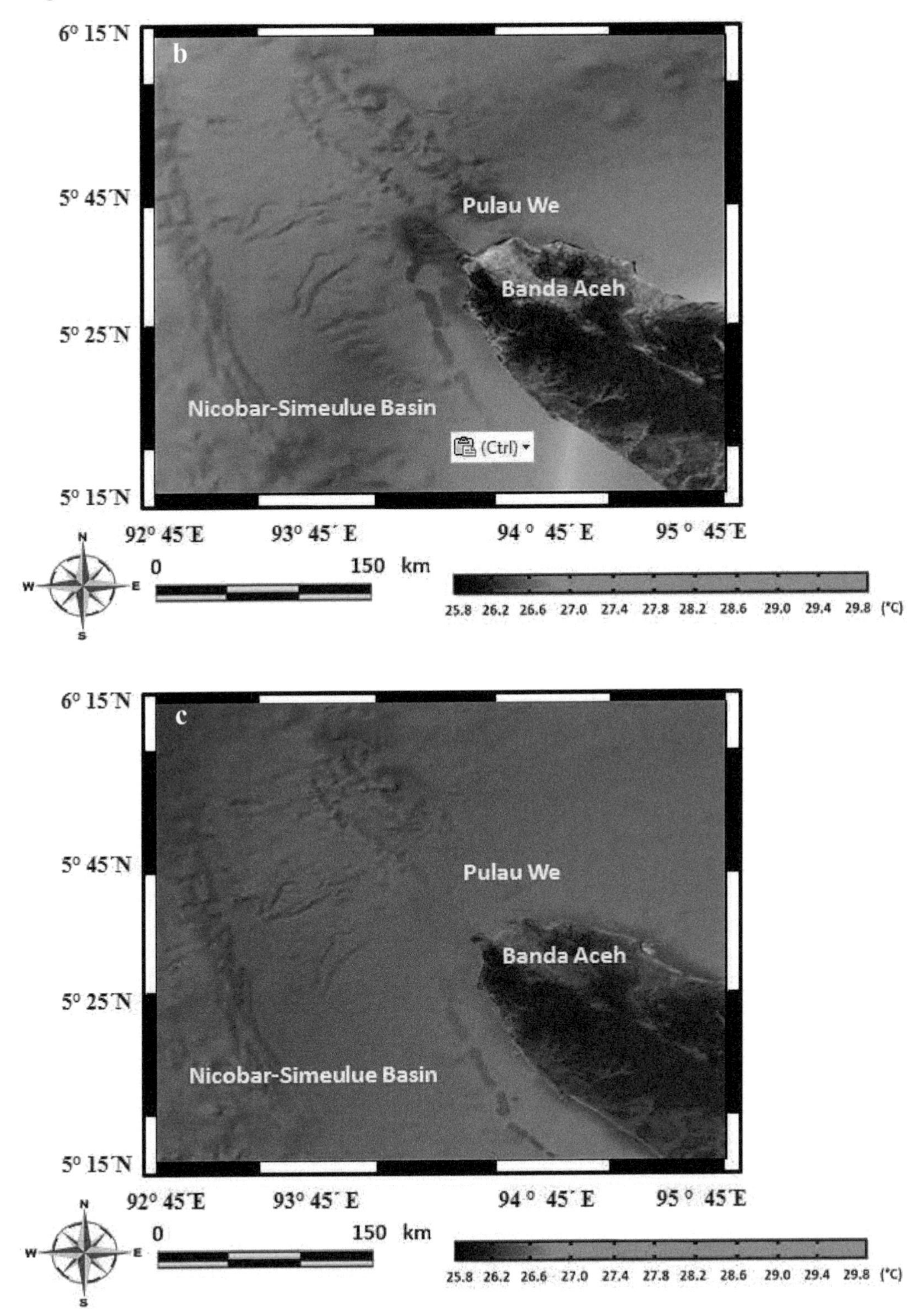

Figure 9.12 SST retrieved from MODIS data (a) pre-tsunami, (b) during the tsunami and (c) post-tsunami.

29.8°C (Figure 9.12a). On 24th December 2004, the SST decreased and ranged between 26.2°C and 27.4°C (Figure 9.12b). Therefore, the dramatical SST decrease occurred on the 27th December, 2004 where SST decreased to 25.8°C (Figure 9.12c). The SST tended to be homogenous along the coastal water of Aceh and Andaman Sea.

It is interesting to note that the SST decreased with a 4°C difference from the 23rd December, 2004 to the 27th December, 2004. In this regard, the cooling of surface water (~ 26.2°C) was observed immediately during and post-tsunami. Consistent with Sarangi (2011) [132], the subsurface water was brought to the surface with ebb tide and brought from coastal to offshore water to a distance of hundreds of kilometers.

9.5 Mechanism of Upwelling by Tsunami

Seismogenic upwelling may cause many changes in the ocean. Ocean waters are changing color or becoming turbid as a result of ocean-floor sediments being carried up vertically and suspended sediment embodies the furthermost manifestation of seismogenic upwelling. Besides, upwelling caused change in the general anomalies in the sea surface temperature (SST) and, subsequently, infrequent weather phenomena may develop. The influx of biogenes in the surface layer regularly washed-out such substances which should be conveyed by explosive growth of phytoplankton [158].

In line with the theory of an incompressible liquid, all the water layer instantaneously overhead the moving part of the sea-floor obtains upright speed and, accordingly, the kinetic energy. On the other hand, the displacement results in a perturbation forming on the water surface which contains the potential energy.

There are two motions which can be generated due to bottom deformation by an earthquake in (i) the tangential direction, and (ii) the vertical direction. The energy transferred to the water layer from the ocean bottom undergoing motion can be computed as the work performed by this force along the path of displacement (Figure 9.13). Consequently, the friction velocity is known to be essentially smaller than the velocity of the average flow. The entire volume of water is dislodged by the displacement due to rapid movement as a result of sudden sea bottom deformation. Hence, the energies are transferred to the water layer by the normal and tangential displacements. Subsequently, it monitors that tangential motions of the ocean bottom can be neglected in the problem of tsunami generation.

Turbulence generation is possible in the near-bottom region owing to the shear instability, when a fault ridge is on the surface (off the bottom), or in the case of horizontal movements of the bottom which induced turbulent flow. Perhaps it is assumed that intense turbulent mixing (Figure 9.14) which is involve the near-bottom region of water of insignificant thickness of the order of the height of the bottom inhomogeneity or of the vertical displacement within the fault.

The Earth's rotation manifested at the tsunami source area, considering residual displacements of the ocean bottom to form at the site, must result in the formation of a certain vertical structure. Usually, bipolar deformation of the ocean bottom occurs at real tsunami sources, therefore, it may be assumed that several vortices structures are formed with different directions of rotation. The part of the energy due to vortices motion is seen to increase quadratically with the horizontal dimension of the source and to decrease as the ocean depth increases. But, in any case, the contribution of this energy does not exceed 1% of the energy of the tsunami wave. Moreover, stratification of the ocean and rotation of the Earth cannot significantly influence the process of tsunami generation by an earthquake. But a small part of the earthquake's energy is transferred both to baroclinic motions and to vortices fields (Figure 9.14). Displacement of the free surface and the flow velocity vector are related to the potential of the flow velocity [159].

In this regard, the turbulence generated due to the vertical displacement can cause shear flow instability. This phenomenon takes place at any depth and may involve the entire thickness of the water column. Under this circumstance, pumping is generated by turbulent energy. The high-power source of turbulence is characterized by a decrease in the maximum temperature gradient and by significant variations in the surface temperature. In both cases, after pumping is switched off, the energy of turbulence pulsations and, consequently, the exchange coefficient in the region of the jump in temperature, rapidly decrease, practically down to zero [160].

The maximum gradient decreases and noticeable changes of surface temperature and of the center-of-mass position occur, which signifies destruction of the thermocline and arrival of depth waters of the

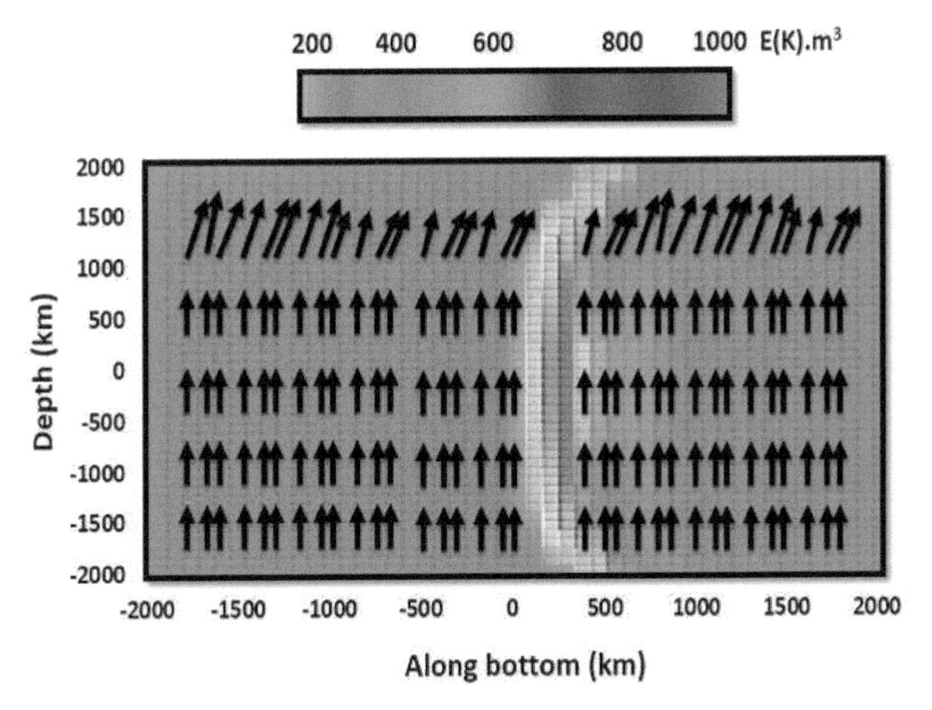

Figure 9.13 Tsunami vertical displacement energy.

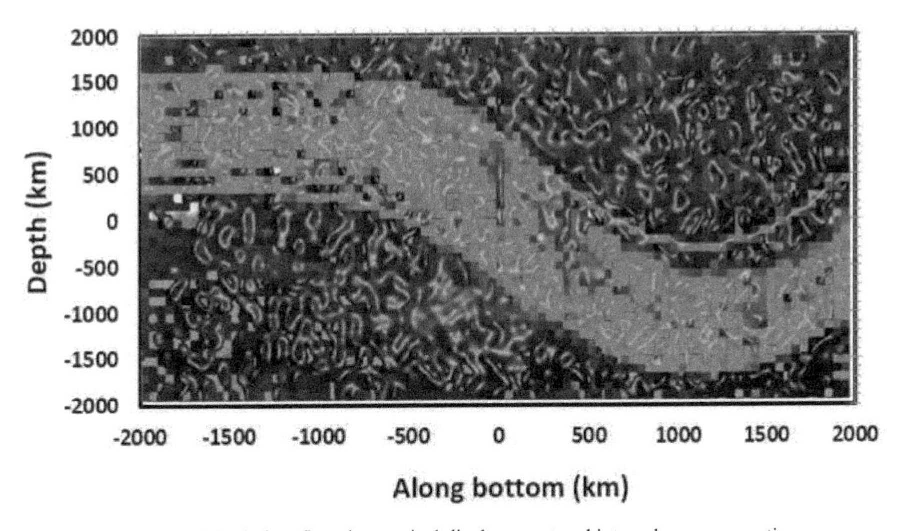

Figure 9.14 Turbulent flow due vertical displacement and internal wave generation.

surface. The maximal temperature gradient increases monotonously with time, no noticeable changes of the surface temperature and of the center of-mass position occur, i.e., the thermocline is not destroyed and it serves an obstacle in the way of depth waters toward the surface. When the water column is mixed until it is practically in a uniform state, the influence of the thermocline depth of the change in the center-of-mass position is insignificant. As to changes in the surface temperature, they are primarily determined by the depth of the thermocline: the thermocline is related to the difference between the density of the surface water compared to the deep layer which is always colder than the surface, where density is based on salinity and temperature [160].

It can be said that strong earthquakes have an adequate reserve of energy for essential transformation of the ocean stratification structure. Tenths of a percent of the energy of an earthquake are sufficient for formation, on the ocean surface, of a temperature anomaly with a characteristic horizontal dimension, measured by hundreds of kilometers and with a temperature deviation of the order of 1°C. Note, that a comparable amount of energy (less than 1% of the earthquake energy) is spent on the formation of tsunami waves. The formation of a temperature anomaly of the ocean surface is most probable in the case of a shallow thermocline and for seismic events, characterized by a persistent process at the source or by a large number of aftershocks. The most striking manifestation of the effect is to be expected in the case of realization of the turbulence generation mechanism with a scale exceeding 10 m [159, 160].

Local variations of the vertical temperature distribution should serve as a source of internal waves even in those cases when temperature variations are insignificant. Internal Waves are ocean waves that propagate underwater along the thermocline. They can have extremely large amplitudes (~ 100 meters) and can readily be seen in optical and radar images from space.

Resulting in water displacement, owing to the vertical component of movement involved through the water column. Under this circumstance, cooler bottom water moves vertically and cools the water column. In this regard, the water salinity expands and rapid vertical movements also convey bottom nutrients through the water column. As an after effect, the salinity and Chl-a concentrations ought to grow. But the displaced waters cannot be exchanged through horizontal flow alongside the coast, and consequently, need to be replaced by the upwelling of subsurface waters. Under this circumstance, plate slips, inflicting subsidence and liberating energy into the water, which causes sufficient displacement to provide upward thrust to a giant tsunami and additionally, the water column above the plate slips [130].

Eventually, the tsunami would possibly have disturbed the sea surface horizontally as well as the vertical structure of the water column and the alternate of water temperature and biomass in the mixed layer. It likely indicates that the deep chlorophyll maxima (DCM) of the water column became disturbed and brought up the excessive chlorophyll concentration to the surface. Hence, the chlorophyll concentration increases from 0.5 to 2.0 mg/m^3 and has been determined with the impact of the vertical and horizontal displacement of the water within the Bay of Bengal and Andaman Sea [132–140].

Chapter 10

Genetic Algorithm for Simulation of Tsunami Impacts on Water Mass Variations using MODIS Satellite Data

On-the-ground measurements are notoriously difficult in the harsh environment due to tsunami, but satellites could help close the gap in measuring water masses. There is no doubt that the tsunami performs a remarkable role to alternate the physical properties of the ocean through mixing turbulent approaches. In continuing with Chapter 9, the tsunami accelerated the water salinity in Aceh coastal waters to approach the maximum value of 38 ppt. Ocean salinity is a keystone of measuring water mass physical properties by compiling the temperature of a water body.

10.1 Water Mass Definition

Specific volumes of water has particular features and retain their properties as they alternate. Consequently, their physical properties may be used to comprehend the source of their generation. In other words, the physical properties, for instance, temperature and salinity, can be exploited to categorize the world ocean into divisions which might be referred to as water masses. In this context, the water masses can be distinguished as a function of temperature and salinity.

Though density is a function of temperature and salinity, it cannot be conducted for the water mass identification. In fact, the two water masses could have a similar density; however, they have unlike temperatures and salinities. It is well known that temperature and salinity can be plotted as coordinated diagram to identify the water masses. This plot is referred to as T-S diagram. In this regard, the temperature and salinity are memory of the ocean. Therefore, they are revised only by dynamic interactions with the atmospheric changes and mixing with other waters. In contrast, both biological and chemical activities cannot impact the water masses formation. Thus, temperature and salinity as the memory of ocean can allow an accurate understanding of the tsunami impacts on the coastal waters. In this view, the unstability of the water column due to tsunami generation, propagation and run-up can be identified from the slope of T-S diagram with reference to sigma-t lines. The water type occurs due to well mixed water mass and might be represented by a single point. This chapter can answer the question of how the tsunami can modify the mixed layer.

Water masses are cornerstone to comprehend the elements of the sea water [268]. Fundamentally, profound sea dissemination is controlled by minor adjustments in seawater thickness which is activated by contrasts in temperature and saltiness, indicated as thermohaline flow. Thermohaline envelops the arrangement and dynamic exchange of particular water masses [269]. These are enormous homogeneous volume of waters which are a function of a particular assortment of temperature and saltiness. Most profound water masses are created at high scopes at the sea surface wherever they achieve their particular drained temperature and saltiness [270].

In any case, temperature is not customarily used to track a water mass since temperature steadily tumbles to the base and therefore uncovers no extrema (greatest or least) rates like oxygen or even saltiness. Actually, thickness contrasts between water masses is emerging power that actuates water developments, generally vertical developments [269]. The stream of the significant profound water gravity caused flow, pulling the denser water masses downwards, dislodging lighter masses upward. An oceanographic water mass is an identifiable waterway with a typical development history which has unmistakable physical properties from encompassing waters. These incorporate temperature, saltiness, compound—isotopic proportions, and other physical properties of the sea water [270].

10.2 Remote Sensing and Water Masses

After more than a decade, both of Terra and Aqua satellites of the MODIS (Moderate Resolution Imaging) are part of NASA's Earth Watching Framework. They have been giving worldwide ocean surface temperatures (SSTs) for over 10 years. In 1978, IR radiometers on the board of satellite platform of Advanced Very High Resolution Radiometer (AVHRR) has been used to estimate SST.

Indeed, Infrared (IR) satellite SST recuperation relies upon estimations taken where the surroundings are reasonably straightforward, in supposed "air windows" in the mid-wave infrared (MWIR, λ = 3.5–4.1 μm) and long-wave, thermal infrared (LWIR, λ = 10–12 μm) unearthly interims [15]. Ocean surface saltiness (SSS) recovery from satellite information is a noteworthy test. To be sure, damaged salts and suspended resources have a noteworthy consequence on the electromagnetic radiation, weakening outdoor the unmistakable spectra run [22].

In this specific situation, the electromagnetic wavelength better than 700 nm is regularly consumed even though the wavelength beneath 300 nm is scattered through non-engrossing particles, for example, zooplankton, suspended residue and broke up salts [149]. Along these lines, Ahn et al. (2008) [15] and Palacios et al. (2009) have decided utilising SSS shaded broke down natural issue (aCDOM) from optical satellite information [12]. Ahn et al. [15] have created robust and appropriate territorial calculations from big *in-situ* estimations of clear and natural optical data (i.e., remote detecting reflectance, Rrs, and retention coefficient of shaded broke down natural issue, aCDOM) to decide saltiness using SeaWiFS data [151].

Currently, there may be three limitations to optical remote sensing technique: (i) strength constraints of the spacecraft permit about two hours of scanning consistently within the day; (ii) cloud covers prevent unique observations, and (iii) moderate information from close to the ocean surface is sensed through the scanner. In this context, retrieving water masses particularly based on SST and SSS is not always feasible, at the same time as SST and SSS are appropriate for direct observation with optical satellite sensors.

This chapter hypothesizes that water mass can also be retrieved by the use of optimization genetic algorithm from MODIS satellite for digital information. In this regard, genetic algorithm (Marghany and Mansor 2016) is taken into consideration to optimize the water mass which is notably based on SST and SSS for efficient synoptic measurements of water mass spatial version in Aceh coastal waters during 2004 tsunami using MODIS satellite data.

10.3 Genetic Algorithm

In this study, genetic Algorithms (GAs) are biological process algorithm that manipulate a population of individuals delineated by fixed-format which strings the water density as function of temperature and salinity. Their acceptance as a method to unravel real-world optimisation issues owes to the speculation of artificial adaptation. An initial population of individuals (solutions) of retrieving water density from MODIS data is generated for the matter domain which then endure evolution by suggesting the reproduction, crossover and mutation of individuals till an appropriate resolution exists.

Therefore, similar to most alternative evolutionary algorithms, genetic algorithm demands that solely the parameters of the matter be quantified. Subsequently, the algorithmic program is smeared to attain a solution which is usually problem-independent. Genetic algorithms typically represent all solutions within the sort of fastened length character strings, analogous to the DNA that's found in living organisms. The rationale for the fastened length character strings is to permit easier manipulation, storage, modelling and implementation of the genetic rule [264].

In this view, binary numbers conjointly afford straightforward conversion to and from the precise solution. Conversely, since there are evidently infinitely several real numbers between 1 and 2, fixed-length strings cause a further weakness for the computer programmer. To unravel this, the real number range should be discretized into a finite variety of constituent real segments, resembling every binary number utilized in the character string. Suppose that the character strings have a length of n = 10. Then the potential values for the character string of water density would be from 0000000000 to 1111111111 [265].

One of the explanations for exploitation of binary numbers is to forbid incorrectly formatted solutions automatically. In binary, it is easier to visualise some characteristics of water density being present (by a 1) or absent (by a 0). This can be also applicable to non-numeric problem domains. There are solely two potential binary values of water density (1 and 0). This suggests that every one of the potential binary values is often created by these two values. Consequently, the binary individuals of water masses 0000000000 and 1111111111 contain all the genetic material attainable, i.e., they span the solution space (Figure 10.1).

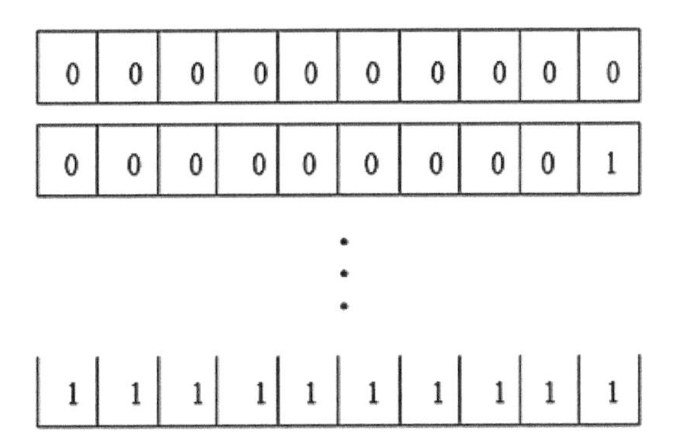

Figure 10.1 Bit-string GA representation of water masses.

10.3.1 Population of Solutions

A collection of potential solutions is unbroken throughout the life cycle of the genetic algorithmic program. This assortment is mostly called the population since it's analogous to a population of living organisms. The population, typically, may be either of fastened or variable size; however, fastened size populations are used a lot often in order that the precise quantity of computer resources will be pre-determined. The population of solutions is held on in main memory or on external storage, counting on the sort of genetic algorithmic program and computer resources accessible.

At the starting of the algorithmic program, a population of solutions is generated indiscriminately. Within the case of the square root problem, a hard and fast variety of 10 character binary strings are generated indiscriminately. This population is then changed through the mechanisms of evolution to result eventually in individuals that are nearer to the solution than these initial random ones (Figure 10.2).

Figure 10.3 represents the initial populations of 1000 for water mass individual generations from MODIS data. The binary number between 0 and 1 is randomly generated and sorted as string in the computer memory. It can also be noticed that they are not sorted in order but randomly which represent only the row of the MODIS data.

Random individuals

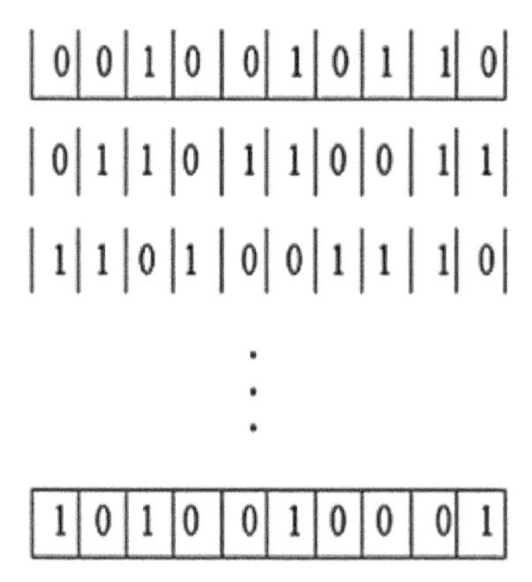

Figure 10.2 Initial populations.

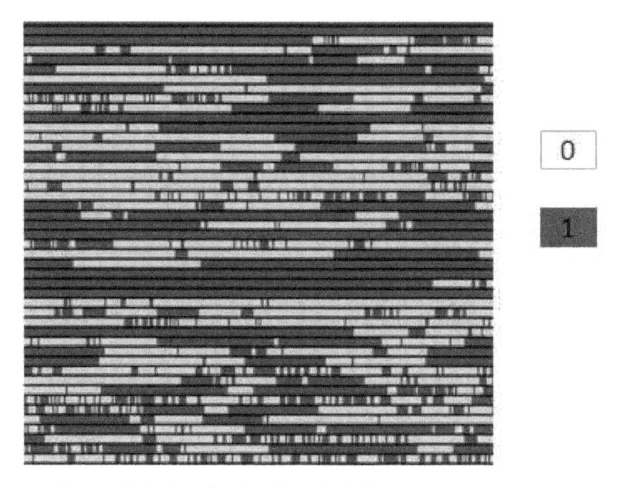

Figure 10.3 Genetic algorithm for initial water mass generation.

10.3.2 Fitness

Therefore, the range of the population is maintained by a widespread fitness sharing function. The satisfying *N* determination is used to determine the severe Pareto solution. In doing so, the blended crossover is used to generate children on fragment identified via two parents and specific parameter. In this optimization, new plan variables of water masses has a weight common as

$$Ch_1 = \varpi * P_1 + (1 - \varpi) * P_2 \tag{10.1}$$

$$Ch_2 = (1 - \varpi) * P_1 + \varpi * P_2 \tag{10.2}$$

where $\varpi = (1 + 2\ell)\_ran_1 - \ell$, Ch_1 and Ch_2 are child 1, 2, P_1 and P_2 are parent 1, 2 which represent programmed scheme variables of the members of the new population and a reproduced pair of the old generation. Therefore, *ran* is random number which is uniform in [0, 1]. When the mutation takes place, Equations 10.1 and 10.2 can be given as follows:

$$Ch_1 = \varpi * P_1 + (1 - \varpi) * P_2 + \alpha(ran_2 - 0.5) \quad (10.3)$$

$$Ch_2 = (1 - \varpi) * P_1 + \varpi * P_2 + \alpha(ran_2 - 0.5) \quad (10.4)$$

where ran_2 is random number which is uniform in [0, 1], and α is set to 5% of the given range of each variable (Figure 10.3).

Subsequently, since the water mass is feature of sea temperature (ST) and sea salinity (SS), its diagram parameters have to be addressed predictably. Else, the computation deviates and infinite population cannot be weighed. Consequently, if set to 0.0, then mutation takes place at a likelihood of 10% [266].

In line with Sivanandam and Deepa (2008) [267], genetic algorithm is commonly a characteristic of the reproducing step which entails the crossover and mutation techniques in MODIS data. In crossover step, the chromosomes interchange genes. A local fitness value results in every gene as given by:

$$f(P_i^j) = |\sigma_t - P_i^j| \quad (10.5)$$

where σ_t is the water density, $f(P_i^j)$ is local fitness value for every gene and P_i^j is probability variation along the i and j, respectively. In this view, the fitness values are changed between 0 and 1 (Figure 10.4) with iteration increments. The highest fitness value of 0.8 does not provide a clear information about σ_t (Figure 10.4a). As the fitness is gradually decreased with iteration increments, the σ_t is being to be estimated (Figure 10.4b). Lowest fitness value indicates the clear σ_t information (Figure 10.4c).

Figure 10.4a shows the fitness of 0.71 and 0.8. In fact, the standardised fitness attempts to restrict the fitnesses to the range of positive real numbers only. The adjusted fitness changes the fitness value so that

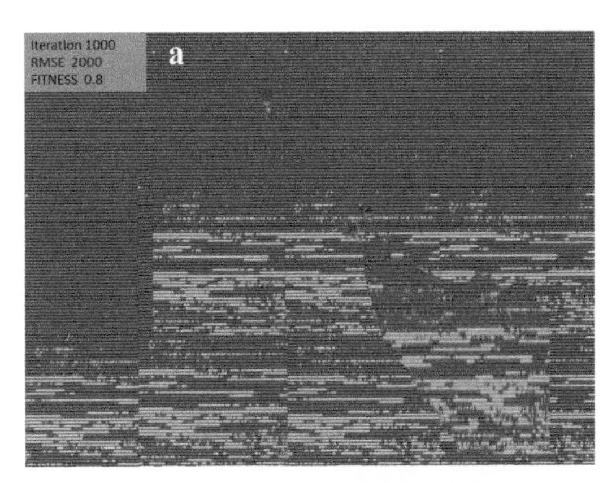

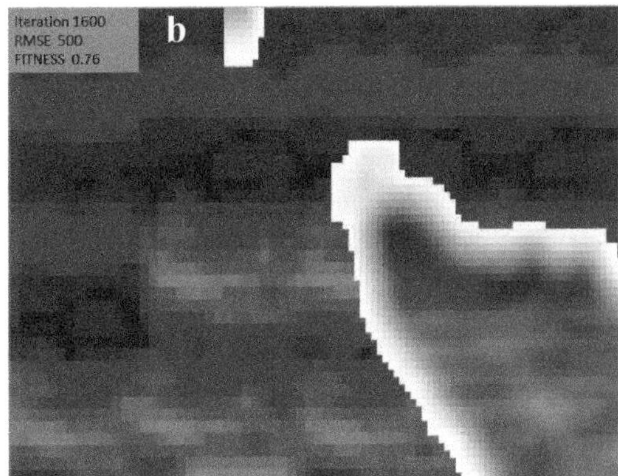

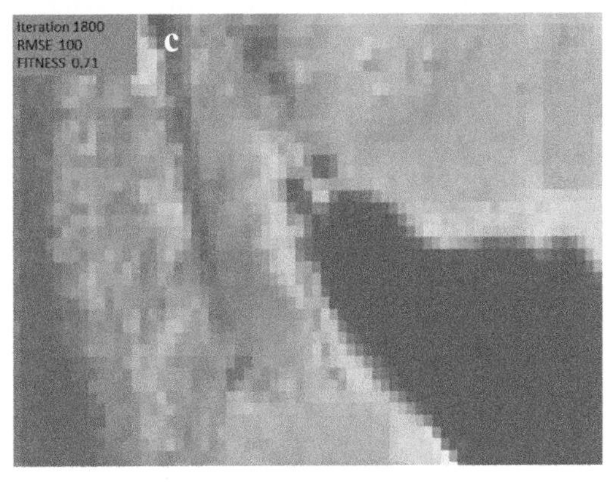

Figure 10.4 Fitness variations with iterations (a) 0.8, (b) 0.76, and (c) 0.71.

it lies strictly within the 0–1 range. Further, Figure 10.4 indicates whether the algorithm is convergent or not. If there is visible convergence and no solution has yet been found, then the algorithm can be extended over more generations. If convergence is not reached, then the parameters of the run can be tweaked to better suit the problem domain.

10.3.3 Cross-over and Mutation

The cross-over and mutation are described in short in the following sections which are difficult to a subsequent step of genetic algorithm [264]. In this understanding, the crossover operator constructs to converge around options with excessive fitness. Thus, the closer the crossover probability is to 1, the quicker is the convergence [265–267].

Then the crossfire between two individuals consists of keeping all individual populations of the first parent which have a local fitness greater than the average local fitness $f(P^j_{av})$ and substitutes the remaining genes by the corresponding ones from the second parent. Hence, the average local fitness is defined by:

$$f(P^j_{av}) = \frac{1}{K}\sum_{i=1}^{K} f(P^j_i) \tag{10.6}$$

Hence, the mutation operator denotes the phenomena of great likelihood in the evolution process. Under this circumstance, the fitness value reduces to 0.53, is a clear feature of Aceh coastal water and σ_t is well identified (Figure 10.5). In addition, this step improves the fitness procedure by showing lowest RMSE value of 2 kg/m^3 as compared to Figure 10.4.

Truly, some useful genetic records involving the chosen population may be required to replace the duration of reproducing step. As a result, mutation operator introduces new genetic facts to the ordinary gene. Generally, the genetic algorithm will take two match individuals and mate them (a process referred to as crossover). The offspring of the mated pair will acquire some of the traits of the parents. The methods of selection, crossover, and mutation are called genetic operators [265].

Figure 10.6 shows the synoptic information retrieved σ_t from MODIS satellite data which is not appropriately simulated. This is because of the existence of heavy cloud covers, pre and post boxing day of 2004, on MODIS data. Most of the information of σ_t, either onshore or offshore, are definitely fuzzy because of the heavy cloud covers throughout the month of December 2004. In fact, the dominant

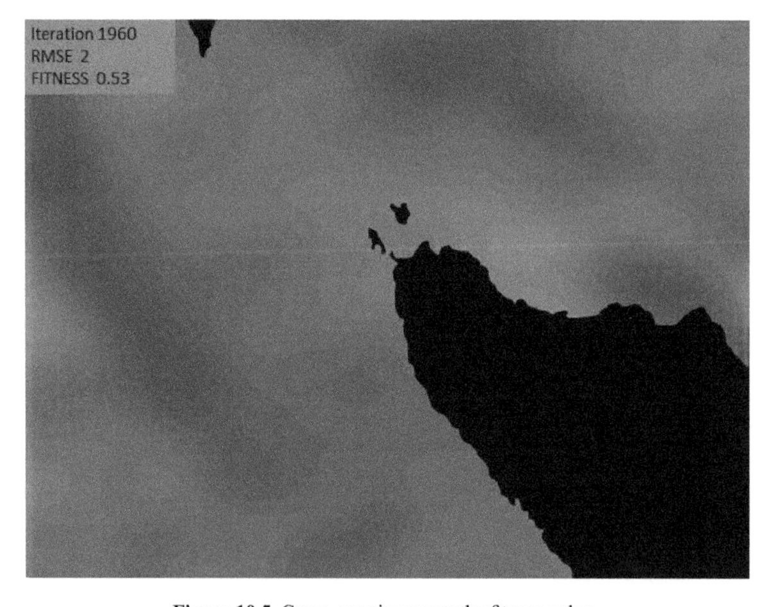

Figure 10.5 Cross-over improves the fitness value.

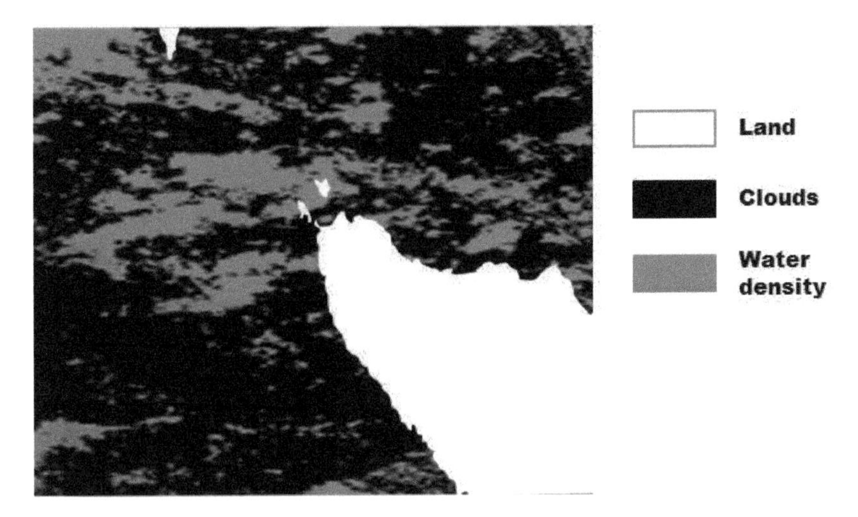

Figure 10.6 σ_t retrieved from MODIS data as function of temperature and salinity.

feature of the tropical zone is heavy cloud covers over the year. However, the retrieved σ_t from MODIS data is because of the potential of visible bands and passive thermal infrared bands. In fact, both visible and thermal infrared bands can deliver the SST and SSS, as explained in Chapter 9. In this context, σ_t is a function of both temperature and salinity.

Generally, the error factors among the SST approximation do not appear to be restricted to atmospherical impacts but nonetheless, conjointly embody SSS and wind speed, affecting the emission of the surface. In this understanding, tsunami turbulent flows exaggerated the SSS that is considered as a significant error factor of SST estimation.

Consequently, genetic algorithmic rule delivers the various pattern of sea surface density, pre-tsunami, during the tsunami and post-tsunami. Figure 10.7a casts σ_t pre-tsunami which is retrieved from MODIS data on 23rd December 2004. The sea surface density ranges from 20.5 to 22.5 kg/m^3. The lower density value of 20.5 kg/m^3 suggests light water layer moving from the north of Sumatra towards the Andaman Sea. On the contrary, σ_t increases to 24.6 kg/m^3 on 26th December 2004 (Figure 10.7b). The isopycnal pattern tends to be irregular as compared to the pre-tsunami. In fact, the tsunami created a turbulent flow on the coastal waters off Aceh that it caused vertical and horizontal mixing. Then, σ_t continued to extend rapidly to 26 kg/m^3 through the post-tsunami period. These results are confirmed by T-S diagram. Prior to tsunami, the two water masses existed along the coastal water of Aceh.

These results are confirmed by the T-S diagram (Figure 10.8). Prior to the tsunami, the two water masses exist along the coastal water of Aceh. These two water masses are characteristic of core water density of 20.5 kg/m^3, the temperature of 27.7°C and water salinity of 29.7 psu (Table 10.1). This light water mass is because of the run-off freshwater from Myanmar rivers, i.e., Chindwin River and Irrawaddy River, beside the Ganges river along Bay of Bengal. This light density water moves from the north of Andaman Sea towards Sumatra to form two water masses. On the contrary, the post-tsunami water mass suggests well mixed water along the Sumutra to Andaman Sea. T-S diagram retrieved from MODIS data shows the post-tsunami water mass is denser than the pre-tsunami. The water mass has core temperature of 25.3°C, salinity of 37.7 psu and σ_t of 25.7 kg/m^3. This dense water mass is formed in the depth of Sunda trench and carried upward to the surface by undersea megathrust earthquake.

Clearly, the genetic algorithm is able to discriminate different pattern of water masses from the surrounding environment. In fact that, genetic algorithm is commonly used to generate high-quality solutions to optimization and search problems by relying on bio-inspired operators such as mutation, crossover and selection procedures on the new populace in sequences of MODIS data. In this aspect, the crossover operator creates the population to acquire optimization options based on excessive fitness. This confirms the work of Sivanandam and Deepa (2008) [267]. In other words, genetic algorithm (GA) is

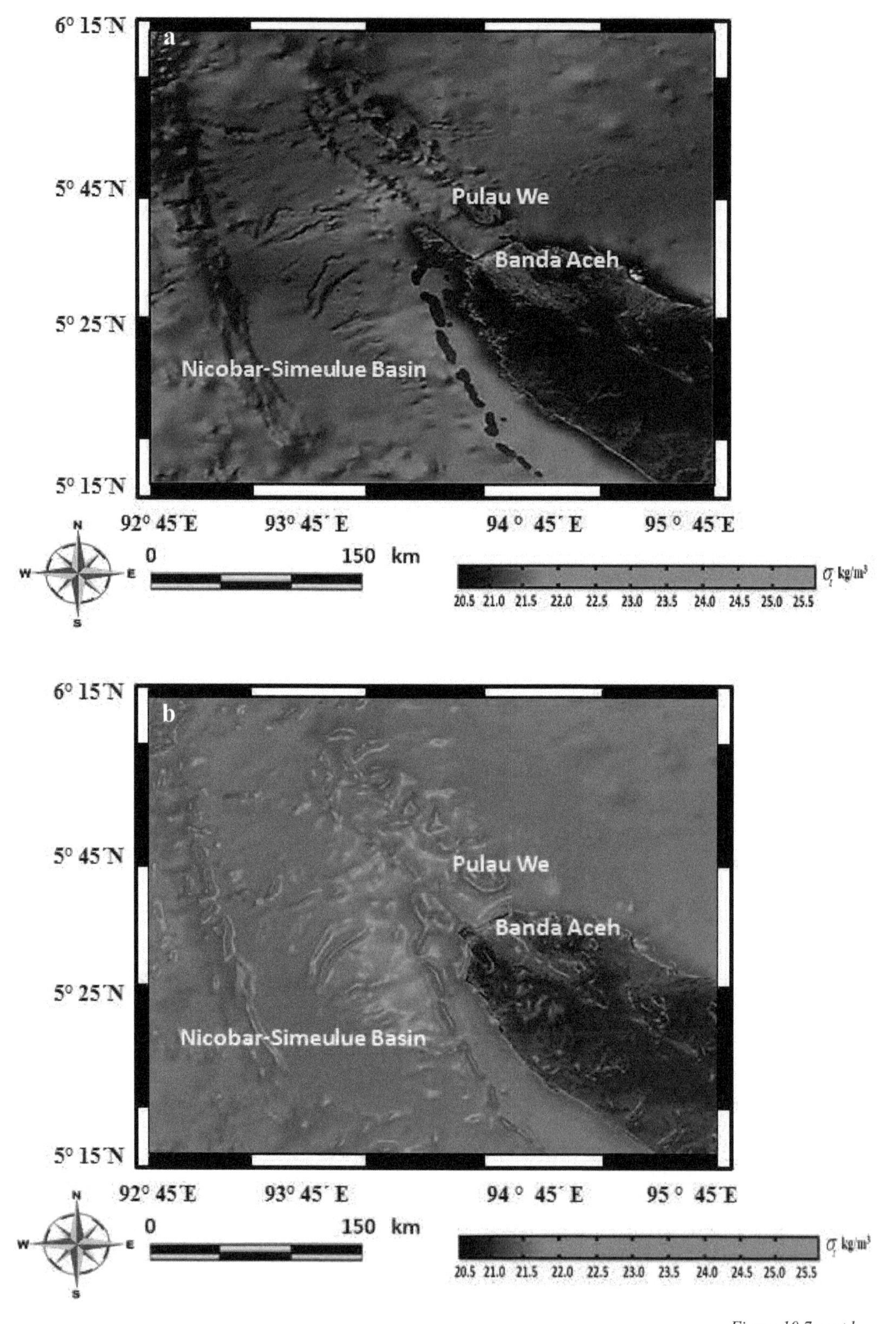

Figure 10.7 contd. ...

...Figure 10.7 contd.

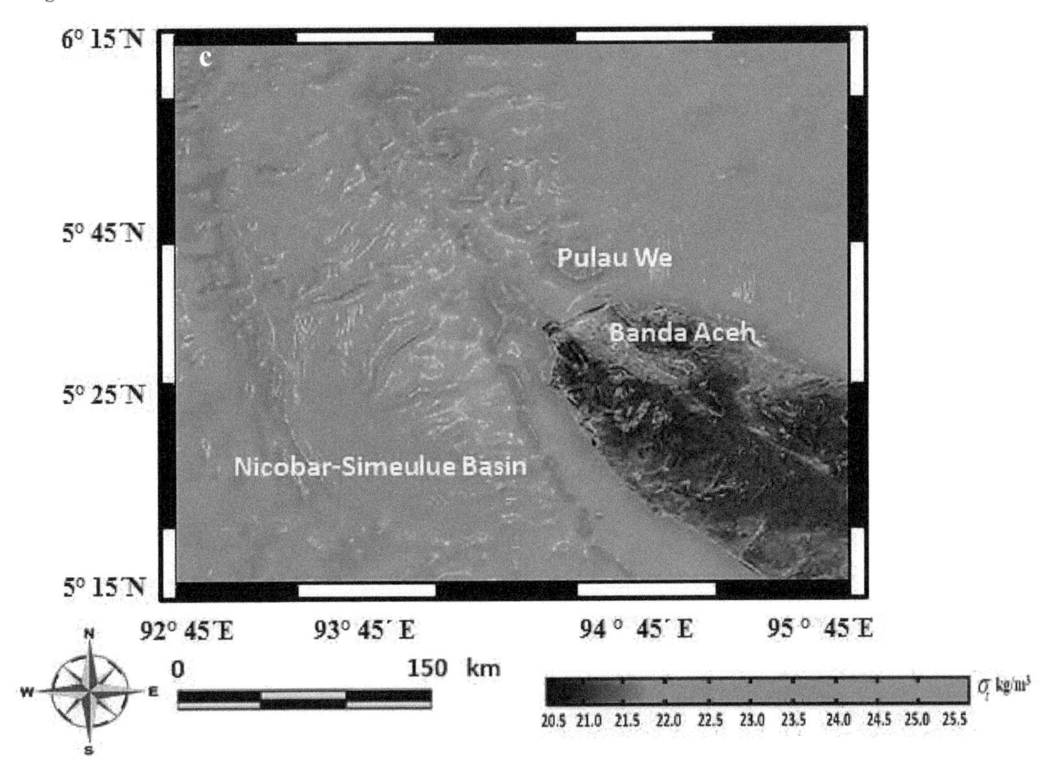

Figure 10.7 Retrieving σ_t from MODIS data using genetic algorithm (a) pre-tsunami, (b) during tsunami, and post-tsunami.

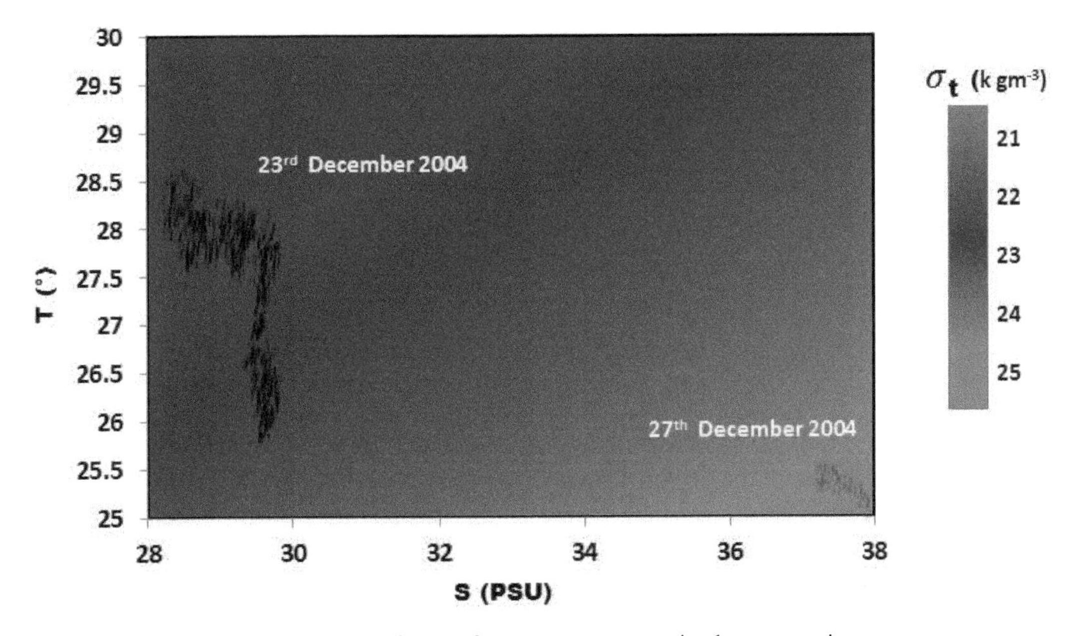

Figure 10.8 T-S diagram of water mass, pre-tsunami and post-tsunami.

Table 10.1 Summary of water mass characteristics, pre tsunami and post-tsunami.

Period	Core Water Mass		
	Temperature °C	Salinity psu	σ_t kg/m^3
Pre-tsunami	27.7°C	29.7	20.5
Post-tsunami	25.3°C	37.7	25.7

successful in producing complicated patterns and performing awkward computations. In addition, fitness characteristic is chosen to determine the similarity of every individual gradient adjustments in MODIS facts, despite the heavy cloud covers. This suggests that genetic algorithm is a notable simulator of ocean water loads in optical facts such as MODIS satellite data.

10.4 Tsunami Causes Water Masses Redistribution

Water masses are created owing to dynamic processes that occur exclusively at the surface. Even the water at the very bottom of the ocean acquired its characteristics—its density, salinity, and temperature structure—at the surface of Earth. In this view, the tropical waters such as Aceh coastal waters, have warm water and less sline because of heavy rain. Besides, the water slowly diffuses upward by approximately 1 cm/day. Further, the surface waters and the deep waters are circulating on very different types of timescales. There is only a limited amount of exchange between both of them—especially from the point of view of deep water turning into surface water.

The equatorial and coastal upwelling zones also contribute to surface-water formation by bringing deep water up. The winddriven downwelling zones bring shallow water down, and the mixing of these systems is occurring especially around the continental margins all over the world all the time in a complex pattern of exchange. However, during and post-tsunami, water masses form due to tsunami effects.

Lastly, the tsunami has changed the coastal water physical characteristics such as water mass due to its effects on water temperature and salinity (Chapter 9). The increment of water density could be due to two factors. First, the upward movement of the meagathrust earthquake has brought dense, deep water with low temperature and high salinity to the surface. Second, it was by sediments which were back washed by the tsunami and added to coastal waters and scattered around the Aceh coastal water by tsunami propagation. In this understanding, high accumulation of sediments increased the water density.

As with every ocean wave, tsunami propagation relies on the Elliptical movement of water masses. Even though the group velocity of the wave is generally high (700 km/h in the open ocean, slows down with shallower bathymetry), the horizontal orbital velocities of water masses accomplishing tsunami propagation are extremely small. The horizontal orbital velocity of a tsunami wave U_h can be estimated by

$$U_h = A\sqrt{gD^{-1}} \tag{10.7}$$

where A is the wave amplitude, g is the gravitational constant and D is the ocean depth. For the Sumatra Boxing Day tsunami in 2004, which had amplitude of 0.7 m in the open ocean with a depth of 4000 m, the horizontal orbital velocity U_h results to 3.5 cm/s. This value clearly indicates the upper bound of expected velocities; smaller (average) tsunamis should be in the range of 1–2 cm/s. These orbital velocities are amplified by shallow bathymetric features and might reach tens of cm/s, or even m/s in coastal areas. Tsunami orbital velocities might be detected by along-track interfero-metric SAR systems and HF surface wave radars.

10.5 Can Water Masses Redistribution Affect Length of Day?

Tsunamis propagate in the sea as gravity waves and since their wavelength is much larger than the sea depth, they are considered as long waves propagating in shallow waters. This plays a great role for redistribution of water mass due to the tsunami propagation. In this view, the effect on the pole motion associated with a tsunami wave propagation is quite different from that produced by a permanent solid mass redistribution because it is the effect of a transient phenomenon due to the water mass redistribution. The rotational theory predicts that a transient phenomenon can be modeled by an excitation function with a delta-like temporal dependence that will give a step-discontinuity.

Generally, the redistribution of water masses across the Indian ocean induced changes in length of the day (LOD). In fact, the tsunami induced rapid rotational pole perturbation, which has accelerated Earth's spin, shortening the length of the 24-hour day by about 6.8 microseconds. However, tsunami magnitude is considered as one hundred times smaller than the detected one. In this context, the LOD distinction persuaded by the water mass redistribution comes out to be as not weighty.

Indeed, the entire consequence is smaller than the uncertainties of recent measurements. In this regard, further studies are required to ensure the reality regarding the impacts of a tsunami on the LOD.

Chapter 11

Three-dimensional Tsunami Wave Simulation from Quickbird Satellite Data

11.1 Introduction

Previous chapters have demonstrated that the tsunami of 26th December, 2004 turned into an extraordinary catastrophe which modified the marine ecosystem and land covers. The mathematical model plays a tremendous role in computing the mechanical generation of disasters. The criterion numerical model is required for forecasting and predicting long-term disaster occurrences. Precise numerical model, consequently, can assist in designing early warning systems and coastal defense structures to prevent the massive devastation of the coastal zones.

Consistent with Paris et al. (2007) [171], at 00:58:53 UTC on Sunday, 26th December 2004, the Indian Ocean earthquake passed off due to an undersea megathrust earthquake with an epicenter off the west coast of Sumatra, Indonesia. The hypocenter of the earthquake turned into approximately 160 km (100 mi) just north of Simeulue island, off the western coast of northern Sumatra. It came at a depth of 30 m in the northern section of the Sunda megathrust [172]. As stated by Marghany and Suffian (2005) [173], Indian Ocean tsunami released the energy of approximately 20×10^{17} Joules, or 475,000 kilotons (475 megatons), or the equivalent of 23,000 Hiroshima bombs. Also, the tsunami of 26th December 2004, was travelling approximately 600 km in 75 mins. Furthermore, Lovholt et al. (2006) [174] stated that splay faults, or secondary "pop up faults", triggered lengthy, narrow components of the sea floor to pop up in seconds. This speed raised up the height and expanded the speed of waves, inflicting destruction on the nearby Indonesian metropolis of Lhoknga [175].

There is no doubt that these walls of water were capable of inflicting massive damages along the coastal lands. The consequent tsunami devastated coastlines around the ocean and killed around 226,000 people, with millions left destitute. The tsunami travelled both east and west away from the fault line, which runs north-south. This is why nearby countries to the North such as low-lying Bangladesh, escaped unscathed, while several more distant countries to the West such as Somalia, suffered considerable damage [173]. Nevertheless, satellite remote sensing data could not work as a warning alarm to prevent the occurrences of the disaster.

Walter et al. (2005) [176] stated that theoretically, sea level anomalies observed by altimetry should reflect tsunami waves. Conversely, observation is difficult, since the additional height is one of the signals of ocean variability. In this regard, scientists just used remote sensing tool to identify the tsunami's impact zones [173, 177, 178, 182].

Kouchi and Yamazaki (2007) [181] have implemented the normalized difference vegetation index (NDVI), soil index (NDSI), and water index (NDWI) to detect tsunami-inundated areas of ASTER images.

Further, they employed Shuttle Radar Topography Mission (SRTM) data to perform geomorphological classification and to determine the extent of tsunami run-up. Consequently, they reported that most of the pixels with NDVI decreased after the tsunami had occurred in the low plain areas.

Ibrahim et al. (2009) [182] have used SPOT-5 satellite data to determine the rate of land changes due to tsunami for Kuala Muda, Kedah, Malaysia. Further, they improved DEM using texture layers by integeration with Brovey transformed band and Maximum Likelihood Classifier. Therefore, Ibrahim et al. [182] reported that SPOT-5 moderate resolution sensor was reliable in identifying land damages due to tsunami. Nevertheless, the above studies were based on conventional methods of generating DEM that caused high level of uncertainties. Walter et al. (2005) [176] stated that theoretically, sea level anomalies observed by altimetry should reflect tsunami waves. Conversely, observation is difficult, since the additional height is one of the signals of ocean variability.

There are few studies that concern the simulation of tsunami run-up from remote sensing data. Therefore, run-up model requires several parameters such frequency, wavelength, wave height, and the beach slope. In addition, run-up model is required for accurate Digital Elevation Models (DEMs) of coastal zones. Salinas et al. (2006) [180], conversely, stated that to approach the run-up and inundation problem, the non-linear shallow water equations must be solved with an appropriate treatment for breaking waves and moving shore lines.

Likewise, they have reported that the complex geometry of the coastal line coupled with arbitrary beach and sea floor profiles, makes solving the shallow water equations a formidable task which only can be approached numerically [178]. Under this circumstance, the standard methods are required to acquire an accurate successive tsunami wave propagation from satellite imagery, and uncertainty might arise due to absence of real time *in situ* measurements.

In this chapter, we address the question of modelling tsunami run-up using remote sensing data without needing to include any *in-situ* measurement data. Two hypothesis examined are: (i) the Quickbird satellite data can be used to detect tsunami spectra; and (ii) finite element scheme of Galerkin method can integrate with tsunami spectra model and fuzzy B-spline to reconstruct 3-D tsunami run-up and inundation zone.

11.2 Theory of Wave Spectra in Optical Remote Sensing Data

Microwave technology has a great potential for wave studies. Unfortunately, no radar instrument was operational during the occurrence of boxing day 2004. The main concept of the wave spectrum retrieving from the optical sensor is based on the light reflection on the sea surface. The mechanism of light interaction with sea surface is mainly understanding how optical sensor can imagine a surface wave. In this understanding, the electromagnetic reflection from a sea surface involves the specular constituent and the diffuse constituent. In view, the specular constituent equals to the incident electromagnetic with respect to the sea surface (Figure 11.1). The diffuse constituent, consequently, is related to the scalar product between the incident light and sea surface. In this view, when the sun is high in the horizon, the sea surface creates a stronger diffuse component. Further, the specular constituent is also known as glitter which is supplementary robust per unit of solid angle. Concretely, these two constituents produce an accurate image of the sea surface. As a rule, the visible range of the electromagnetic spectrum depicts the specular reflection of visible sunlight on the manifold facets of the sea surface.

Let Θ be the sun angel elevation and Φ be the viewing incidence with respect to the verticle. Then the specular reflection is formulated as:

$$\Theta + \Phi = 90^{\circ} \tag{11.1}$$

Equation 11.1 demonstrates that the specular reflection fluctuates with the swell slope α symmetrically around the still water specular reflection direction. In other words, the unidirectional specular reflection does not exist when the swell is restrained by the sinusoidal transfer (Figure 11.2). Glints, therefore, are created on image due to the geometry positions between sensor, the surface sea slope and the Sun azimuth

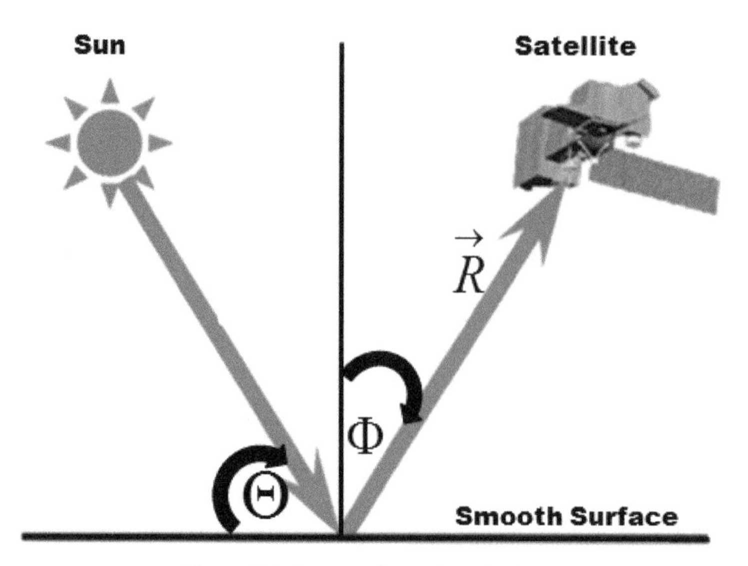

Figure 11.1 Concept of specular refraction.

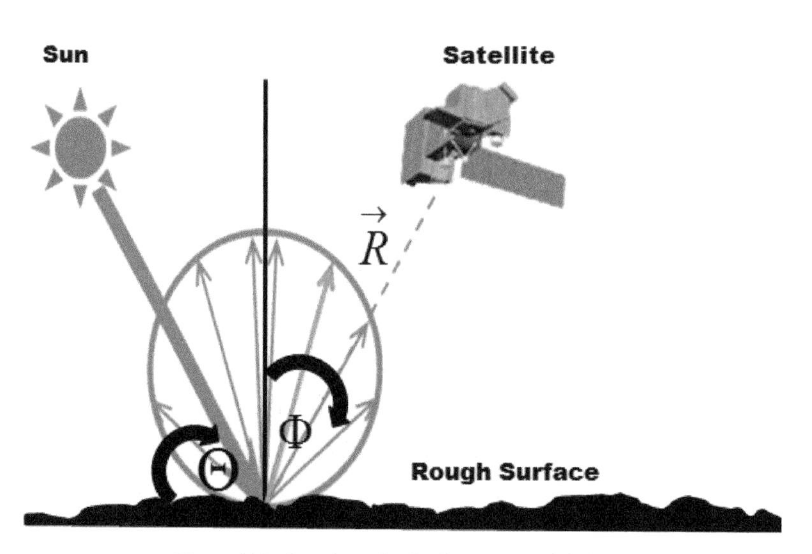

Figure 11.2 Specular reflection by wave modulation.

and elevation. Under these circumstances, if the strong sun light reflection on the sea surface is higher than 1 m, then the glint also occurs. Furthermore, the glint is reconstructed on the sea surface slope in altered sites. The mathematical circumstance to acquire accurate sea surface image is given by

$$60^{o} < \Theta + \Phi < 120^{o} \tag{11.2}$$

Equation 11.2 explains that there are two components that influence the image quality. The main parameters are $\Theta + \Phi$ value, in addition to wave slope, wavelength and angular dispersion. The second component involves the weather conditions which includes cloud covers and visibility.

When the $\Theta + \Phi$ is less than 60°, the acquired image cannot be excluded. The optical sensors cannot detect sea surface less than 1 m. On the contrary, excellent satellite image must be obtained with sea surface higher than 2 m. Further, the strong angular dispersion occurs with wind sea in the short period of 5 s and 40 m wavelength. When the sea surface slope is small due to a decrease of the ratio of wave height per wavelength, the signal to noise ratio decreases and sea surface cannot be detected accurately in the optical sensor.

Both the solar zenith angle and the direction of reflection are influenced by the glitter patter. In this regard, the gilter is variably polarized when it is viewed in the forward scatter. For instance, it is highly polarized in the range of 35° to 45°, and cascades once again to approximately 20% at $\Theta = 90°$. On the contrary, in the path of extreme radiance, the imitated energy is unpolarized at $\Theta = 0°$.

At the near-vertical angles, the impact of wind and waves in a modulation of the smooth sea surface is trivial. The reflectivity of the smooth sea surface, however, can be increased as the wind speed increases when the Sun azimuth is near the horizontal of the sea surface. The increase of reflectivity concerning the horizon is owing to the wide-ranging variation of intensity in conjunction with azimuth angle that depicts dielectric of the sea water. In this regard, the smooth sea surface would appear to be very dark near nadir. Nevertheless, any wave propagation rapidly diminishes the reflectivity at large incidence angles. Under this circumstance, the roughened sea seems much darker in the direction of the horizon than the smooth one.

11.2.1 Kirchhoff Approximation for Sea Surface Reflection

Kirchhoff approximation model is a keystone to understanding the reflection in the sea surface. It is used when the incident light wavelength is shorter than horizontal roughness scale which is a function of the large average radius of curvature. In this view, the Green's theorem as a core of understanding of the Kirchhoff approximation (KA) describes the reflection on the sea surface as tangential fields on the surface.

Let $R_{sea}(\omega, \boldsymbol{r}_r|\boldsymbol{r}_s)$ be the sea surface reflectivity, and ω be angular frequency where the Sun zenith at $\boldsymbol{r}_s = (\boldsymbol{x}_s, z_s)$ and the satellite sensor at $\boldsymbol{r}_r = (\boldsymbol{x}_r, z_r)$; so the sea surface reflectivity which is recorded by single sensors is formulated as [271]:

$$R_{sea}(\omega, \boldsymbol{r}_r|\boldsymbol{r}_s) = -\int_{S_{fs}} [G(\omega, \boldsymbol{r}_r|\boldsymbol{r}_{fs})[\nabla P(\omega, \boldsymbol{r}_{fs}|\boldsymbol{r}_s).\boldsymbol{n}_{fs}]]\, dS_{fs}, \tag{11.3}$$

being, $G(\omega, \boldsymbol{r}_{fs}|\boldsymbol{r}_r)$ is the free-space Green functions with sources at $\boldsymbol{r}_{fs} = (\boldsymbol{x}_{fs}, z_{fz})$ on the sea surface and sensor at $\boldsymbol{r}_r$. Additionally, $\nabla P(\omega, \boldsymbol{r}_{fs}|\boldsymbol{r}_s)$ and $\boldsymbol{n}_{fs}$ are the pressure gradient and the normal vector at the sea surface S_{fs}, respectively.

The statistical parameters of sea surface are required from remote sensing to compute the plane wave reflection coefficient matrix. This procedure is considered the statistical parameters of sea surface height owing to the difficulties to obtain surface height from remote sensing data. According to Thorsos (1987), a Gaussian distribution can be used to describe sea surface height. In this regard, zero mean and standard deviation of σ, the coherent plane wave reflection coefficient matrix at the mean sea level based on the KA is determined from:

$$\hat{R}(\omega, \mathrm{K}_r|\mathrm{K}_S) = \left\langle \hat{R}(\omega, \mathrm{K}_r|\mathrm{K}_S) \right\rangle = e^{\{-2[k_z^S\sigma]^2\}\hat{R}_{coef}^{Flat}(\omega, \mathrm{K}_r|\mathrm{K}_S)} \tag{11.4}$$

$\langle\ \rangle$ being an anticipation operator, K_r is the wavenumber of the sensor while K_s is incident sun radiation. In addition, $\hat{R}_{coef}^{Flat}$ is a finite length flat sea surface plane wave reflection coefficient matrix, and k_z^s is the vertical wavenumber for sun radiation.

In this regard, Kirchhoff approximation considers the quasi-specular reflection; however, it ignores polarization. The limitation of geometrical optics leads the incident wavelength to be zero. Under this circumstance, Kirchhoff approximation is precise. If the sun and sensor points are far from the reflection of the sea surface, the absolute amplitudes of the reflection are very nearly the same. Consequently, the small-perturbation method (SPM) is used if the wavelength is larger than both the standard deviation and the correlation length of surface heights. However, SPM neither explicates for long-scale features in the surface spectrum nor for specular scattering. Both approximations are considered rough surface which has to be either large or small compared with the incidence wavelength.

For a sea surface, the slopes are usually small excluding for steep breaking waves, which constitute a distinctly small percentage and are most effective at robust and really sturdy wind speeds. The small-slope approximation is the consequence of a Taylor expansion with reverence to the influences of surfaces slopes. Finally, the strength of the KA is retraced in expressions of the specular and non-specular reflections from the rough sea surfaces.

11.3 QuickBird and Kalutara, Sri Lanka

The satellite QuickBird data are used to extract the information on tsunami wave spectra using 2-D Fourier transform (FFT) (Figure 11.3). Digital globe QuickBird black and white merchandise permits superior visual evaluation primarily based on 61-centimetre resolution (at nadir) and 11-bit gathered facts depth. The panchromatic sensor collects facts at the visible and near-infrared wavelengths and has a bandwidth of 450–900 nm. This data is acquired from the southwestern coast of Sri Lanka which is just south of the city of Colombo in a resort area known as Kalutara. This data was acquired shortly post the moment of tsunami impact at 10:20 a.m. local time, slightly less than four hours after the earthquake. Moreover, the panchromatic QuickBird satellite data reveals massive turbulent flow along coastal water of Kalutara (Figure 11.3). In fact, the tsunami wave is curved around the southern part of Sri Lanka and struck the west coast and triggered strong turbulent flow. This turbulent flow spelled out the name of Allah in Arabic [173].

The Kalutara coastline is a resort town located approximately 40 km south of Colombo in Sri Lanka. The coastline is planted by coconut trees. Further, the 38-meter long Kalutara Bridge was built at the mouth of the Kalu Ganga River and serves as a major link between the country's western of Kalutara. Kalutara district is a district in Western Province, Sri Lanka. Its area is 1,606 km^2. The district was hit by the tsunami generated by the 2004 Indian Ocean earthquake. This area is located between 6° 30′ 36″ N to 6° 35′ 24″ and 79° 55′ 48″ E to 80° 00′ 36″ E. The coastline is dominated by muddy sediment due to the existence of several rivers such as Kalu Ganga River (Figure 11.4).

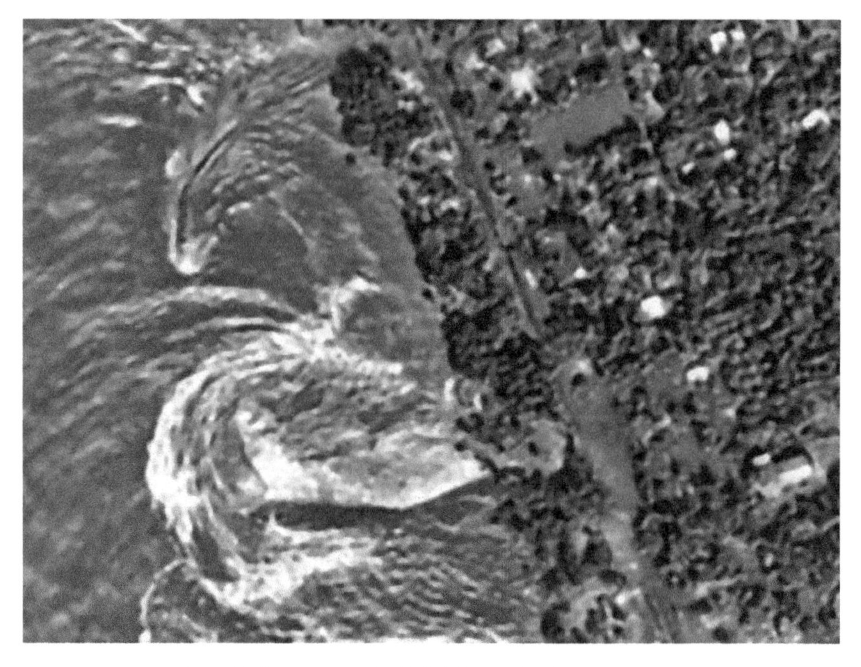

Figure 11.3 Panchromatic QuickBird with 61 cm resolution.

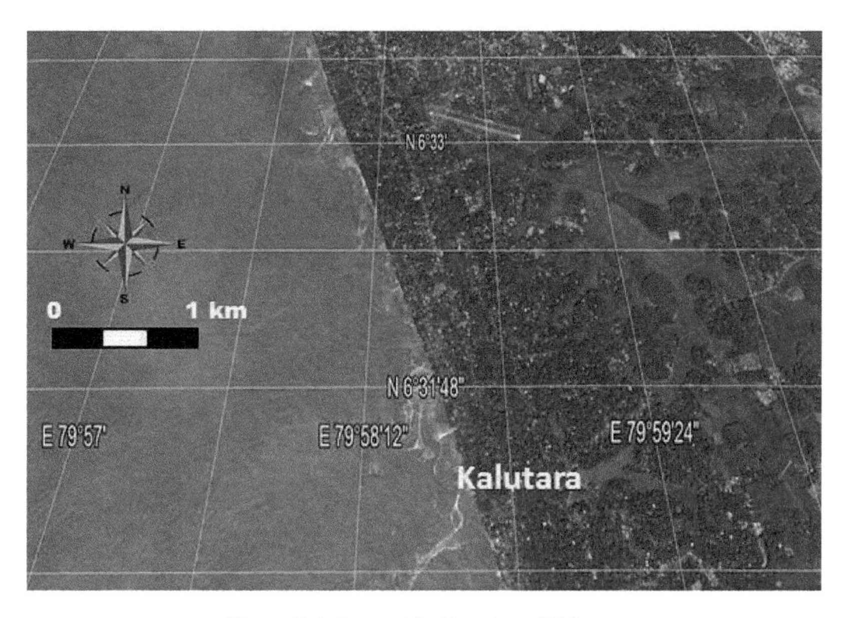

Figure 11.4 Geographical location of Kalutara.

11.4 Wave Spectra Estimation from QuickBird Satellite Data

Since the wave modifies its direction and wavelength as it propagates, the two dimensional (2-D) Discrete Fourier transform (DFT) was used to derive the wave quantity spectra from QuickBird satellite data. First, select a window kernel length of 512 × 512 pixels and lines with the pixel size identical to ΔX. In fact, it is impossible to acquire wave spectra parameters by using kernel window size less than 512 × 512 pixels and lines (Figure 11.5). According to Populus et al. (1991) [183], 2-DFFT transfers the 2-D QuickBird satellite data into frequency spectra domain (k_x and k_y). The frequency spectra domain is used to estimate the variation of wavelength variation along the spectra frequency domain as wavelength is inversely proportional to frequency domain as discussed below.

Following Populus et al. [183] and Marghany (2001) [184], let $X(m_1, m_2)$ represent the digital count of the pixel at (m_1, m_2) which is used to perform DFT, which is given by:

$$F(k_x,k_y) = N^{-2} \sum_{m_2=0}^{N-1} \left[\sum_{m_{11}=0}^{N-1} X(m_1,m_2).e^{-ik_x.m_1.\Delta X} \right] .e^{-ik_y.m_2.\Delta X} \tag{11.5}$$

where, n_1 and n_2 = 1,2,3,.......,N and k_x and k_y are the wave numbers in the x and y directions, respectively. Following Marghany (2004) [185], the Gaussian algorithm has been applied to remove the noise from the image and smooth the spectral peak into normal distribution curve. The wavelength has been estimated by using autocorrelation algorithm. The autocorrelation algorithm has been implemented in the mid row and the mid column. In a two dimensional wave number spectrum, the spectral peak (C) is located at X and Y, respectively (C_x, C_y) of a $N x N$ image spectrum which has the wavelength λ and the wave direction θ [183]:

$$\lambda = \frac{\Delta x.lag(pixels)}{(k_x^{\ 2} + k_y^{\ 2})^{0.5}} \tag{11.6}$$

$$\theta = \tan^{-1}(\frac{k_y}{k_x}) \tag{11.7}$$

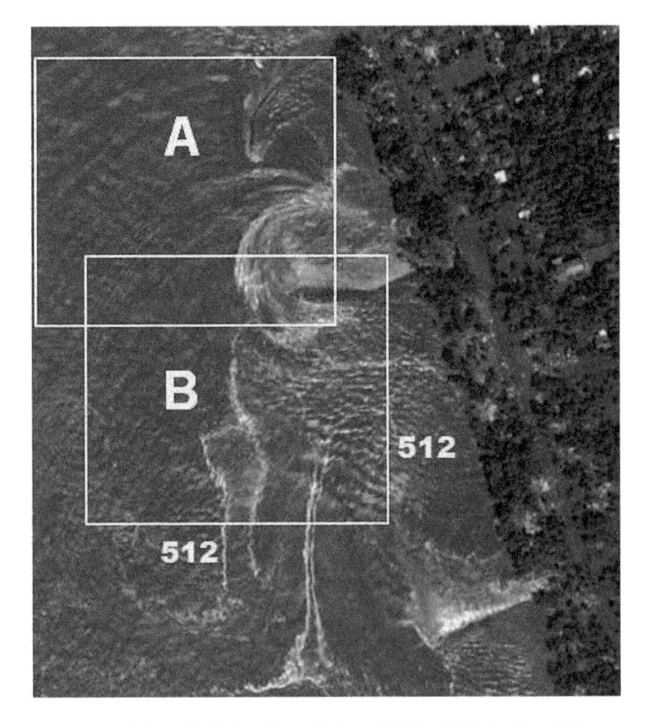

Figure 11.5 Selected window kernel size of 512 × 512 pixels and lines.

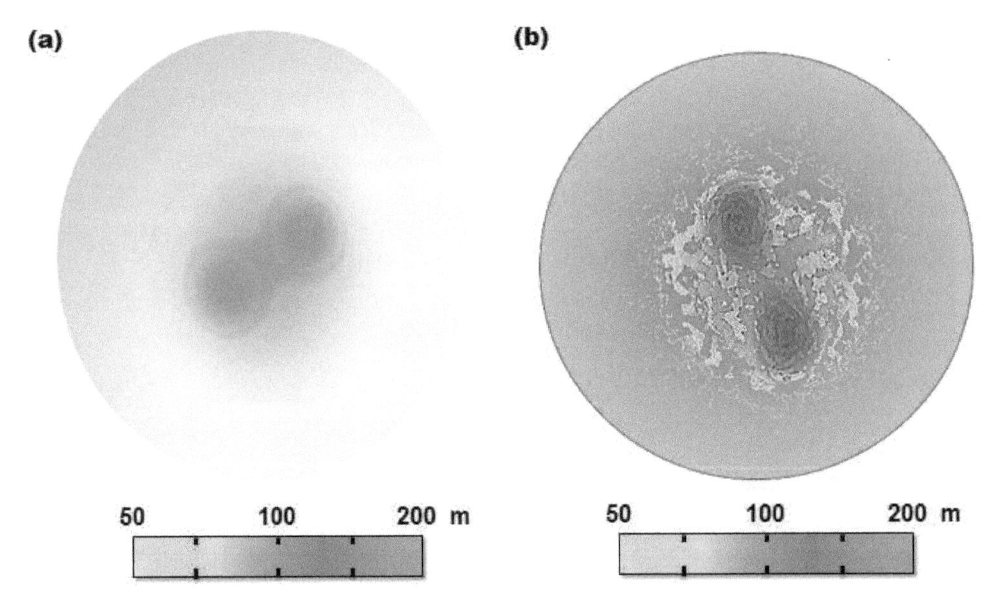

Figure 11.6 Spectra are derived by 2-DFFT in (a) region A and (b) region B.

Figures 11.6a and 11.6b reveal the specific patterns of tsunami wave spectra alongside the coastal water. Figure 11.6a depicts that the tsunami wave spectra direction of 150° is in the direction of the shoreline whilst Figure 11.6b indicates tsunami propagation is in the direction of 70°. It is thrilling to discover the dominant wavelength is between 50 and 200 m. The changes of wave path from region A

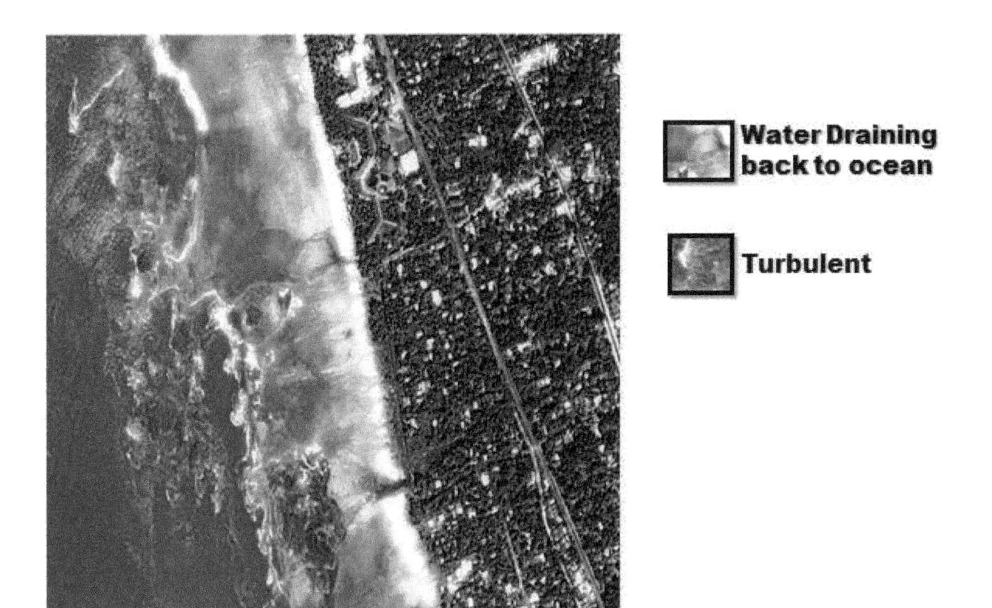

Figure 11.7 Turbulent flow along coastline because of tsunami.

to B are because of diffraction. Consistent with Marghany and Suffian (2005) [173], the tsunami waves have diffracted around Sri Lankan Island after which they moved perpendicular to the Kalutara coast and unfolded inland, inflicting massive flooding.

Figure 11.7 shows that the water drained back into the ocean and it built two barriers along Kalutara coastline. As successive tsunami passed along the two large barriers, the wave is then diffracted. Therefore, another barrier blocked part of the wave and allowed the rest to pass and generate large eddy with radius of 150 m behind the barrier. This indicates that the successive tsunami waves that hit the Kalutara coastline have changed the coastal zone morphology pattern.

11.5 Numerical Model of Tsunami Run-up

Then water elevation η caused by tsunami wave can be modeled using solitary wave. In this regard, solitary wave consists of a complex spectrum of frequencies which is uniquely described by the height-to-depth ratio $\boldsymbol{H} = H/d$, where H is height and d is depth, according to Gedik et al. (2005) [186]. The modelling of η is adapted using solitary wave model as follows:

$$\eta(x, 0) = H \sec h^2 (0.75H)^2 (x - X_1) \tag{11.8}$$

where x and X_1 are pixel locations of tsunami wave propagation in QuickBird satellite data, where $x = X_1$ at in time t = 0. The estimation of the tsunami height H can then be done by using the formula adopted from Synolakis (1987) [187] as follows:

$$\lambda = \frac{2d}{\sqrt{0.75H}} \operatorname{arcosh} \frac{1}{\sqrt{0.05}} \tag{11.9}$$

where λ is the wavelength extracted from QuickBird satellite data using Equation 11.6, and d is the water depth which ranged between 5 to 200 m along the coastal waters of Sri-Lanka [188, 189]. The bathymetry information was acquired directly on USGS site [190]. Equation 11.9 is used to estimate tsunami wave height from QuickBird satellite data which is then substituted in Equation 11.8 to calculate tsunami wave elevation η.

In the coastal zone, the wavelength of the incident tsunami becomes shorter and the amplitude becomes larger as the tsunami propagates into shallower water [191]. Therefore, in Cartesian coordinate system, the equation of tsunami wave propagation can be written as follows

$$\frac{\delta\eta}{\delta t}+\frac{\delta}{\delta x}(du)+\frac{\delta}{\delta}(dv)=0 \tag{11.10}$$

$$\frac{\delta u}{\delta t}+g\frac{\delta\eta}{\delta x}=0 \tag{11.11}$$

$$\frac{\delta v}{\delta t}+g\frac{\delta\eta}{\delta y}=0 \tag{11.12}$$

where η is the tsunami elevation which has been estimated from the Equation 11.10, u and v are the tsunami velocity components in the x and y directions, g is the gravity acceleration and h is the water depth which is obtained from USGS [190]. Then, the run-up (R) is modeled using the equation adopted from Gedik et al. (2005) [186]

$$\frac{R}{d}=1.25\left(\frac{\pi}{2\beta}\right)^{0.2}\left(\frac{H}{d}\right)^{1.25}\left(\frac{H}{\lambda}\right)^{-0.15} \tag{11.13}$$

where β the inclination angle of the plane beach, λ is the wavelength modelled from QuickBird satellite data using Equation 11.8, d is the water depth, and H is tsunami height modelled using Equation 11.9.

11.6 Fuzzy B-spline Method for 3-D Run-up Simulation

The fuzzy B-splines (FBS) are introduced allowing fuzzy numbers instead of intervals in the definition of the frequency domain of B-spline. According to Marghany et al. (2011) [192], a fuzzy number is defined using interval analysis. There are two basic notions that we combined together: confidence interval and presumption level. A confidence interval is a real values interval which provides the sharpest enclosing range for tsunami wave spectra propagation in spatial domain. Following Marghany et al. [192], an assumption level μ-level is an estimated truth value in the [0, 1] interval on our knowledge level of the tsunami wave spectra [195]. The 0 value which corresponds to minimum knowledge of tsunami frequency spectra, and 1 to the maximum variation in tsunami frequency spectra was retrieved from Quickbird imagery. A fuzzy number is then prearranged in the confidence interval set, each one related to an assumption level $\mu \in [0, 1]$. Moreover, the following must hold for each pair of confidence intervals which define a number: $\mu \succ \mu' \Rightarrow \omega \succ \omega'$.

Let us consider a function $f: \omega \rightarrow \omega'$, of N fuzzy variables, $\varpi_1, \omega_2,\ldots,\omega_n$ where ω_n are the global minimum and maximum values wave angular frequency derived by 2-DFFT. Based on the spatial variation of the digital number, the fuzzy B-spline algorithm is used to compute the function f.

Following the studies conducted by Marghany (2012) [194], let $\omega(i,j)$ be the depth value at location i,j in the region D where i is the horizontal and j is the vertical coordinates of a grid of m times n rectangular cells. Let N be the set of eight neighbouring cells. The input variables of the fuzzy are the amplitude differences of digital number intensity ω defined by [193–197]:

$$\Delta\omega_N = \omega_i - \omega_0,\ N = 1,\ldots\ldots,8 \tag{11.14}$$

where the d_i, $N = 1, 8$ values are the neighbouring cells of the actually processed cell ω_0 along the horizontal coordinate i. To estimate the fuzzy number of digital value belonging to wave spectra frequency ω_j which is located along the vertical coordinate j, we estimated the membership function values μ and μ' of the fuzzy variables ω_i and ω_j, respectively, by the following equations as described by Rövid et al. (2004) [207]

$$\mu = \max\{\min\{m_{pl}(\Delta d_i) : d_i \in N_i\}; N = 1....,9\} \tag{11.15}$$

$$\mu' = \max\{\min\{m_{LNl}(\Delta d_i) : d_i \in N_i\}; N = 1....,9\} \tag{11.16}$$

where m_{pl} and m_{LNl} correspond to the membership functions of fuzzy sets. From Equations 11.15 and 11.16, one can estimate the fuzzy number of water tsunami flow ω_j

$$\omega_j = \omega_i + (L-1)\Delta\mu \tag{11.17}$$

where $\Delta\mu$ is $\mu - \mu'$ and $L = \{\omega_1,.........,\omega_N\}$. Equations 11.16 and 11.17 represent turbulent flow in 2-D. In order to reconstruct fuzzy values of turbulent flow in 3D, fuzzy number of turbulent flow in z coordinate is estimated by the following equation proposed by Russo (1998) [208],

$$\omega_z = \Delta\mu MAX\{m_{LA}\left|\omega_{i-1,j} - \omega_{i,j}\right|, m_{LA}\left|\omega_{i,j-1} - \omega_{i,j}\right|\} \tag{11.18}$$

where ω_z fuzzy tsunami wave propogation flow values over water depth z which is function of i and j coordinates, i.e., $\omega_z = F(\omega_i, \omega_j)$. Fuzzy number F_O for tsunami wave spectra propagation in i, j and z coordinates then can be given by

$$F_O = \{\min(\omega_{z_0},.........., \omega_{z\Omega}), \max(\omega_{z_0},.........., \omega_{z\Omega})\} \tag{11.19}$$

where $\Omega = 1, 2, 3, 4$, and the fuzzy number of turbulent flow F_O is then defined by B-spline in order to reconstruct 3D of tsunami wave spectra. In doing so, B-spline functions including the knot positions, and set of control points are constructed. The requirements for B-spline surface are set of control points, set of weights and three sets of knot vectors and are parameterized in the p and q directions [208].

A fuzzy B-spline surface $S(p,q)$ is described as a linear combination of basis functions in two topological parameters p and q. Let $R = r_0; \ldots; r_m$ be a non decreasing sequence of the real numbers. The r_i is called knots and R is the knot vector. The interval r_i and r_{i+1} is called knot span. According to Anile et al. (1995) [196], the *Pth*-degree *(order P + 1)* piecewise polynomial function B-spline basis function, denoted by $\beta_{i,P}(r)$, is given by

$$\beta_{i,1}(r) = \begin{cases} 1 & \text{If } r_i \le r \le r_{i+1}, \\ 0 & \text{Otherwise;} \end{cases} \tag{11.20}$$

$$\beta_{i,P}(r) = \frac{r - r_i}{r_{i+P-1} - r_i}\beta_{i,P-1}(r) + \frac{r_{i+P} - r}{r_{i+P} - r_{i+1}}\beta_{i+1,P-1}(r) \quad \text{for } P > 1 \tag{11.21}$$

To exercise more shape controllability over the surface, and invariance to perspective transformations, fuzzy B-spline is introduced. Besides having the control point as in the B-spline, fuzzy B-spline also provides a set of weight parameters $w_{i,j}$ that exert more local shape controllability to achieve projective invariance. Following Fuchs et al. (1977) [205] and Russo (1998) [208], fuzzy B-spline surface, that is composed of $(O \times M)$, i.e., O and M are the element vectors that belong to knot p and q patches, respectively, is given by

$$S(p,q) = \frac{\sum_{i=0}^{M}\sum_{j=0}^{O} F_O C_{ij}\beta_{i,4}(p)\beta_{j,4}(q)w_{i,j}}{\sum_{m=0}^{M}\sum_{l=0}^{O}\beta_{m,4}(p)\beta_{l,4}(q)w_{ml}} = \sum_{i=0}^{M}\sum_{j=0}^{O} F_o C_{ij} S_{ij}(p,q) \tag{11.22}$$

$\beta_{i,4}(p)$ and $\beta_{j,4}(q)$ are two basis B-spline functions, $\{C_{ij}\}$ are the bidirectional controls net and $\{w_{ij}\}$ are the weights. The curve points $S(p,q)$ are affected by $\{w_{ij}\}$ in case of $p \in [r_i, r_{i+P+1}]$ and $q \in [r_j, r_{j+P'+1}]$, where P and P' are the degree of the two B-spline basis functions constituted by the B-spline surface. Two sets of knot vectors are $knot\,p = [0,0,0,0,1,2,3,\ldots\ldots,O,O,O,O]$, and $knot\,q = [0,0,0,0,1,2,3,\ldots.,M,M,M,M]$, respectively. Fourth order B-spline basis are used $\beta_{j,4}(.)$ to ensure continuity of the tangents and curvatures on the whole surface topology including at the patches boundaries [193, 208].

11.7 Galerkin Finite Element

In this chapter, a mesh-less method is applied because it does not require a time-consuming mesh generation for modeling the computational domain. The element-free Galerkin method (EFGM) proposed by Hiroshi et al. (1994) [198] is one of the practical mesh-less method. Then, the EFGM is used to solve the shallow water wave equations. For the EFGM, state value of arbitrary evaluation points in the domain is solved by a kind of Moving Least Square Method (MLSM) from state value of nodal, which exists in the circle of which the center is the evaluation point. The radius of circle is called 'domain of influence'. In this regard, the distance between evaluation point and every node in "the domain of influence" is a function of MLSM estimation (Figure 11.8).

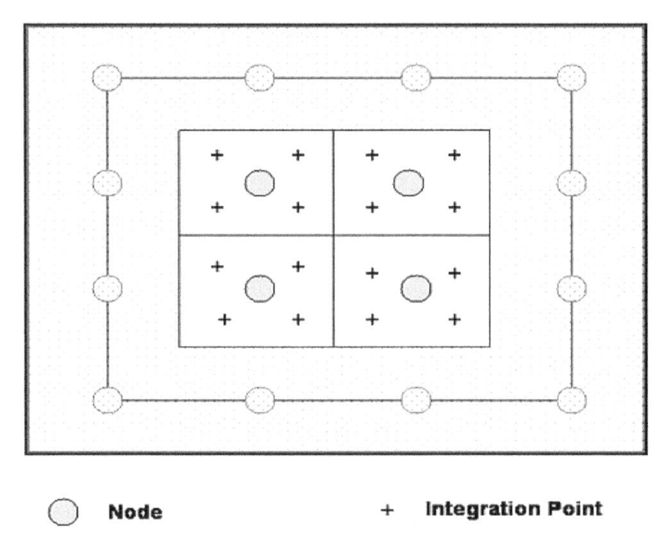

Figure 11.8 Finite element grid used in this study.

11.7.1 Moving Least Square Method (MLSM)

The moving least-squares (MLS) method can be used in function approximation. The so-called MLS method, which reconstructs the continuous function from an arbitrary set of particles via the calculation of a weighted Least Square (LS) around the evaluation point, is used to obtain the local approximate derivatives. The MLS is generally more accurate than the grid-based numerical methods near irregular surfaces [199]. The unknown function, in MLSM is expressed as follows,

$$\phi^h(x) = \sum_{j}^{N} p_j(x)a_j(x) \equiv \{p(x)\}^T\{a(x)\} \tag{11.23}$$

where p_j is linear basis including in the space coordinates, a_j is undetermined coefficient and N is number term used at expansion. For example, p is expressed as follows [199],

$$\{p(x,y,z)\}^T = \{1, x, y, z\} \qquad (N=4) \tag{11.24}$$

$$\{p(x,y,z)\}^T = \{1, x, y, z, x^2, xyz, y^2\} \qquad (N=7) \tag{11.25}$$

Following Cinegoski et al. (1988) [200] and Belyschko et al. (1994) [201], the coefficient $a_j(x)$ is obtained by minimizing the performance function J as follows,

$$J = \sum_i^N w(x-x_i)\{\phi_i^h(x) - \phi_i\}^2 = \sum_i^N w(x-x_i)\{\{p(x)\}^T\{a(x)\} - \phi_i\}^T, \tag{11.26}$$

where N is the number of nodes within neighborhood of x and x_i is the space coordinates of an arbitrary node x_i within neighborhood of x. $\sum_i^N w(x-x_i)$ is weighting function that depends on the distance between point x and node x_i. Finally $\phi_i^T(x)$ is interpolation function and is defined as [199],

$$\phi_i^T(x) = \{p(x)\}^T[\sum_i^N w(x-x_i)p(x_i)p^T(x_i)]^{-1}[w(x-x_1)p(x_1), \\ w(x-x_2)p(x_1), \ldots\ldots\ldots\ldots, w(x-x_n)p(x_n)] \tag{11.27}$$

11.7.2 Weight Function

The weight function value (w) is changed according to distance and nodes in the domain of effect, then a smoothly approximate curved line is given. For the reason of this characteristic, approximated value is changed according to the movement of current along the distance. It is very important to select a weight function w in the MLSM. There is a large range for selecting the weight function. The basic characteristics are described as follows: (i) number of weight function is positive; and (ii) weight function is defined as the function of distance between the two points [199–202].

Following Goto and Ogawa (1992) [191] and Singh (2004) [199], Equations 11.6 to 11.28 are multiplied by weight functions for elevations $w(r)_\eta$, and velocity components $w(r)_{u,v}$ where $r = \|\eta - \eta_i, u - u_i, v - v_j\|$ over the domain V as follows,

$$\int_V w(r)_\eta \frac{\delta\eta}{\delta t} dV + \int_V w(r)_\eta \frac{\delta}{\delta x}(hu) dV + \int_V w(r)_\eta \frac{\delta}{\delta x}(hv) dV = 0 \tag{11.28}$$

$$\int_V w(r)_u \frac{\delta u}{\delta t} dV + g\int_V w(r)_u \frac{\delta\eta}{\delta x} dV = 0 \tag{11.29}$$

$$\int_V w(r)_v \frac{\delta v}{\delta t} dV + g\int_V w(r)_v \frac{\delta v}{\delta y} dV = 0 \tag{11.30}$$

As tsunami water elevation is derived from Equation 11.8, velocity components and their weighting function are integrated in each three node triangular elements with the interpolation function $\phi_i^T(x)$ that is obtained from Equation 11.27.

11.7.3 3-D Waves and Run-up Study Case: Kalutara Coastline, Sri Lanka

The coastal bathymetry is governed by water depth of 1000 to 3000 m which surrounds the Sri Lankan Island (Figure 11.9). However, the coastal zone is dominated by Digital Elevation Model (DEM) that ranges between 5 to 15 m above the sea level (Figure 11.10). The main feature of this coastal zone is the existence of the high sand dunes. They are scattered along stretches of the southeast and the far northeast coast which were very effective in withstanding the tsunami wave [173, 203, 204].

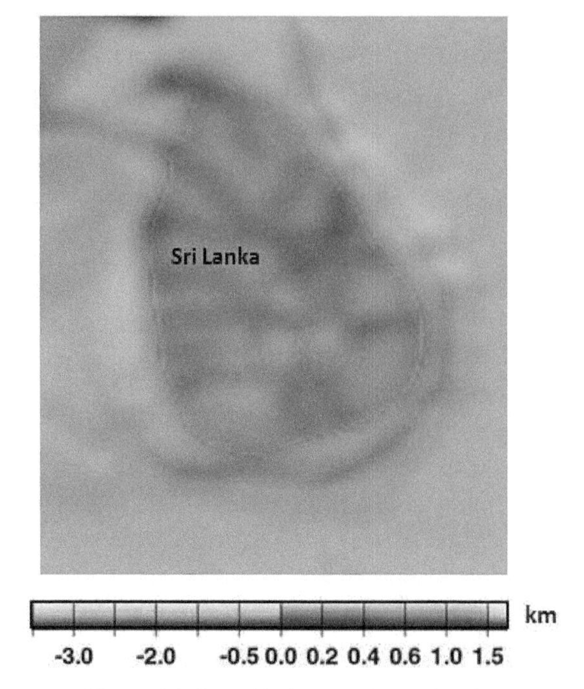

Figure 11.9 Coastal bathymetry along Sri Lanka.

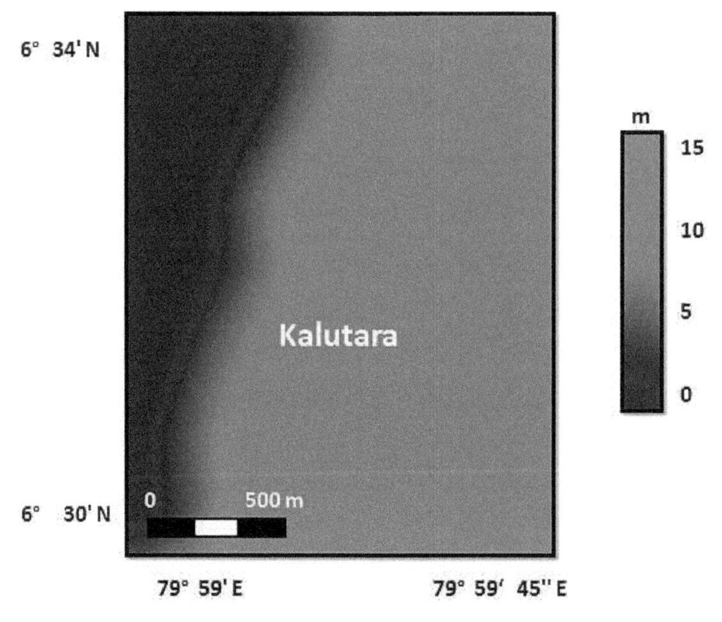

Figure 11.10 Kalutara DEM.

Figure 11.11 suggests the 3-D tsunami wave propagations constructed by using the use of fuzzy B-spline. It is fascinating to determine the clear pattern of tsunami wave heights which are ranged between 3 and 6 m. The maximum wave peak of 6 m was occurred due to the wave breaking. The maximum wave height is shown throughout an eddy movement while the waves have turned into 4 to 6 m height inland.

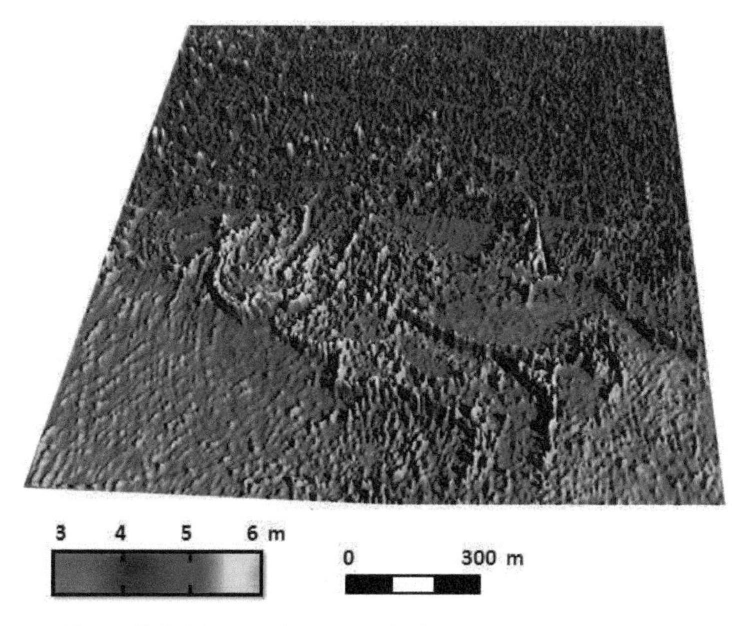

Figure 11.11 3-D tsunami pattern north of Kalutara using fuzzy B-spline.

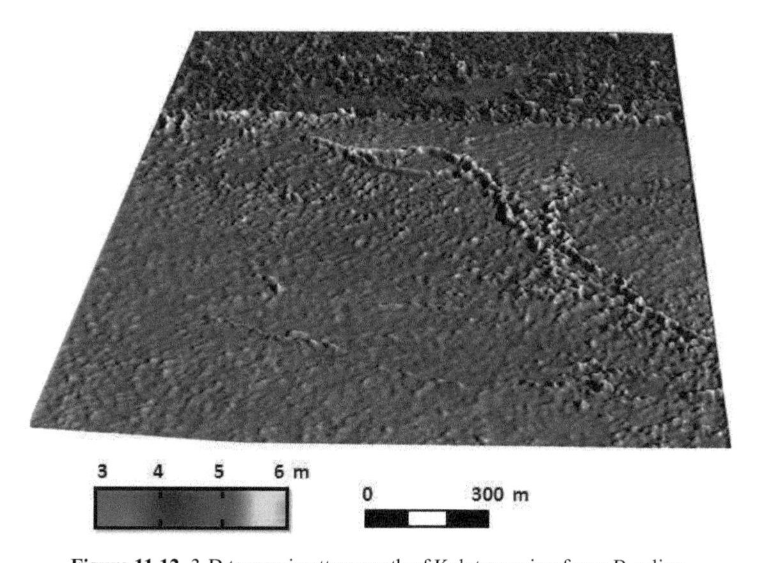

Figure 11.12 3-D tsunami pattern south of Kalutara using fuzzy B-spline.

Further, Figure 11.11 indicates 3-D wave diffraction along Kalutara shoreline. This suggests a turbulent water movement because of the combination of wave diffraction, refraction, reflection and longshore current movements between the two jetties.

Taken together, these were able to cause a pattern which spelled out, approximately, the pattern of the Arabic word for Allah and El-Gbar as shown Figure 11.12. Figure 11.12 shows that the run-up is ranged between 4 and 6 m. The minimum runup was observed inland while the regions were closed to the coastline dominated by run-up of 6 m (Figures 11.11 and Figure 11.12). It is obvious that the mechanism of run-up was accompanied by convergence zone (Figure 11.12).

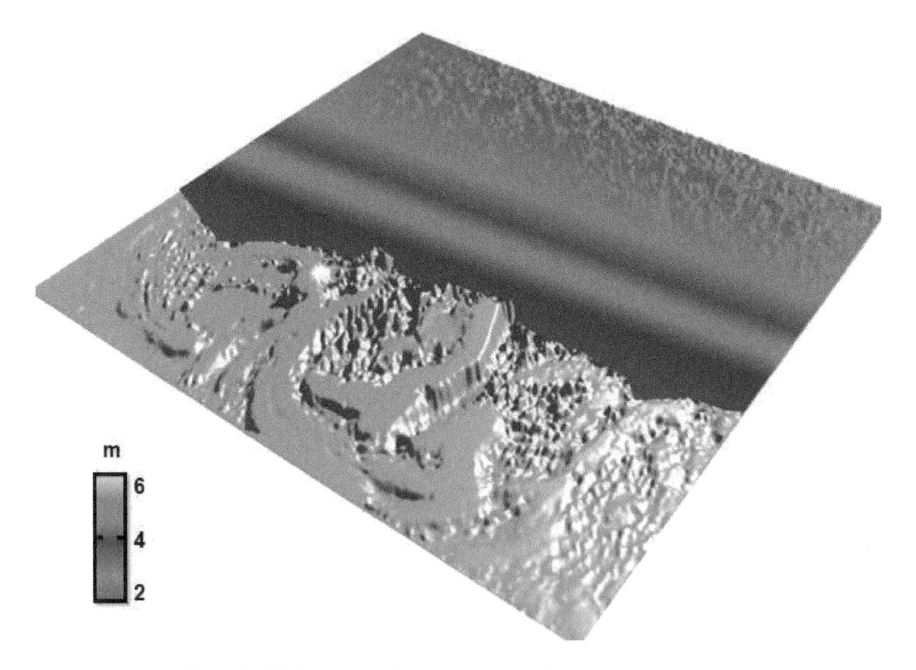

Figure 11.13 3-D tsunami run-up pattern using EFGM algorithm.

Figure 11.13 reveals the 3-D tsunami run-up which is derived by EFGM algorithm. The fuzzy B-spline which results in Figures 11.10 to 11.12 agree with output of EFGM algorithm. Both EFGM algorithm and fuzzy B-spline agreed that wave run-up is ranged between 2 and 6 m wave height.

Clearly, EFGM is able to reconstruct 3-D visualization of tsunami propagation and run-up in QuickBird satellite data. High resolution of 0.6 m is relieving EFGM algorithm to perform accurate visualization. In this regard, high resolution of QuickBird is able to capture a fine information about tsunami morphology propagation.

Further, EFGM is able to track fine information about tsunami causing rough turbulent flow along the coastal water of Kalutara. EFGM based mesh-less method, therefore, does not need a time-consuming mesh generation for modeling the computational domain of tsunami propagation in QuickBird satellite data. Further, Moving Least Square Method (MLSM) is used to acquire state value of arbitrary evaluation points in the domain for the EFGM.

On other hand, the moving least-squares (MLS) method can be used in function approximation which reconstructs the continuous function from an arbitrary set of particles by calculating a weighted Least Square (LS) around the evaluation point. Consistent with Hiroshi et al. (1994) [198], the MLS is more precise than the grid-based numerical methods near irregular surfaces such as tsunami wave propagation. Finally, the weight function in MLS is flexible to track the changes of the different parameters in every three node triangular element with the integration of linear interpolation function.

11.7.4 3-D Whirlpools and Solitary Waves

The whirlpools are best viewed at high or low tide which varies daily. Nevertheless, the rushing tsunami can generate whirlpools in shallow waters. In this regard, the tsunami formations are captivating, creating swirls and ripples that are rather hypnotic to watch. It is interesting that whirlpools are followed by tsunami. In this view, a tsunami can be twisted like a corkscrew around its direction of travel. This is similar to the unusual quantum concept which allows wave particles to whirl around in a vortex, even when no external force is implemented to the wave particles.

When a tsunami is twisted, waves at the central axis cancel each other out forming a bright core (Figure 11.14). As the tsunami particles spin around the axis, they carry orbital angular momentum that can rotate its particles and create massive turbulent flow. The similarities arise because the twisting angular momentum of the water particle interacts with their forward motion in the same way that intrinsic angular momentum (spin) interacts with the motion of the tsunami. In the view of quantum mechanism, this is known as spin-orbit coupling. In this understanding, an interaction of tsunami particle's spin with its motion in a potential is considered as a relativistic interaction.

This reinforces the prominent theory of wave-particle duality, which declares that all particles have a wave correlated with them. Further, it proposes that the convenient claims of tsunami generating vortices could be reproduced at much shorter wavelengths (Figure 11.14). In this view, the generated vorticity is robust against perturbations.

In other words, whirlpools occur because of the interface between running water and the geology of the coastline and seafloor. As the tsunami proceeds up the continental slope, the wave period remains constant while the velocity decreases, and as the wavelength decreases to 50 m, the height increases from 1 to 4 m. The whirlpool is located 443 m away from the shoreline. Its diameter is 165 m (Figure 11.15). Under these circumstances, a complicated train of waves is formed along the coastal waters. Figures 11.14 and 11.15 indicate a lot of water that is being pushed around, and it interacts with the shape and the bathymetry, near the coastline.

As long as the large whirlpool is developed, the massive turbulent flow must be created due to shallow bathymetry of less than 20 m. Consequently, the edge waves are generated too upon the reflection from the coastline boundary. These edge waves occur post-tsunami run-up which is concomitant to reflected offshore wave energy. In this view, the edge waves travel back and forth, parallel to the coastline (Figure 11.15). In this context, edge waves have length much larger than the water depth. Figure 11.16 shows the fuzzy b-spline classification of the edge waves using a fuzzy-Bspline algorithm. The vorticity pattern indicates that the upstream is essentially irrational while the downstream is tended to be extremely turbulent.

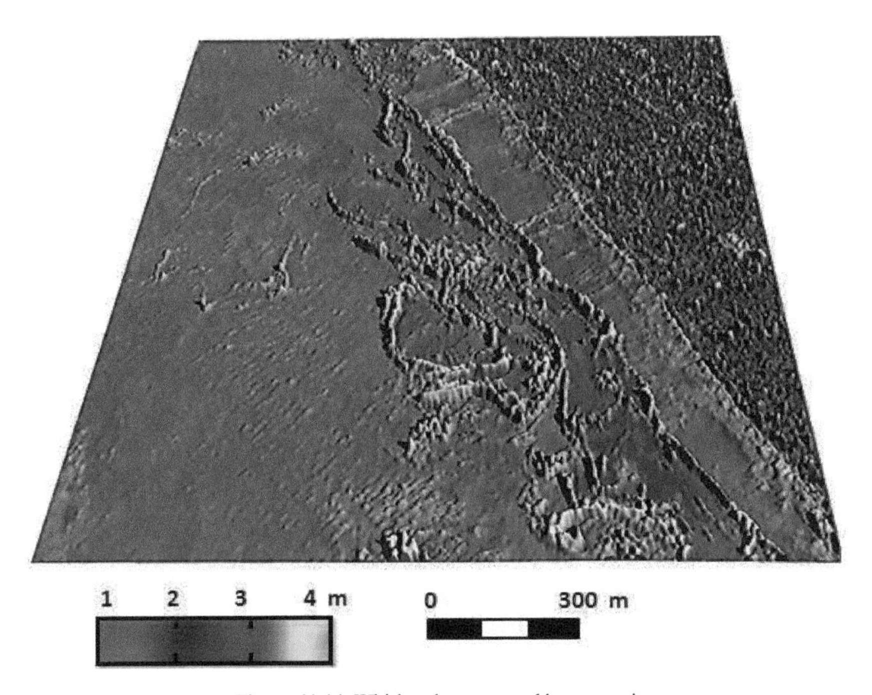

Figure 11.14 Whirlpools are caused by tsunami.

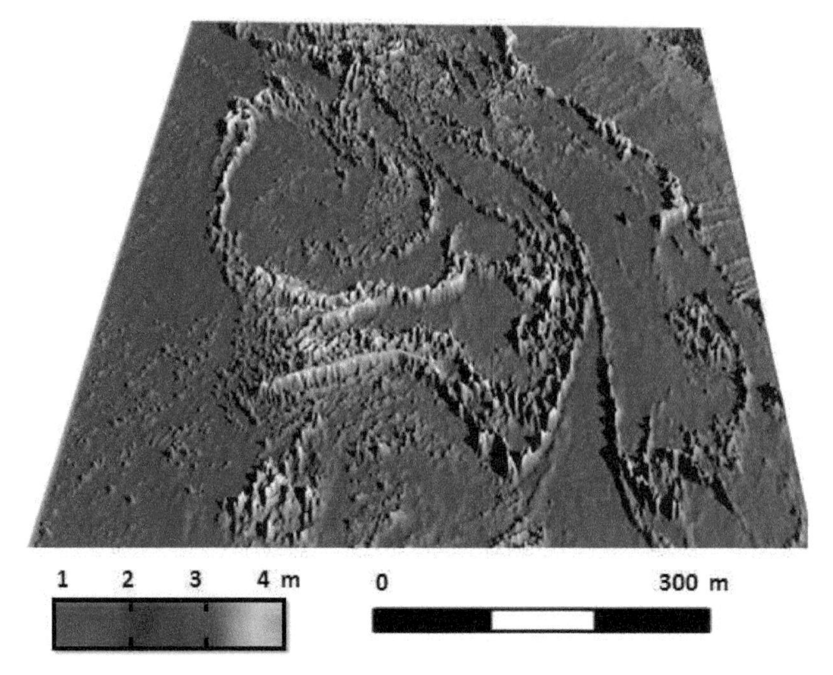

Figure 11.15 3-D large whirlpool simulated by fuzzy-B-spline.

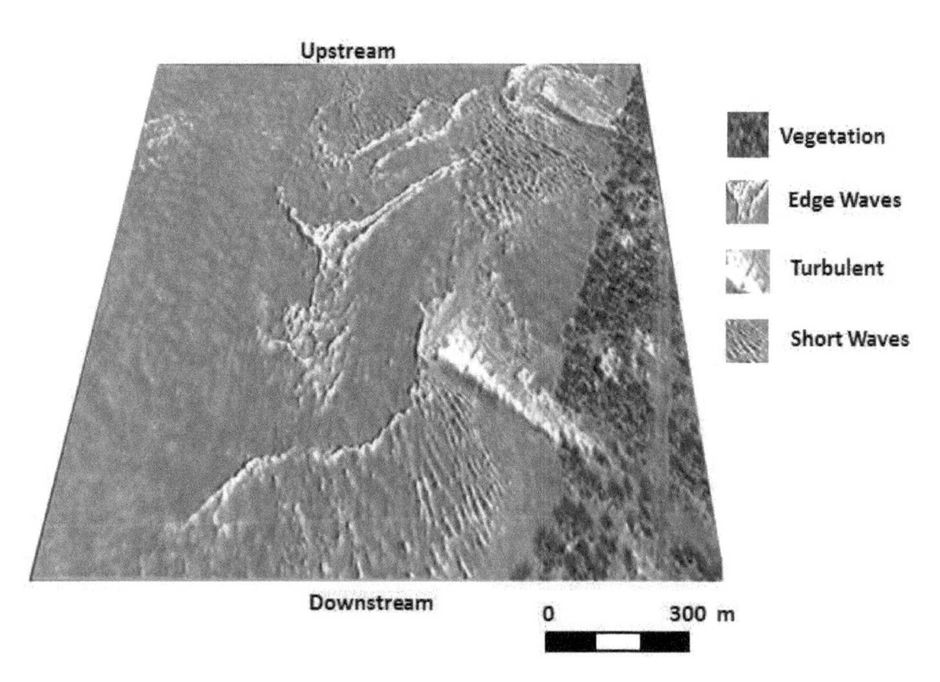

Figure 11.16 Edge waves using fuzzy B-spline classifier.

Fuzzy B-spline approximation of 3rd order provides 3-D images which were virtually free of visible artifacts. This is due to the fact that each operation on a fuzzy number becomes a sequence of corresponding operations on the respective μ-levels, and the multiple occurrences of the same fuzzy parameters are evaluated as a result of the function on fuzzy variables [194–197].

It is very easy to distinguish between small and long waves. Typically, in computer graphics, two objective quality definitions for fuzzy B-splines were used: triangle-based criteria and edge-based criteria. Triangle-based criteria follow the rule of maximization or minimization, respectively, of the angles of each triangle [197] which prefers short triangles with obtuse angles. The so-called max-min angle criterion prefers short triangles with obtuse angles. This finding confirms those of the studies of Keppel (1975) [206], Anile (1997) [195], and Marghany et al. (2010) [193].

In addition, fuzzy B-spline algorithm produced 3D tsunami turbulent flow reconstruction without existence of any information regarding water or topography elevation. In fact, fuzzy B-spline algorithm is able to keep track of uncertainty and provide tools for representing spatially clustered gradient flow points [208].

Chapter 12

Four-dimensional Hologram Interferometry of Tsunami Waves from Quickbird Satellite Data

12.1 Introduction

Hologram and four-dimensional (4-D) are required theoretical thoughts prior to their implementation in 4-D tsunami reconstruction from optical satellite data. In this regard, hologram techniques can be used to acquire three-dimensional (3-D) of tsunami wave which is encoded in two-dimensional (2-D) wave. Therefore, the main question which arises is how is 4-D encoded in 3-D? Besides, how hologram interferometry can be derived from incoherent satellite optical data such as QuickBird data? In fact, QuickBird data are considered due to reflection of sun radiation, which is recorded by the sensor. Under this circumstance, the information recorded in optical satellite data is just the amplitude of electromagnetic wave reflection. However, it is impossible to retrieve the hologram fringes from QuickBird data or other optical remote sensing due to an absence of phase information. Therefore, it may be possible to implement the hologram interferometry from optical satellite data by considering the procedures of incoherent hologram.

Sooner or later, holograms could be theoretically transferred electronically through an advanced display device in homes and business places. In 1947, Dennis Gabor invented the theory of hologram. Consequently, the hologram becomes possible because of the advancement in laser technology. However, it is far from the pattern manufactured on a photosensitive medium that has been exposed through holography after which photographics evolved. The photosensitive medium, hence exposed and so developed, is likewise known as a holograph.

12.2 Physics of Hologram

12.2.1 Definition of Hologram

It is worth mentioning that the origin of a hologram lies in Greek. In this view, the Greek word 'holos' denotes a whole while 'gramma' means a message. It is pronounced as HOL-o-gram which is a 3-D image, formed by the photographic projection. In contrast to 3-D or virtual fact on a dimensional computer display, a hologram is a virtual three-dimensional and unfastened-status photograph that doesn't simulate the spatial intensity or require a unique viewing tool.

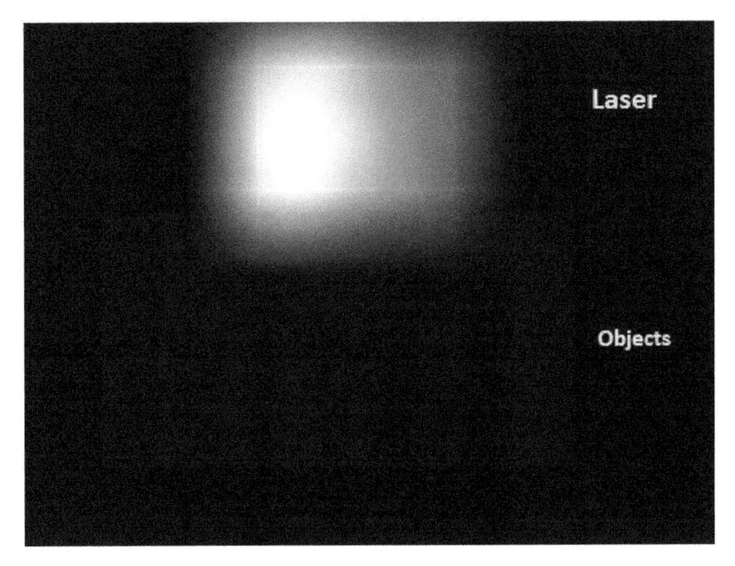

Figure 12.1 3-D Hologram of chess pieces is created by a laser illumination.

The coherent light is the keystone of hologram technique. In the view of physics, when laser illuminates the object, the image is recorded. Indeed, if the light is exposed to the film, it is reflected from the object and to a direct beam of laser. In this regard, the interference of laser beam or coherent light on the film illuminates the object and creates a 3-D image (Figure 12.1).

12.2.2 LASER

Acronym of LASER involves many terms for each of its letters. The full form of LASER, for instance, alone stands for **L**ight **A**mplification by **S**timulated **E**mission of **R**adiation. Therefore, the spectral radiation of the visible light is not a harmful radiation [40].

Laser light is a superior sort of illumination. Foremost, it is recognized as black-and-white, i.e., monochromatic, or one solitary wavelength or color. Subsequently, it is coherent. In other words, the electromagnetic waves are travelling out of the spout in an identical well-ordered and harmonized routine (Figure 12.2). The laser, therefore, is extra intensive and influential owing to its coherence.

In contrast, the sunlight seems to be white. Conversely, it is without a doubt made up of each color of light, consisting of hues our eyes cannot realize. That's why sunlight can be detached into its constituent hues through a prism, or in the case of a rainbow, through airborne water vapor. Covered in daylight is infrared light, which we cannot see; nevertheless, we feel it as heat. It is also comprised of an invisible ultraviolet light. In fact, there are many colorings, or wavelengths of light in daylight.

Lasers converge all their supremacy in a trivial spot. The coherent light can be conveyed over abundant superior spaces than incoherent light. The coherent light, for instance, can be transferred hundreds, even thousands of miles through fiber optic cables starved of considerable defeat. Correspondingly, laser light is inclined to infiltrate simply to the zeniths and move along them, for energy-stirring influences in the zeniths, even at a space from the site being considered as the coherent light.

12.2.3 Hologram and Holographic

Holography is the technique of physics to create holograms. Classically, a hologram is an accurate taping of a light field, rather than a photograph which evolved with the aid of a lens. Theoretically, holography

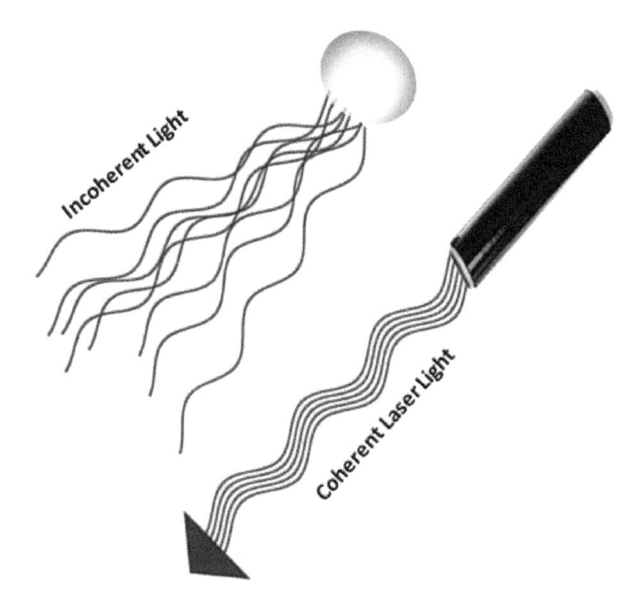

Figure 12.2 Difference between light and laser.

Figure 12.3 3-D hologram image.

entails the use of laser light for instructive purposes and for viewing the completed hologram. Therefore, assessing under ideal circumstances, a holographic photograph is visually indistinct from the actual item, if the hologram and the item are ignited impartially as they have been at the time of verifying.

Consequently, it is far more expensive to create an exhibition of a whole 3-D photograph of the holograph which is deprived of the aid of unique optical devices (Figure 12.3). The hologram is not a photograph and it is often incoherent whilst it is visible underneath the diffuse ambient light. It is a coding of the electromagnetic field as an interfering pattern of reputedly arbitrary variability inside the obscurity, density, or shallow profile of the photogenic medium. While it should be struck, the interference configuration diffuses the light into an imitation of the particular mild subject and the items that had been in it appears to nevertheless be there, displaying visible intensity cues inclusive of parallax and perspective that exchange realistically with any trade in the relative function of the observer [211].

12.2.4 Duality of Hologram and Universe

The main part of quantum theory is based on the duality between waves and particles. Evidently, light is a wave which consists of wavelength and regulates its color. Furthermore, light can intermingle with each other as wave to make up coherent, for instance, lasers. Consequently, light is made up of discrete particles which are known as photons. From the point of view of particle-wave duality, light is considered as quantum which has an asymmetry between its wave phases and particles (Figure 12.4). From the point of view of quantum, it does not require discrimination between wave phases and particles. Both quantum objects have the duality of their particle and wave aspects [209].

Credibly, the holographic theory is acknowledged as the utmost authoritative duality. This principle is often misrepresented as the idea that the universe is actually a hologram; nevertheless, there is a duality between a volume of space and the surface enclosing that volume. The holographic principle states that all the information contained within a region of space can be determined by the information on the surface containing it. Mathematically, this means the volume of space can be embodied as a hologram of the surface, hence the idea's name.

Generally, the holographic theory develops the impression of volumes and surfaces, and it covers everything from black holes to cosmology. With a black hole, for instance, we can't observe the interior because the gravity near a black hole is too strong for light to escape. But the black hole has a "surface" known as the event horizon, and we could observe everything outside of that. So, from information near the event horizon, we can understand the interior of a black hole.

The main question is whether that implies that the universe is a hologram? Not really. There may be a duality: ours is a universe with hologram properties or a hologram with universe properties (Figure 11.5). Theoretical physicists and astrophysicists investigating irregularities in the cosmic microwave heritage (the 'afterglow' of the Big Bang) have discovered that there is considerable evidence of holographic universe. In this regard, the theory of cosmic inflation can explain a holographic rationalization of the universe. Within the mathematical formalism, you don't need to differentiate between one of the alternatives, which is what offers the holographic precept its strength.

Clearly, a simple clarification of hologram universe can be, as an instance, brought by looking at a replicate. This may truly provide an explanation for the duality which presents the real world and the reflection of the mirror. This indicates our image within the replicate is as real as us. In other words, it does not mean a ghost image of the real world in a mirror, but a bodily reflected version of the real world. In this understanding, the duality of hologram can be described as the mirror universe [209].

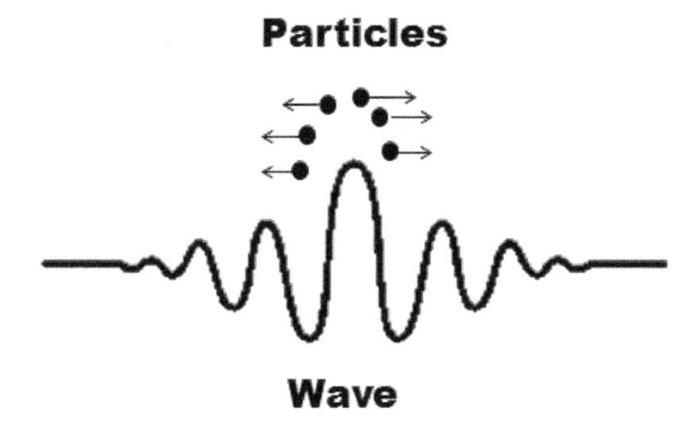

Figure 12.4 Wave-particle duality.

Figure 12.5 Simple concept of universe hologram.

12.3 How Holography Works?

12.3.1 Reflection and Transmission Holograms

Basically, there are two sorts of holograms: (i) reflection holograms; and (ii) transmission holograms. In this manner, reflection holograms sculpt images by reflecting beams of light off the surface of the hologram (Figure 12.6). This sort of hologram produces extraordinary features, nonetheless it is extremely costly to reconstruct. There are, nevertheless, countless concerns, for instance, 3-D color registration, equalization between different colors and attaining the precise color to achieve high quality 3-D hologram [211].

Taping reflection holograms can be accomplished as follows. The laser affords an exceedingly coherent resource of light. The light strikes the beam splitter, which is a semi-reflecting plate that breaks the beam into two beams: (i) an object beam and (ii) a reference beam. The object beam is extended by a beam transmitter (expanding lens) and the light is reflected off the object and is positioned on the photographic plate. Then, the reference beam is also amplified by a beam splitter and the light reflects off a mirror and shines on the photographic plate. Finally, the reference and object beams encounter at the photographic plate and generate the interference pattern. This accounts for the amplitude and phase of the resultant wave [210].

Consequently, reconstructing reflection holograms are used to reconstruct the object wavefront. The reconstruction beam is positioned at the same angle as the illuminating beam that was used during the recording phase. Then, the virtual image appears behind the hologram at the same position as the object (Figure 12.7).

Transmission holograms form images by transmitting a beam of light through the hologram. This type of hologram is more commonly seen since they can be inexpensively mass-produced. Embossed holograms, such as those found on credit cards, are transmission holograms with a mirrored backing (Figure 12.8).

Consistent with Caulfield (2012) [211], three steps are involved in recording transmission holograms. First, as with reflection holograms, a laser is used to provide a highly coherent source of light. A beam splitter and beam spreaders are also used in the recording of transmission holograms. After the object beam passes through the beam spreader, the light is reflected off a mirror and onto the object. The object beam is then reflected onto the photographic plate. Third, the reference beam is also reflected off a mirror and shines on the photographic plate [212]. Consequently, the incoming object and reference beams create a resultant wave. The amplitude and phase of the resultant wave are recorded onto the photographic plate as an interference pattern (Figure 12.9).

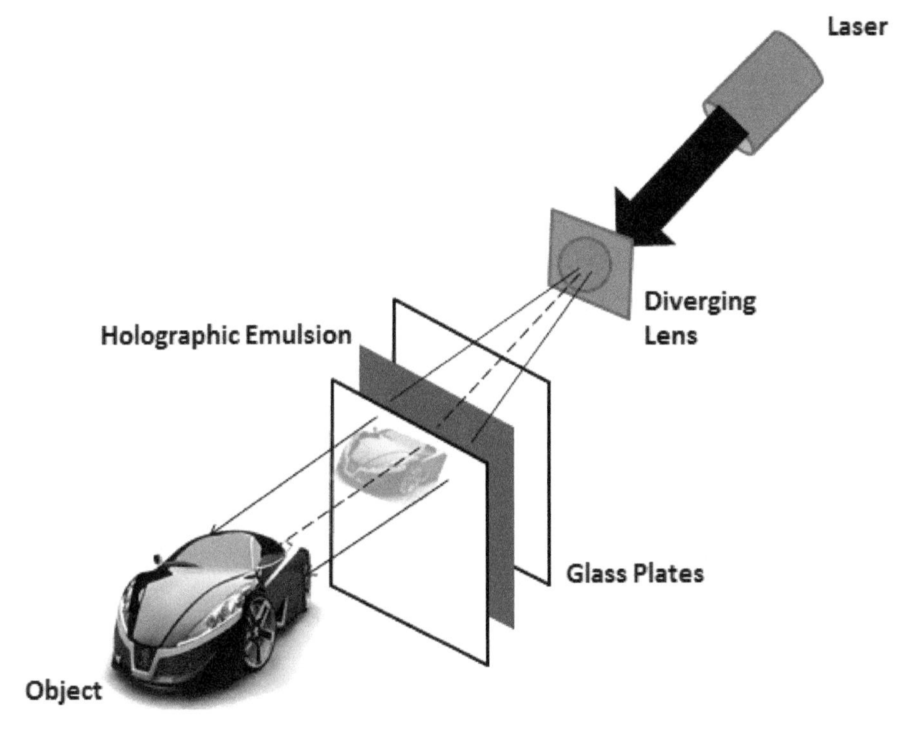

Figure 12.6 Reflection hologram.

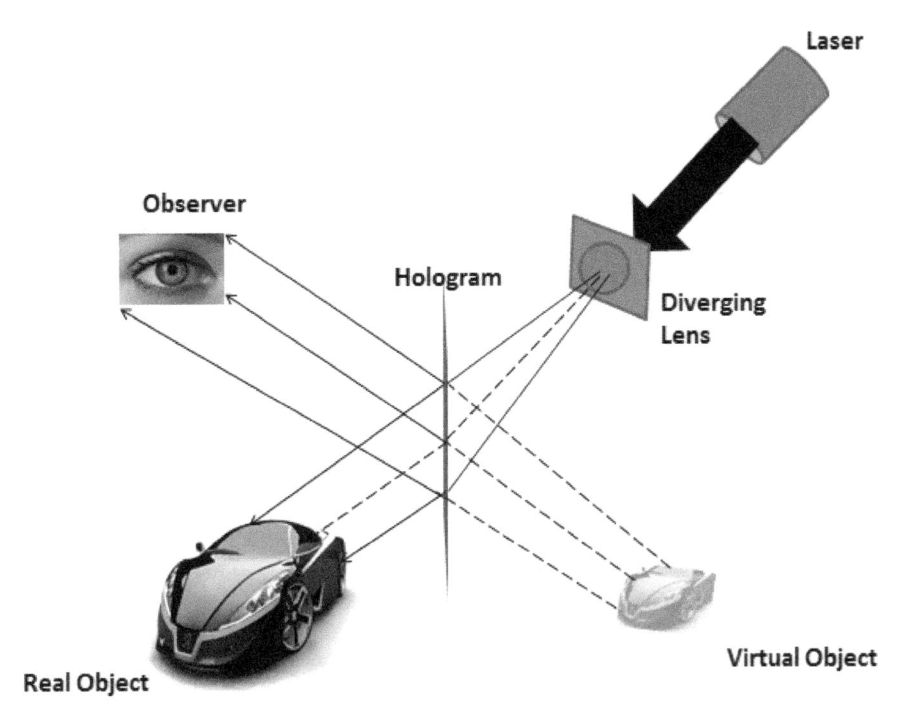

Figure 12.7 Reconstructing reflection light.

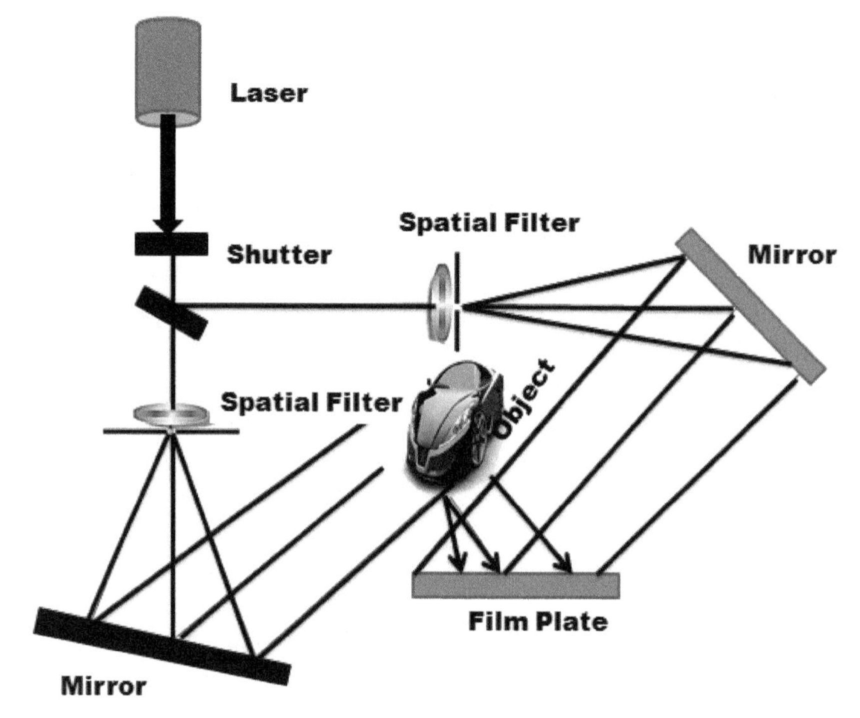

Figure 12.8 Transmission hologram.

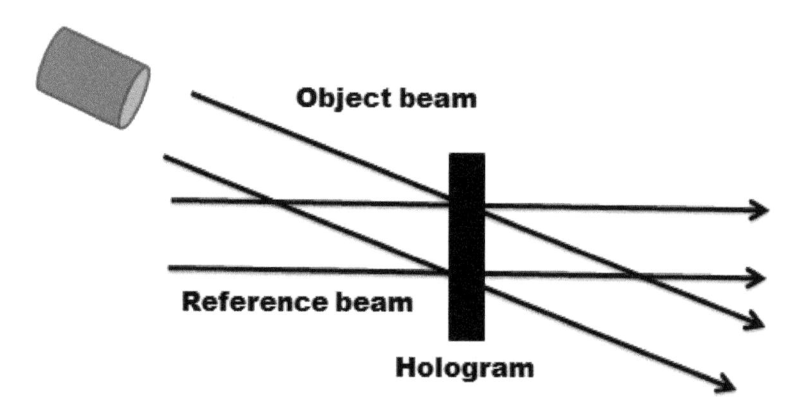

Figure 12.9 Transmission image recording.

Collier (2013) [212] stated that a reconstruction beam is used to illuminate the hologram and is positioned at the same angle as the reference beam that was used during the recording phase. When the reconstruction beam is placed at the right angle, three beams of light will pass through the hologram. Consequently, Hariharan (1996) [213] agreed that an un-diffracted beam (zeroth order) will pass directly through the hologram but will not produce an image. Thus a second beam forms the primary (virtual) image (first order) that is diffracted at the same angle as the incoming object beam that was used during recording. Finally, a third beam forms the secondary (real) image (first order) (Figure 12.10).

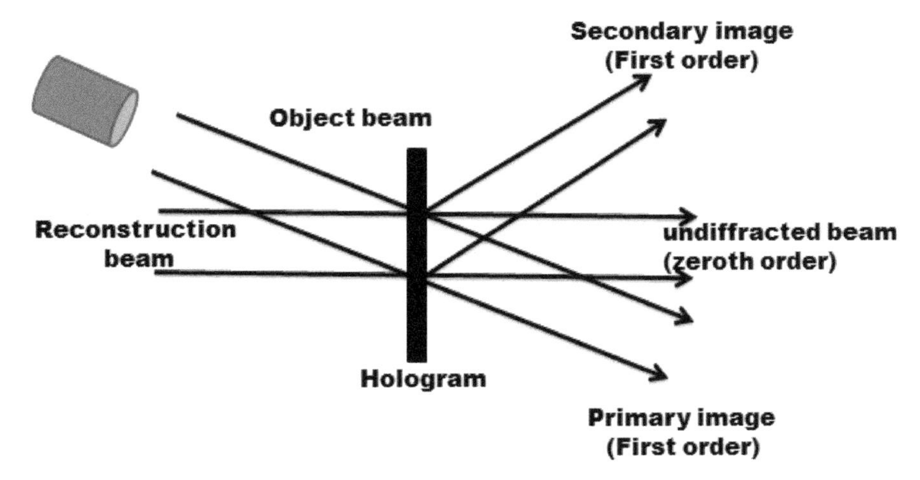

Figure 12.10 Transmission image reconstruction.

12.3.2 Capturing the Fringes

The hologram records the interference between the light waves in the reference and object beams. In this regard, holograms are created by the light-sensitive emulsion. In this view, if the two similar wave peaks interfere, they strengthen each other which is known as a constructive interference [215]. On the contrary, the destructive interference occurs if a peak alliances with a trough, and cancels one another out (Figure 12.11).

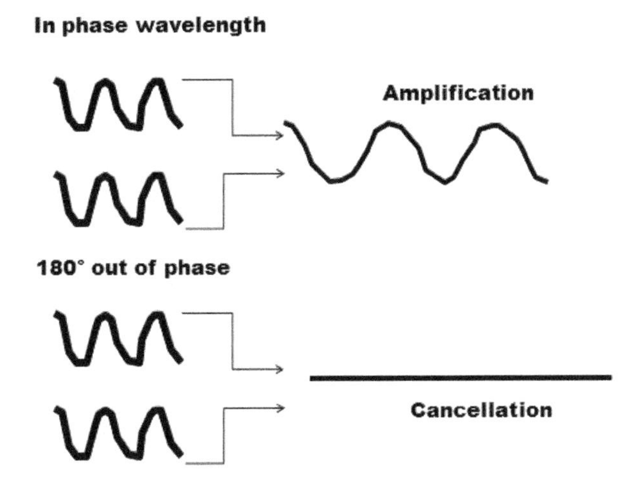

Figure 12.11 Constructive and destructive interferences.

12.3.3 Coherent and Incoherent Holography

Laser as coherent light is the basis of creating coherent holography. In fact, laser is a high energy spatially and temporally coherent source. In contrast, this restriction can be solved by incoherent holography. In this understanding, each scene point can split into two beams. In this view, the method successfully generates a reference for individual scene.

Hong and Kim (2013) [214], therefore, stated that holographic imaging is extremely recognized for recovering both the amplitude and phase figures of a real object. Rationally, nevertheless, the practice of coherent illumination engaged the comprehension of feasible holographic imaging. Besides, it also circumscribed the extensive practice of holographic imaging. The possibility of incoherent holography was implemented with the development of digital electronic devices and computer science over the past decade. Self-interference is the most important technique to achieve Incoherent holography. In this regard, it splits the light from the object into two pathways and permits those beams from the same object point to interfere with each other. Nevertheless, the three-dimensional incoherence of the light from the object washes out the fringe of the recorded intensity image. As a substitute, the complex hologram can be computationally retrieved from three or four phase-shifted images. The completely incoherent imaging process of these techniques has widened the application of holographic imaging [40].

12.3.4 Hologram Classifications

The hologram can be classified into volume and rainbow holograms. When the object is larger than the light wavelength, it is recorded as volume hologram. Under this circumstance, diffraction of light from the hologram is feasible solely as Bragg diffraction, i.e., the light needs to have the proper wavelength (color) and consequently, the wave should have the proper phase. Volume holograms are known as thick holograms or Bragg holograms [209].

Rainbow holograms are invented to be observed beneath white light-weight radiance, as opposed to the laser light-weight that was needed before this. The rainbow holography is a recording method which uses a horizontal slit to eliminate vertical parallax within the output image and greatly reduces the spectral blur whereas conserving 3-D for many observers. A viewer moving up or down before a rainbow holograph records dynamic spectral colors instead of absolutely different vertical perspectives [210].

Because of perspective, as effects are reproduced on one axis solely, the object can seem multifariously stretched or pressed if the holograph is not viewed at an optimum distance. In this view, this distortion might expire unobserved if there's not abundant depth. However, it is often severe once the space of the object from the plane of the holograph is extremely substantial. Under this circumstance, stereopsis and horizontal motion parallax, two relatively powerful cues to depth, are preserved [215].

In a holograph, the two intersecting light wavefronts develop a shape of 3-D hyperboloids that seem like hyperbolas revolving around one or more focal points. In this regard, an observer might be able to read a lot of concern rounded shapes in wolfram mathematics world.

12.4 Four-dimensional

In physics and mathematics, the dimension of a mathematical space is confidentially delineated, for instance, the minimal range of coordinates required to become aware of any point contained by means of it. The purpose of the classical mechanics is to describe how bodies change their position in space with time. In this regard, a four-dimensional space or 4D space is a mathematical extension of the concept of three-dimensional or 3D space. 4-D of space-time involves activities which are not sincerely termed, spatially and temporally. In this regard, any object can be power in the space of four and even higher dimensions. The key venture is to modify rationality tactics to renovate such high-dimensional stuffs.

The Fourth Dimension was dissected with the aid of scientists, psychologists, mathematicians and physicists, later in the 1800s. Indeed, scientists have utilized 4-D principles to explicate roughly on the universe. At an early stage, scientists created 4-D from 3-D with the aid of spinning 3-D about its image or itself. Scientists, therefore, have explained time as a dimension, besides the 3-D. Nevertheless, scientifically, this concept is no longer precise. The fourth Dimension axis goes by using the Z, Y and Z. The 4-D object has 4 quintessential units: width, length heights and 4-D which is W. A hypercube, for instance, has a length, width, top and a fourth dimension that is perpendicular to all three of the different units. Consequently, 4-D is exploring the internal objects of 3-D.

Hitherto developed remote sensing technology disables the implementation of n-dimensional and is simply strained to simulate three-dimensional of any object on the bottom. However, n-dimensional is a remarkable theme for mathematicians and physicians. String theory, M-theory, and Supergravity are the most accepted n-dimensional theories. String theory, consequently, planned that the universe is formed in multiple dimensions: (i) height, (ii) width, and (iii) length compose three-dimensional space; and (iv) time contributes to the completeness of 4 discernible dimensions. String theories, notwithstanding, continued the probability of ten dimensions—the remaining six that human capability cannot depict precisely.

Thus, the supergravity theory fuses the theories of supersymmetry and Einstein's theory of relativity. Further, Supergravity theory has contributed a vital part in substantiating the fundamentals of the eleventh dimension. To close, M-theory conjoins the five various string theories (along with a later discarded challenge to fuse Einstein's theory of relativity and quantum physics known as 11-D-Supergravity) into one theory [37, 51].

M-theory, hypothetically, can assume an assembly of developing a unified theory of all of the elementary potencies of nature. Challenges to contact M-theory to investigate usually converge on compactifying (i.e., theory with relevance to one of its space-time continuum dimensions) and its additional dimensions to construct candidate models of our multidimensional world (Figure 11.10). However, the geometry of four-dimensional (4-D) space is more convoluted than that of three-dimensional space owing to the extraordinary degree of freedom.

Consequently, 4-D comprises of 4-polytopes that are made of polyhedral (Figure 12.13). Additionally, 4-D conjointly contains half a dozen lenticular regular 4-polytopes that are the analogues of the platonic solids. Hence, the restful conditions for reliability spawn an additional 58 lenticular unvarying 4-polytopes, analogous to the thirteen semi-regular Archimedean solids in three dimensions [37, 51]. Consequently, as 3-D beings, it cannot transfer freely in time; however, in 4-D, it might be realizable. In this context, 4-D can distinguish three-dimensional and it cannot be sent to Euclidean space that suggests that fourth-dimension is an abstraction. Therefore, 4-D is often generated algebraically, by utilizing the rules of vectors and analytical geometry to a space in 4-D [38, 50, 51].

A vector as an example, with four components (a 4-tuple), may be investigated to perspective as an object in 4-D. Space may be a Euclidean space that incorporates a metric and norm. Further, all directions are handled because of the same: the intercalary dimension is incoherent from the opposite three. Consequently, the 4-D space has an additional axis, orthogonal to the opposite three, that is typically tagged as compared to 3-D. An edge on the W axis may be known as specified. In this regard, space, a tool referred to as dimensional analogy, is often used. The dimensional analogy is the study of how $(n - 1)$ dimensions correlate to n-dimensions, and then supposing how n dimensions would draw a parallel to $(n + 1)$ dimensions [38, 51].

Tesseract is an ideal example of the 4-D. To grasp a hypercube, a square and a cube are defined in terms of a lower dimension (Figure 12.14). A hypercube, consequently, by analogy, is an infinite variety of a cube retained in an exceeding direction perpendicular to the prevailing three dimensions. Because of the definite fact that we cannot delineate the fourth dimension, it is imaginary and does not exist easily in the 3D space. A "true" cube has no length within the time.

The W-axis is the 4th dimension which is added to the three-dimensional coordinates of X, Y, and Z. In other words, W is the fourth lines which are holding the two cubes located rigorously on the W-axis. These W-lines are perpendicular to all the three axes X, Y and Z. In this understanding, the three axes, X, Y, and Z cannot be represented as components of the fourth dimensional segments.

On the other hand, the n-dimensional establishment in the real world is still in a preliminary step. This is because of the main limitation of human eyes to visualize objects in 2-D which can convert in 3-D image by the assistance of computer graphic algorithms. Though there are great advances in remote sensing technology which can provide 3-D maps through LIDAR, TanDEM TerraSAR–X, etc., the human eyes are restricted to view 2-D or 3-D objects and cannot be viewed in n-dimensional. In free space universe, there are parallel coordinates which can view the universe in n-dimensional. One of the accurate remote sensing technique for 3-D visualization is interferometry synthetic aperture radar technique.

Yet, the performance of the interferometric phase estimation suffers seriously from poor image coregistration. Interferogram filtering algorithms such as adaptive contoured window, pivoting mean filtering, pivoting median filtering, and adaptive phase noise filtering are the main methods of the

Figure 12.12 Four-dimensional which is a subset of 3 sphere.

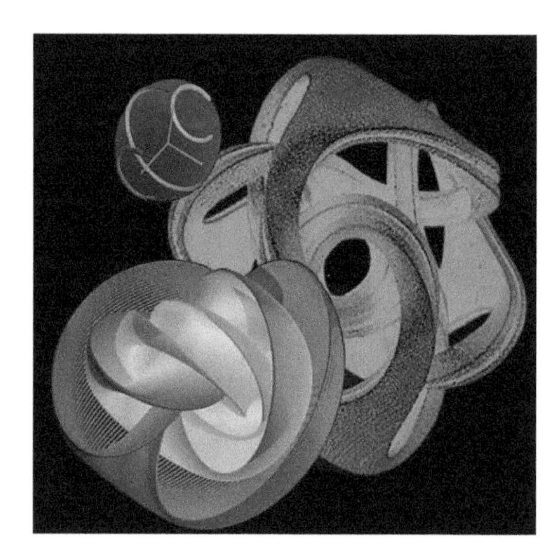

Figure 12.13 Four-dimensional of 4-polytopes.

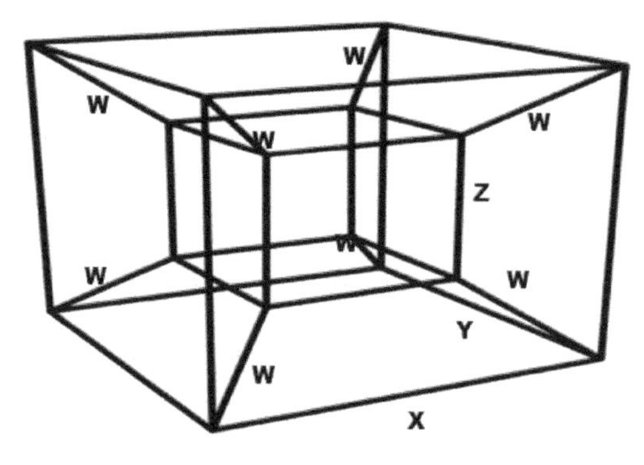

Figure 12.14 Tesseract parallel to four axes of space.

conventional InSAR interferometric phase estimation [52]. Recently, Marghany (2014) [48] implemented hologram interferometry for shoreline change. Marghany (2011) [49] introduced a new technique to reconstruct 4-D from hologram interferometry for optical remote sensing and ENVISAT ASAR data. However, these studies do not explain clearly the mathematical formula used to reconstruct 4-D. In fact, 4-D is required to formulate mathematical protocols to be implemented in several applications [51].

The main question is how to reconstruct 4-D from 3-D? This chapter postulates that 4-D can be implied from 3-D phase unwrapping of optical hologram interferometry. As a matter of fact, optical interferometry is a powerful approach to measure shifts of the stability of electromagnetic wavelength spectra. One significant limitation of common interferometric methods is that they require specular reflectors. This limitation can be removed by utilizing holography, allowing very small motions of arbitrary, diffusely reflecting, objects to be detected.

The main novelty of this chapter is to derive a new formula for 4-D hologram interferometry phase unwrapping Hybrid Genetic Algorithm (HGA). The main objective is to reconstruct fourth-dimension of tsunami impacts using high resolution QuickBird satellite data by optimization of 4-D hologram interferometry.

12.5 Mathematical Model for Retrieving 4-D using Hologram Interferometry

12.5.1 Hologram Interferometry to Reconstruct Fourth-dimensional of Tsunami Wave

The multiple viewpoint projection (MVP) holography is implemented to achieve incoherent hologram interferometry to acquire 4-D tsunami wave propagation in QuickBird data. Consistent with Shaked and Rosen (2008) [216], MVP holograms are created by first acquiring multiple projections of a 3D scene from various perspective viewpoints, and then digitally processing the acquired projections to yield the digital hologram of the scene. Under this circumstance, Fourier, Fresnel, image or other types of holograms can be generated. With this comprehension, the superposition of these waves at any point in space can be simulated to acquire the interference pattern which is involved in hologram generation.

Computer generated holograms do not compel definite targets to produce the hologram which provides information of the light scattered or diffracted off the object. In this regard, mathematical procedures can be involved to present these incoherent hologram procedures. Under these circumstances, the light transmission and reflection properties of the object are no longer a problem for the ideal object wave which can be computed mathematically. Consequently, a computer created holographic image can be simulated by being numerically based on the physical phenomena of light diffraction and interference.

Following Marghany (2015) [52], assume that I_1 and I_2 are the different acquisition times of two optical satellite data, for instance, *QuickBird*. Consequently, $I_1 \in E_1$ and $I_2 \in E_2$, where $E_1 \notin E_2$ or $E_1 \neq E_2$ as E is electromagnetic spectra which presents in two *QuickBird* satellite data (Figure 12.15).

Both electromagnetic spectra waves interfere at the surface of point in space or time. Hence, their amplitudes will add as a vector. If one of these is a plane wave pointing in the z direction and the other is a spherical wave, then they can be given by the following expressions:

$$\begin{aligned} I(x,y) &= \left| E(x,y)\, e^{(i\phi_O(x,y))} + r(x,y)\, e^{(i\phi_R(x,y))} \right| \\ &= (E_2(x,y)\, e^{(i\phi_R(x,y))})(r(x,y)\, e^{(i\phi_R(x,y))})^* + \\ &(E_1(x,y)\, e^{(i\phi_O(x,y))})\,(E(x,y)\, e^{(i\phi_O(x,y))})^* \\ &+ (E_1(x,y)\, e^{(i\phi_O(x,y))})(E_2(x,y)\, e^{(i\phi_R(x,y))})^* + \\ &(E_2(x,y)\, e^{(i\phi_R(x,y))})\,(E_1(x,y)\, e^{(i\phi_O(x,y))})^* \end{aligned} \tag{12.1}$$

where $\boldsymbol{E}$ is the complex amplitude of the object wave with real amplitude E_l and phase ϕO, $\mathbf{E}_2$ is the complex amplitude of the reference wave with real amplitude E_2 and phase ϕR and * denotes the conjugate complex. The phase ϕ is derived from the image intensity information by using phase demodulation procedures. In this step, let A be the amplitude image and I be the image intensity, then the low pass filter L for A and I can then be expressed as

$$L_f = A_0\,(i,j) + I_0\,(i,j) \tag{12.2}$$

Then determine component L_1 of low pass filter is given by:

$$L_1 = I_0\,(i,j) - L_f \tag{12.3}$$

$I_0\,(i, j)$ being obtained from L_f post the rejection of the residual, low-frequency term by using notch filter H. The notch filter is derived from the low pass filter and will thus be given by

$$H = 2L_f - 1 \tag{12.4}$$

Then L_1 and H are multiplied by the sine and cosine functions whose spatial frequencies equal f_0 and f_H. The spatial frequency f_s is calculated using:

$$f_s = f_0 + f_H \tag{12.5}$$

$$I_{fc}(i,j) = I_0\,(i,\, i)\cos(2\pi f_0 i) \tag{12.6}$$

$$I_{fs}(i,j) = I_0\,(i,\, i)\sin(2\pi f_0 i) \tag{12.7}$$

$$H_c(i,j) = H(i,\, i)\cos(2\pi f_H i) \tag{12.8}$$

$$H_s(i,j) = H(i,\, i)\sin(2\pi f_H i) \tag{12.9}$$

Then phase map can be determined mathematically from Equations 12.6 to 12.9 as follows:

$$\phi_I = \tan^{-1}\left[\frac{I_{fs}(i,j)}{I_{fc}(i,j)}\right] \tag{12.10}$$

$$\phi_H = \tan^{-1}\left[\frac{H_s(i,j)}{H_c(i,j)}\right] \tag{12.11}$$

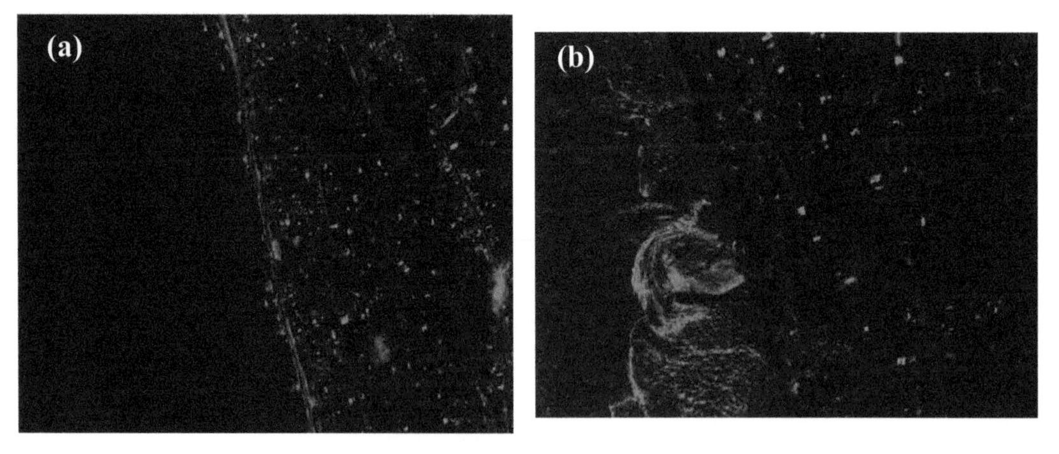

Figure 12.15 Quickbird satellite data, pre and post tsunami, along Kalutara (a) January 1, 2004, and (b) December 26, 2004.

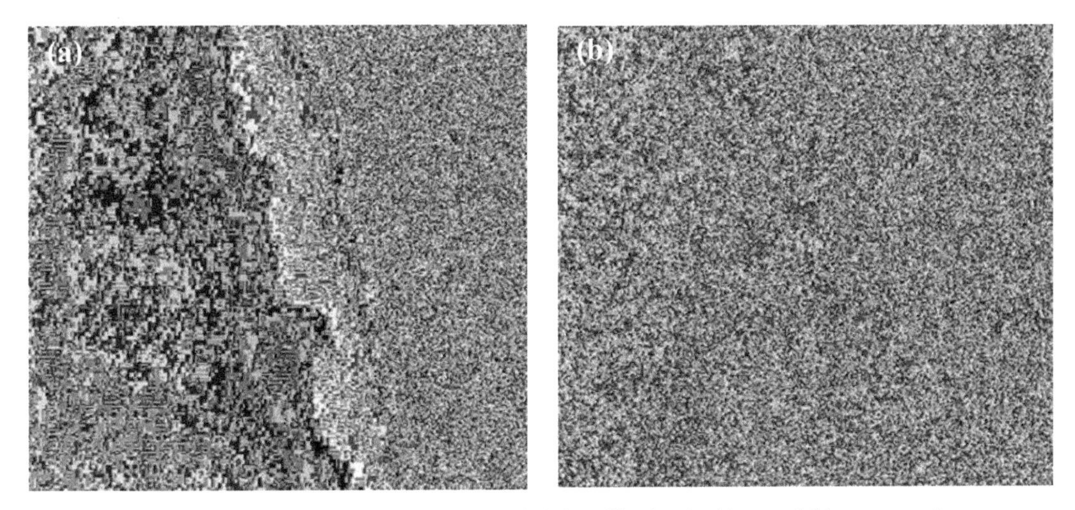

Figure 12.16 Phase image generated for QuickBird satellite data for (a) pre and (b) post tsunami.

In this regard, ϕ_I can be used to extend the non-ambiguity range of ϕ_H. Consequently, ϕ_H indicates higher sensitivity than ϕ_I against height (z) variations. The phase maps are estimated for both QuickBird satellite data, pre-tsunami and post-tsunami, in order to obtain phase changes between two periods (Figure 12.16). Then 2-D phase unwrapping $\Phi(i, j)$ can be obtained by

$$\Phi(i,j) = \phi_H(i,j) + 2\pi INT\left[\frac{N\phi_I(i,j) - \phi_H(i,j)}{2\pi}\right] \tag{12.12}$$

where *INT* denotes rounding to the nearest integer.

12.5.2 Fourier Computer Generated Hologram

The Fourier spectral distribution of two QuickBird satellite data, pre and post tsunami, is created by a Fourier transform hologram. The expression of master image and slave image are expressed by:

$$I_M(x, y) = \eta_M(x, y)e^{[j\phi(x, y)]} \tag{12.13}$$

$$I_s(x, y) = \eta_s(x, y)e^{[j2\pi\alpha x]} \tag{12.14}$$

where η_M and η_s are amplitude images of QuickBird data, pre and post tsunami. η_M is $\sin\alpha\lambda_M^{-1}$ and η_s is $\cos\alpha\lambda_s^{-1}$ where λ is selected wavelength band for both QuickBird data, α is zenith angle, ϕ is phase. In the Fourier domain, the complex amplitude spreading of the hologram can be conveyed as:

$$F(u, v) = FT\{I(x, y)\} \tag{12.15}$$

Then the fabricated computer generated hologram is achieved by using Burch's encoding algorithm ($B(x, y)$) which is given by

$$B(x, y) = 0.5\{1 + N(\eta)\cos[\phi_I(i, j) - \phi_H(i, j)\} \tag{12.16}$$

where $N(\eta)$ is normalized amplitude. Figure 12.17 shows the amplitude and phase which are generated by computer from QuickBird satellite data, pre and post tsunami. In this regard, computer generated holograms do not involve real targets to create the hologram on condition that the electromagnetic intensity of the images could be epitomized precisely in the spatial frequency domain signals by using 2-DFFT.

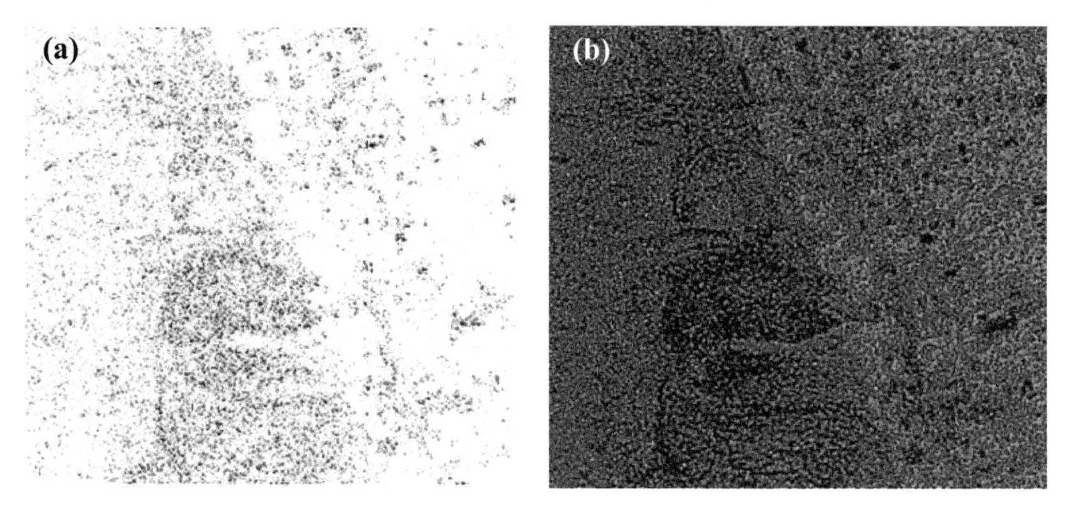

Figure 12.17 FFT computer generation (a) amplitude and (b) phase images of QuickBird satellite data.

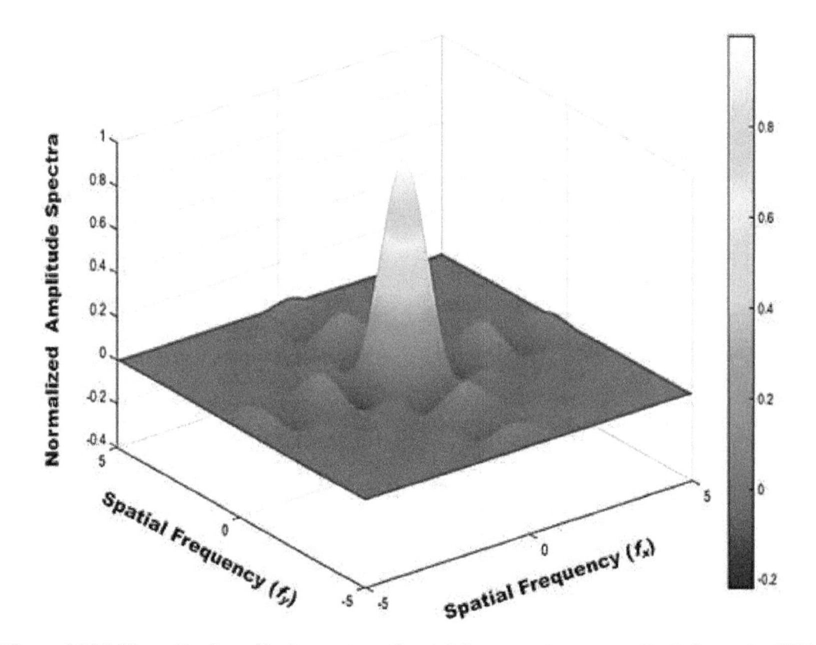

Figure 12.18 Normalized amplitude spectra estimated for generating an amplitude image by FFT.

The Fourier Transforms of both pre and post tsunami images have complex values and cannot be imaged. Consequently, the square of the absolute values of the amplitudes of both images is imaged. Fourier transform, when discretized with periodic sampling, is only the Fourier series representation of the 2D object (Figure 12.18). In this regard, the matrix representing both images is real. The matrix of both images becomes complex when *FFT* transformed [51].

12.5.3 2-D Hologram

Figure 12.19 shows the 2-D hologram derived by 2-DFFT. It is interesting to find that the computer generated hologram based on the 2-DFFT algorithm is able to track coastal water changes due to turbulent flow which

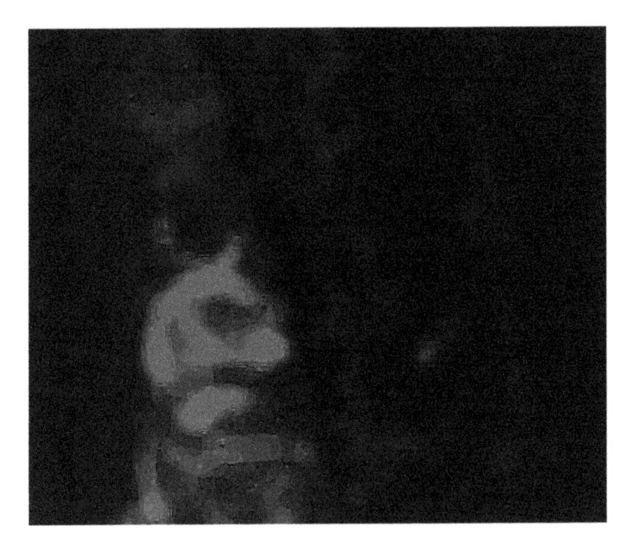

Figure 12.19 Computer generated hologram.

is generated by tsunami wave propagation. This is due to fact that QuickBird satellite has multispectral sensors with 0.61 m–0.72 m and 2.44 m–2.88 m resolution, respectively, and depend on the off-nadir viewing angle (0–25 degrees). The sensor, therefore, has coverage of 16.5 km in the across-track direction. In addition, the along-track and cross-track capabilities provide good stereo geometry and a high revisit frequency of 1–3.5 days. These characteristics of the images enable easy management and observation of the Earth, especially for disaster observation and mapping such as tsunami disaster. Nevertheless, this method is not able to reconstruct the rest of the information which existed in the image. In fact, amplitude information about recording wavefront is lost.

12.5.4 4-D Phase Unwrapping

The 4-D wrapped phase is generated by the following equation:

$$\Psi_w(x,y,z,t) = \mathrm{mod}\left[\eta \sin\left(\frac{(x+y+z+t)2\pi}{\lambda}\right) + G(\sigma)\right] \tag{12.17}$$

where x, y, z, t are the 4-D coordinate space of estimated wrapped phase Φ_w, mod the modulo operator wrapping the data within the range of $0 \leq \Psi_w \leq 2\pi$ and $G(\sigma)$ is an additive Gaussian noise function with standard deviation σ. Then the wrapped phase Ψ_w is related to correct phase by using the Laplacian-based algorithm. Under this circumstance, the Laplacian of the real phase ϕ is formulated in terms of the wrapped phase, which is given below:

$$\nabla^2\phi = \cos\Psi_w \nabla^2(\sin\Psi_w) - \sin\Psi_w \nabla^2(\cos\Psi_w) \tag{12.18}$$

where ∇^2 is the Laplace operator and by using an inverse of ∇^{-2}, the real phase ϕ' can be estimated in 4-D by compiling Fourier-domain forward and ∇^{-2} as follows [217],

$$\begin{aligned}\phi'(i,j,k,t) = {} & FCT^{-1}\left[\frac{FCT[\cos\Psi_w(FCT^{-1}[(P^2+Q^2+R^2+S^2)FCT(\sin\Psi_w)])}{P^2+Q^2+R^2+S^2}\right] \\ & -FCT^{-1}\left[\frac{FCT[\sin\Psi_w(FCT^{-1}[(P^2+Q^2+R^2+S^2)FCT(\cos\Psi_w)])}{P^2+Q^2+R^2+S^2}\right]\end{aligned} \tag{12.19}$$

where FCT^{-1} and FCT are inverse and forward cosine or sine transforms, respectively, in 4-D of time-domain coordinates (i, j, k, t) and the Fourier-domain coordinates $(P^2 + Q^2 + R^2 + S^2)$. Equation 11.19 involves three types of operations: forward and inverse cosine transforms, trigonometric operations, and the masking expression $(P^2 + Q^2 + R^2 + S^2)$.

Following Marghany (2014) [51], the relative optical phase difference can be associated to a physical displacement through the sensitivity vector found in the hologram interferometry in two satellite data which can be expressed in 4-D as:

$$\begin{pmatrix} \Delta\Phi_1 \\ \Delta\Phi_2 \\ \Delta\Phi_3 \\ \Delta\Phi_4 \end{pmatrix} = \frac{2\pi}{\lambda} \begin{pmatrix} \vec{d}_{1i} & \vec{d}_{1j} & \vec{d}_{1k} & \vec{d}_{1p} \\ \vec{d}_{2i} & \vec{d}_{2j} & \vec{d}_{2k} & \vec{d}_{2t} \\ \vec{d}_{3i} & \vec{d}_{3j} & \vec{d}_{3k} & \vec{d}_{3t} \\ \vec{d}_{4i} & \vec{d}_{4j} & \vec{d}_{4k} & \vec{d}_{4t} \end{pmatrix} \begin{pmatrix} P \\ Q \\ R \\ S \end{pmatrix} \tag{12.20}$$

where d is the displacement along orthogonal components of P, Q, R, and S, in i, j, k, and t, respectively. Consistent with Marghany [51], the phase unwrapping can mathematically be extended into fourth-dimensional as:

$$\begin{aligned} &\sum_{i,j,k,t} W^x_{i,j,k,t} \left|\Delta\phi^x_{i,j,k,t} - \Delta\psi^x_{i,j,k,t}\right|^L + \sum_{i,j,k,t} W^y_{i,j,k,t} \left|\Delta\phi^y_{i,j,k,t} - \Delta\psi^y_{i,j,k,t}\right|^L \\ &+ \sum_{i,j,k,t} W^z_{i,j,k,t} \left|\Delta\phi^z_{i,j,k,t} - \Delta\psi^z_{i,j,k,t}\right|^L + \sum_{i,j,k,p} W^z_{i,j,k,p} \left|\Delta\phi^t_{i,j,k,t} - \Delta\psi^t_{i,j,k,t}\right|^L \end{aligned} \tag{12.21}$$

where $\Delta\phi$ and $\Delta\psi$ are the unwrapped and wrapped phase differences in x, y, z, t respectively, and W represents the user-defined weights. The summations are carried out in both x, y, z, and t directions over all i, j, k, and t respectively [51]. The phase unwrapping problem can be solved by L^2 norm second differences reliability criterion. Consequently, L^2 can be extended into 4-D as:

$$\text{Re}(O) = \sum_{n=1}^{40} \sqrt{(O - n_+)^2 + (O - n_-)^2} \tag{12.22}$$

where $\text{Re}(O)$ is the reliability of the pixel O, n_+ a neighbor on the 4-D hypercube of neighbors to O, and n_- the opposite neighbor. In four dimensions, each voxel has 80 neighbors, resulting in a reliable criterion that is the sum of 40 measurements.

12.5.5 Hybrid Genetic Algorithm (HGA)

The HGA algorithm relies on estimating the parameters of the nth order-polynomial to approximate the unwrapped surface solution from the wrapped phase data (Figure 12.20) [44, 51]. Any optimisation problem using a genetic algorithm (GA) requires the problem to be coded into the GA syntax form, which is the chromosome form. In this problem, the chromosome consists of a number of genes where every gene corresponds to a coefficient in the nth-order surface fitting polynomial. This can be extended into 4-D as follows [51],

$$f = n \rightarrow \sum_{t=0}^{n} \sum_{k=0}^{n} \sum_{j=0}^{n} \sum_{i=0}^{n} a_{i,j,k,t} \Delta\phi^x_{i,j,k,t} \Delta\phi^y{}_{i,j,k,t} \, t\Delta\phi^z_{i,j,k,t} \Delta\phi^t_{i,j,k,t} \tag{12.23}$$

where $a[0....n]$ are the parameter coefficients which are retrieved by the genetic algorithm to an approximated unwrapped phase. Further, i, j, k and t are indices of the pixel location in the unwrapped phase in 4-D, respectively, and n is the number of coefficients.

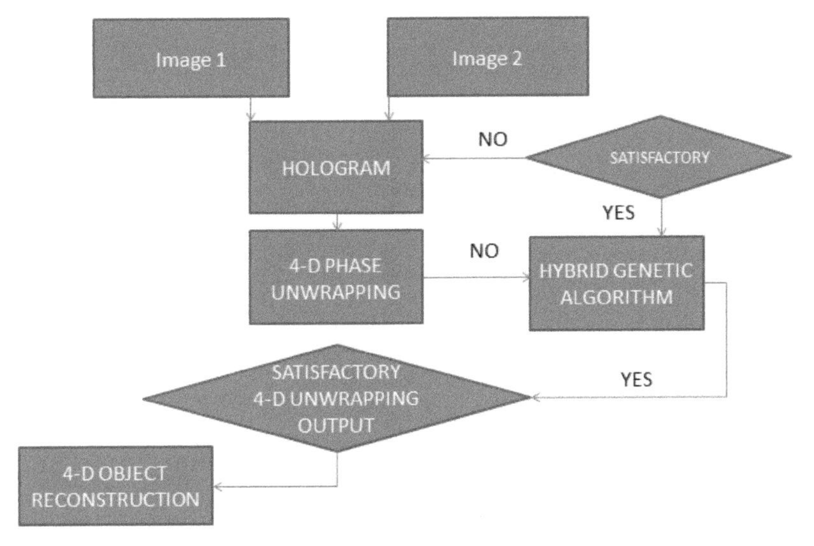

Figure 12.20 4-D Portocalusing holographic interferometry.

12.5.5.1 *Initial Solution and Population*

Following Karout (2007) [44], the initial solution is approximated using a 'polynomial Surface-fitting weighted least-square multiple regression' method. The initial population is then generated based on the initial solution. In doing so, every a_g in each chromosome in the population, a small number relying on the accuracy of the gene that is added or subtracted to the value of the gene is given by [50, 51],

$$a_g = a_g + (\pm 1)\{10^{[\log(a_g)+\Re]}\} \tag{12.24}$$

where a_g is the coefficient parameter stored in gene g, and $\Re$ is a random number generated between the values in 4-D.

12.5.5.2 *Record Pareto Optimal Solutions*

Calculate the objective values of chromosomes in the population and record the Pareto optimal solutions.

Definition: Pareto Optimal Solutions

Let ϕ_i, $\phi_{i,j}$, $\phi_{i,j,k}$, $\phi_{i,j,k,t} \in F$, and F is a feasible region in 4-D coordinate. ϕ_0 is called the Pareto optimal solution in the minimization problem of 4-D phase unwrapping, if the following conditions are satisfied.

1. If $f(\phi_1)$ is said to be partially greater than $f(\phi_2)$, i.e., $f_i(\phi_1) \geq f_i(\phi_2)$, $\forall N = 1,2,\ldots,n$ and $f_i(\phi_1) > f_i(\phi_2)$, $\exists N = 1,2,\ldots,n$, then ϕ_1 is said to be dominated by ϕ_2.
2. If there is no $\phi \in F$ s.t. ϕ dominates, ϕ_0, then ϕ_0 is the Pareto optimal solution.

12.5.5.3 *Fitness Evaluation*

In this step, the quality of the solution is evaluated at every generation to determine the global optimum solution to the parameter estimation phase unwrapping problem [53, 55]. Therefore, the genes on the chosen chromosomes are used to evaluate the approximated phase unwrapping value at four-dimension coordinates of *i*, *j*, *k*, and *t*. Then, the obtained phase is subtracted from the contiguous pixel approximated phase value to retrieve the approximated unwrapped phase solution gradient. It is then subtracted from the gradient of the wrapped phase in the *i*, *j*, *k* and *t* direction [51, 54].

12.5.5.4 Crossover and Mutation

Following Haupt and Haupt (2004) [43], the two point greedy continuous crossovers are implemented in crossover operator. Therefore, crossover is less a problem than the mutation operator [53]. Thus, Mutation operator concerns deliberate changes to a gene at random, to keep variation in genes and to increase the probability of not falling into a local minimum solution. It involves exploring the search space for new better solutions [43]. This proposed operator uses a greedy technique which ensures only the best fit chromosome is allowed to propagate to the next generation [42, 44, 51].

12.5.5.5 Phase Matching

The accurate 4-D phase unwrapping can be obtained by phase matching algorithm which is suggested by Schwarz (2004) [57]. Consistent with Schwarz [57], the phase matching algorithm is matched to the phase of wrapped phase with unwrapped phase by the given equation:

$$\psi_{i,j,k,t} = \Delta\phi_{i,j,k,t} + 2\pi\rho\left[\frac{1}{2\pi}\left(\Delta\hat{\phi}_{i,j,k,t} - \Delta\phi_{i,j,k,t}\right)\right] \tag{12.25}$$

where $\psi_{i,j,k,p}$ is the phase matched unwrapped phase, *i*, *j*, and *k* are the pixel positions in the quality phase map, $\Delta\phi_{i,j,k,t}$ is the given wrapped phase, $\Delta\hat{\phi}_{i,j,k,t}$ is the approximated unwrapped phase, $p[.]$ is a rounding function which is defined by $\rho[t]=\lfloor t+\frac{1}{2}\rfloor$ for $t \geq 0$ and $\rho[t]=\lfloor t-\frac{1}{2}\rfloor$ for $t < 0$ and are *i*, *j*, *k* and *p* the pixel positions in *x* and *y*, *z*, *w* directions, respectively [11, 51, 53, 57].

12.6 4-D Hologram Visualization of QuickBird

The comparison between 2-D phase unwrapping, 3-D phase unwrapping, and 4-D phase unwrapping are shown in Figure 12.21. The hologram interferometry fringe patterns are more vibrant by using 4-D phase unwrapping as compared to other phase unwrapping dimensions, i.e., 2-D and 3-D. Particularly, the complete cycle of hologram interferometry fringe patterns is certain with the 4-D phase unwrapping algorithm. In this regard, 4-D holographic interferometry fringes are produced by using Hybrid Genetic Algorithm based on Pareto Optimal Solutions. It is interesting to find that the proposed algorithm has produced clear feature detection of infrastructures. In fact, the proposed algorithm has minimized the error in interferogram cycle due to the low coherence in water and vegetation zones and along the coastline due to tsunami impact. This could be improvement of such previous works of Hussein et al. (2005) [42], Karout (2007) [44] and Marghany (2012, 2013, 2014, 2011, 2003) [45–50].

Figure 12.21d shows 4-D hologram interferometry, which is generated by HGA. It is interesting to find the complete cycle of the fringe pattern. This leads to coastal deformation and run-up tsunami wave height. The maximum run-up wave height is 7 m which causes coastal erosion of the –60 m/month and coastal deposition of 10 m/month. In fact, the earthquake and the tsunami, which struck the coastline of the Indian Ocean and Sumatra, Indonesia on 26 December 2004, had caused massive changes in the shorelines of many countries especially in Banda Aceh, Indonesia and Kalatura, Sri Lanka. Scientists investigating the damage in Aceh found evidence that the wave reached a height of 80 feet (24 m) when coming ashore along large stretches of the coastline, rising to 100 feet (30 m) in some areas when travelling inland. This deadliest disaster had caused massive destruction to the environment and the world ecosystems, apart from killing million of people all over the world.

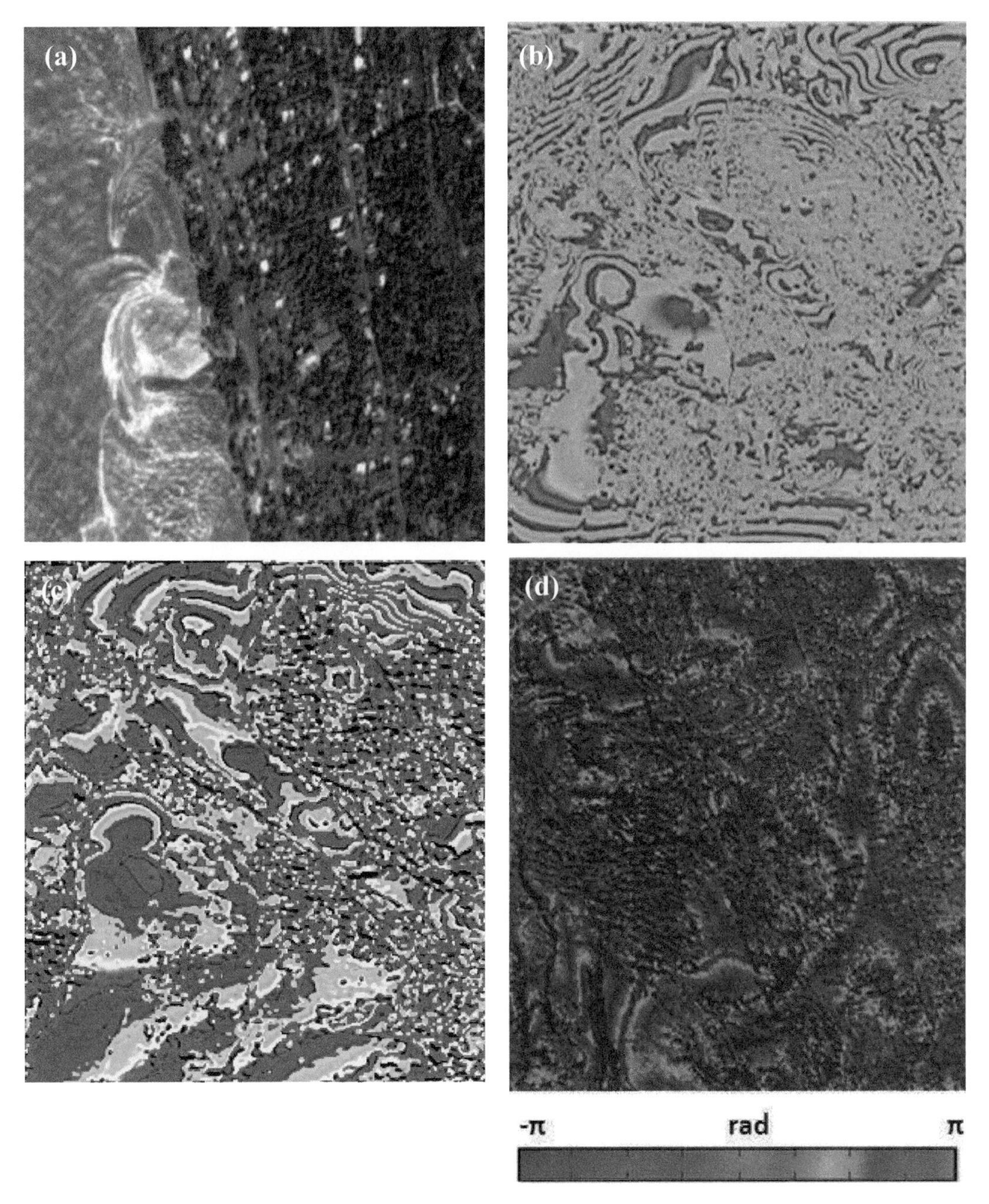

Figure 12.21 Phase unwrapping (a) original data, (b) 2-D, (c) 3-D and (d) 4-D.

Figure 12.22b shows the 4-D visualization derived from hologram interferometry. Obviously, the 4-D visualization distinguishes between infrastructures and buildings. However, Figure 12.22a illustrates unclear 3-D features as compared to 4-D. In this regard, coding 3-D of QuickBird data into 4-D can visualize many information about urban features, i.e., building, infrastructures, and roads in spite of the great damages caused by tsunami. The fourth-dimension represents the tsunami velocity along the coastal zone. The maximum run-up height of 6 m corresponds to the maximum wave propagation of 8 m/s. The maximum wave speed of 8 m/s reduces inland to 3 m/s due to the impact of land covers.

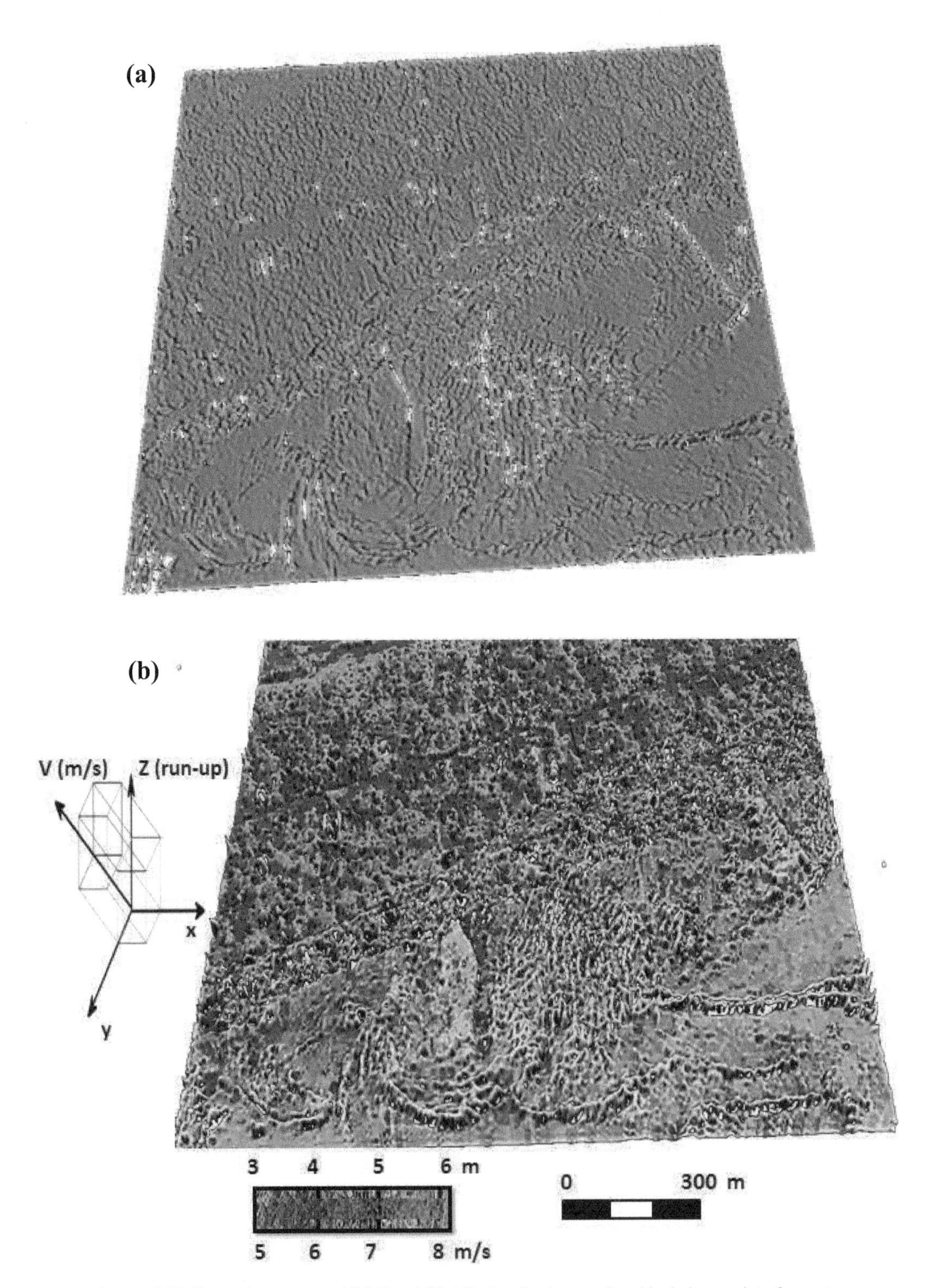

Figure 12.22 Comparison between (a) 3-D and (b) 4-D visualization, produced by hologram interferometry.

The hologram interferometry is considered as a deterministic algorithm which is described here to optimize a triangulation locally between two different points. It could precisely be expressed as a completely three dimensional image verified on a very high-resolution. The hologram archives every slight detail around the object's position in space besides the texture and surface features of the object. All the optical characteristics of the target are preserved, and to an extent, enhanced. This corresponds to the feature of deterministic strategies of finding only sub-optimal solutions usually. This confirms that optical interferometry is a robust means for evaluating displacements of the order of a wavelength of electromagnetic spectra. Therefore, involving a hologram can assist in recovering the matter of specular reflection, besides permitting tiny changes in random diffusely replication, object modifications can be identified accurately [59].

The visualization of the infrastructures is sharp by 4-D phase unwrapping based Hybrid Genetic Algorithm. These results deliver an excellent geometry for 4-D reconstruction using 4-D phase unwrapping of hologram interferometry. This study improves the work done by Marghany (2003) [50]. Furthermore, adding fourth coordinate p in a mathematical formula of 3-D HI shaped an excellent 4-D object visualization.

HGA assists to deliver 4-D image and not only discriminates the individual relaxed deformation but also sorts out impeding objects in 4-D. Generally, HGA corresponds to the phase of the wrapped phase with guessed unwrapped phase to verify the best representation of the unwrapped phase. Consistent with Karout (2007) [44], and Saravana et al. (2003) [55], a genetic algorithm is used to approximate the coefficient of an nth-order polynomial which is considered as the best estimate for the unwrapped phase map. In this regard, it minimizes the variance between the unwrapped phase gradient and the wrapped phase gradient. The genetic algorithm uses an initial solution to speed convergence. The initial solution is achieved by unwrapping using a simple unwrapping algorithm and estimating the parameters of the polynomial using weighted least squares multiple regression. Furthermore, 4-D results obtained with the hybrid genetic algorithm exceed the performance of the 4-D phase unwrapping of hologram interferometry. In fact, hybrid genetic algorithm reduces the number of miss-unwrapped cells by creating the best population members which are equivalent quantities of misalliance unwrapped cells.

The 4-D phase unwrapping methods outlined are applicable to reconstruct 4-D visualization of tsunami effects using QuickBird data. The involvement of Hybrid genetic algorithm to optimize modification formula of hologram interferometry can permit the QuickBird data to be coded into 4-D. The results show that the varieties of 2-D objects in QuickBird data can visualize in 4-D. The fine structures of roads, urban buildings and infrastructures are well visualized in 4-D. In conclusion, the modification and optimization of hologram interferometry formula hold excellent promise for 4-D object visualization of such optical satellite data of QuickBird.

The 4-D hologram interferometry delivers excellent explanations about the generation of solitary wave by the tsunami. When a tsunami travels in shallow water, it can, perhaps frequently, be guessed by a solitary wave. Further, its amplitude turns gradually into higher and sharper peaks and the trough grows into longer and flatter peaks. A solitary wave, consequently, propagates in shoaling water, ultimately becomes unsteady and then breaks. It was also distinguished that as the slope enlarged, the breaking point stirred closer to the shoreline.

Finally, hologram achieved a real 4-D imagery floating in space (Figure 12.22b). Further, the hologram feature is presented in Figure 12.22b which contains bright pixels, bold and pure saturated colors to create a natural look. This is clear with tsunami wave turbulence propagation. 4-D hologram algorithm using the HGA smooth appearance of objects rather than a grainy, fuzzy image is difficult to see, as in Figure 12.22a.

12.7 4-D and Relativity

In Einsteinian theories of relativity, space-time is 4-D since a fourth dimension that coincides with time is attached to the 3-D space. In this view, the space-time is 4-D, not the space alone. The concept of space-time is developed to be 4-D as a function of the Lorentz Transformations. Under this circumstance, the physical phenomena are transferred from an inertial reference frame (x,y,z) for instance, to another (x',y',z',t).

This is clearly remarked in Figure 12.22 where the 3-D tsunami propagation space frame is transferred to tsunami space-velocity frame. In the event of inertial reference frames, the space-time element 4-D is non-linear and it can be termed by a tensor with 16 coefficients that become 10 for reasons of symmetry.

What nevertheless can be declared with certainty is that a tsunami space 3-D exists and that we observed and measured it as a run-up through three coordinates expressed in metres. Likewise, we can articulate that a velocity 1-D exists and we measure it in meter per seconds. However, the space-time unit does not exist and it is only described by tensor and not necessarily be described as a physical reality. In this understanding, the space-time dimensions are presented by the tensor, but it does not prove a precise reconstruction of 4-D.

It can claim that the tsunamis are not 3-D, moving through time; they are themselves 4-D and extended through time. The time, therefore, determines the growth of a tsunami. In this context, tsunami development is a function of space and time. Thus, if the Lorentz transformation is implemented, it will deal with the space and time concepts of identical equilibrium up to the constant rate.

Figure 12.23 reveals that with different viewing angles, the different features can visualize in 4-D. The 4-D of solitary wave propagation is clearly noticed at 0° viewing angles with scattered short waves. At 90° and 360° viewing angles, the breaking of the solitary waves is a most clear feature with coastal water and inland too. However, the turbulent movements along coastal water and inland with extremely sharp crests are clearly distinguished at 180° viewing angles.

In point view of hybercube, the solitary and edge waves are clearly distinguished. This is well recognized in Figure 12.24. This is agreed with different viewing angles as every side of hypercube shows different features of solitary and edge wave propagation. More specifically, as the solitary and edge waves are extended up and down, right and left, fourth dimension, which corresponds to tesseracts, extends. Lastly, as a result, they develop a cubic mass of identical trivial tesseracts, and when the tesseract is prearranged in space world, it performs on the surfaces enclosing the upward and the right and left dimensions. From the point view of duality, the local solitary variables turn into extend on one side of the hypercube and

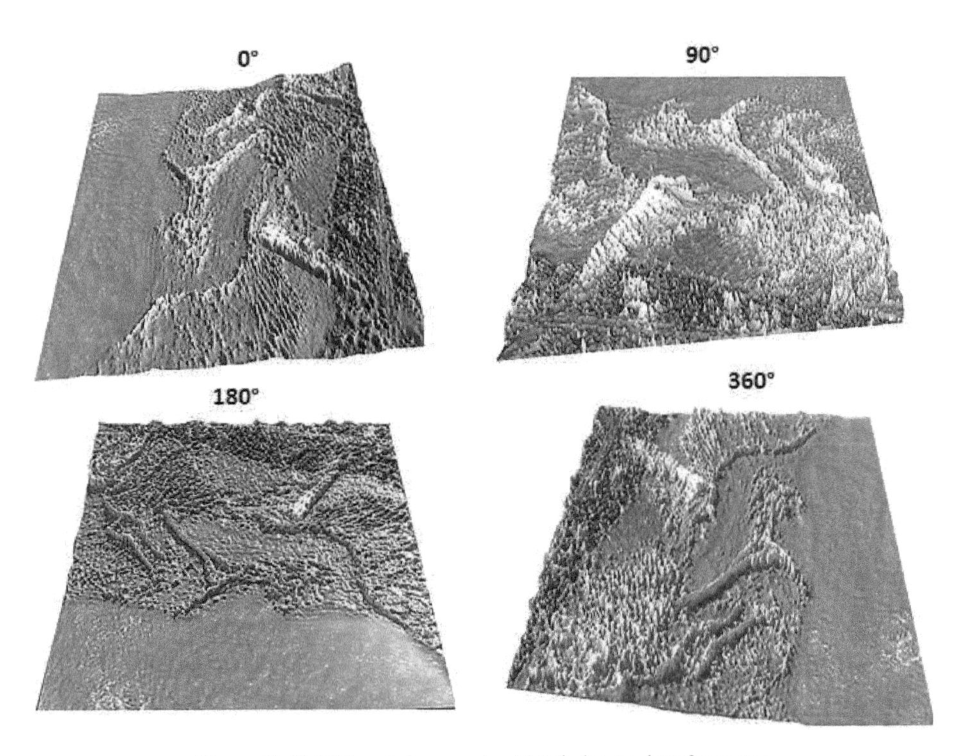

Figure 12.23 Different view angels of 4-D hologram interferometry.

Figure 12.24 Hypercube of solitary wave propagation.

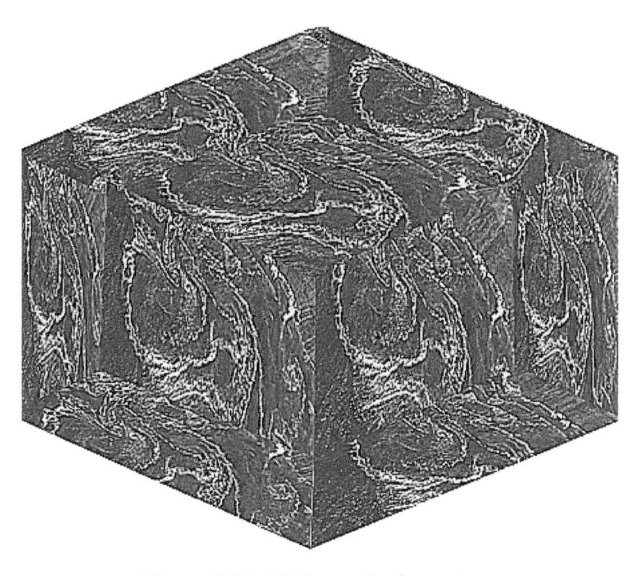

Figure 12.25 4-D hypercube for vortices.

topological objects on the other side. In this view, the solitary features are considered as the unchain of dual connection (Figure 12.24).

In 4-D, the vortices convert extended features, for instance, lines or surfaces as they are remarked in Figure 12.25. With the understanding that vortices, solitary wave and edge waves are vague from each other, it indicates a strong coupling constrain of the vortices. In other words, the duality generates excellent dual gauge fields arbitrary exchanges between individual vortex, solitary, and edge wave sources. In this view, dual gauge shown in Figures 12.21 to 12.25 is because of the hologram modelling. In this respect, holographic duality is known as gauge. Lastly, the boundary system is referred to as a "hologram" of the bulk system.

The different viewing angles spin a 3-D image, length into width and its width into depth. By an artless rotation, any of the three spatial dimensions can be swapped. At present, if velocity or time is the fourth dimension, at that moment it is conceivable to create "spins" which renovate the space into time

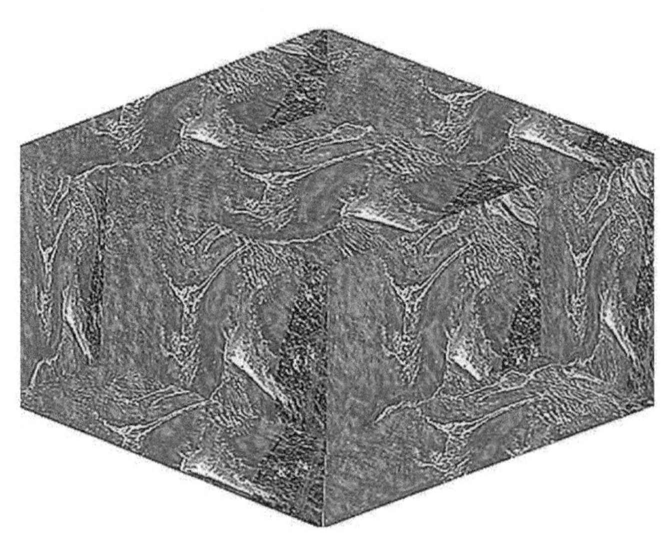

Figure 12.26 4-D hypercube of the solitary wave.

or velocity and vice versa. These four-dimensional spins are precisely the alterations of space and time or velocity, which are claimed by special relativity. In this understanding, space and time have assorted in a fundamental rule, and ruled by relativity. The meaning of time or velocity as being the fourth dimension is that time and space can spin into each other in specific mathematic rules.

Consistent with the special relativity, solitary waves can shatter at changed velocity rates, depending on how fast one is moving as a function of water depth. Velocity being the fourth dimension means that time is basically connected with a transfer in the x,y,z "space" of the coastal water. Though, if time is the fourth dimension, then space and time can rotate into each other and solitary waves have different velocity rates of coastal water and inland depending on how fast they move. More importantly, matter and energy are not connected in a 3-D world. In contrast, 4-D have a duality which connects one to each, for instance, space-time.

Another indication of the nature of four-dimensional space can be extended by deliberating the problem of the array of duality. In this respect, the 4-D generates the sharpest features, for instance, the sharp edge of solitary waves (Figure 12.26). It is clearly understood that the nature of 4-D space is sort of space which represents four qualities. On the contrary, 3-D presents the positions of three qualities. The duality, in this manner, is also established between a hologram and hypercube.

Thus, we may indicate the nature of four-dimensional space by saying that it is a kind of space which would give positions representative of four qualities, as three dimensional space gives positions representative of three qualities.

Chapter 13

Principles of Synthetic Aperture Radar

Previous chapters delivered comprehensive understanding of the utilization of the optical remote sensing for monitoring the tsunami wave impacts on the coastal waters. This chapter critically evaluates the potential of synthetic aperture radar's (SAR) existing theories for observing the effects of tsunami waves on the coastal waters. Unquestionably, SAR techniques are intensively used for monitoring and modelling the sea surface physical properties, for instance, wave spectra, current movements, and bathymetry mapping. The main question which can of course be raised is how SAR can monitor the boxing day effects, although there is no any availability of SAR data during 26th December 2004?

13.1 Principles

The microwave is part of the electromagnetic spectrum, which ranges from the wavelength λ of 1 mm to 1 m, corresponding to signal frequencies f equals 300 GHz and 300 MHz ($\lambda f = c$, with the speed of light c), respectively. Unlike the visible spectrum, the wavelength is extremely larger and intermingles with reasonably dissimilar targets contrasted to passive spectra. In this view, the microwave signal has extremely lower energy than passive spectra to induce molecular resonance. In contrast, passive spectra of visible and near infrared, for instance, exposes their radiation patterns into the atmosphere and soil chemical structures due to their high energy. The microwave domain, conversely, still have adequate high energy to model resonant spin of definite dipole molecules consistent with the frequency which is a function of the changing of signal electric field.

In this sense, the object appears dark because it is smaller than the radar wavelength which does not imitate abundant energy. On the contrary, short-wavelength radar can discriminate minor variants of irregularity than long-wavelength radar. In other words, object roughness fluctuates with wavelength. Therefore, because of the diffuse reflection, the incidence angle does not have a significant role in rough surface.

SAR sensors are reasonably delicate to object's physical properties, for instance, permittivity ε, surface roughness, morphology, and geometry. Active microwave wavelengths tend to be divided into typical wavelength regions or bands (Table 13.1).

Table 13.1 Microwave bands and their physical characteristics.

Bands	Wavelength (cm)	Frequency (GHz)
P-band	30–100	1.0–0.3
L-band	15–30	2.0–1.0
S-band	7.50–15	4.0–2.0
C-band	3.9–7.50	8.0–4.0
X-band	2.40–3.75	12.5–8.0
K-band	0.75–2.40	40–12.5 (rarely employed)

In microwave sensors, therefore, frequency (in Hertz) can also be used to describe a band range. It is worth mentioning that satellite active microwave sensors do not acquire multispectral microwave images. Conversely, they attain data for only a single band of wavelength/frequency. Nevertheless, recent airborne sensors can acquire multi-frequency bands, for instance, NASA's Jet Propulsion Laboratory AIRSAR system; C, L and P-band.

13.2 Radio Detecting and Ranging

With intending to understand the mechanics of synthetic aperture radar (SAR) data image of the tsunami impacts, the principal characteristics of SAR must be considered. The active microwave Earth observation is also known as RADAR (RAdio Detection And Ranging). The term radio used because the first radar used long wavelengths of radiation (1 to 10 m) that fell on the radio band of the electromagnetic spectrum. The term radar is commonly used for all active microwave systems [234].

The radar configuration consists of antenna, transmitter and receiver. The transmitter generates the energy to provide a radar beam and transmits it to the antenna and therefore to the target. The transmitted signal is a short burst rapidly repeated to give the pulsed signal. These pulsed signals travelling at 300000 km/s strike the targets in view where some of the energy is absorbed, some is reflected away and some refracted or backscattered to receiver [235]. The receiver is usually integrated with the transmitter. The receiver receives the pulsed returned signal. It determines the signal strength, and relates signal to the transmission. To calculate target distance, these data are processed into a form suitable for recording. The return signal, consequently, is much lower than the transmitted signal since not all the signal is returned. This signal will provide the grey scale tone of the image [236].

This type of microwave imaging is termed active because they transmit their own microwave energy (pulses) at a particular wavelength (single frequency) for a particular duration of time, known as pulsed coherent radar. Figure 13.1 shows the general concept of radar system. In this context, the following sections are concerned with the fundamental concepts of SAR image data [237].

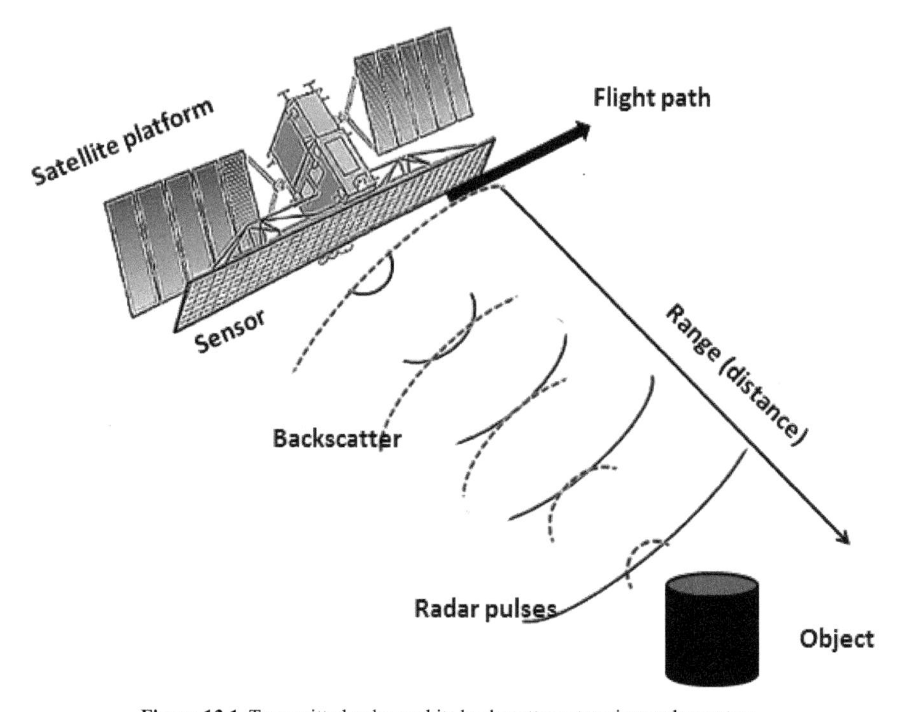

Figure 13.1 Transmitted pulse and its backscatter return in a radar system.

13.3 Synthetic Aperture Radar and Radar Resolutions

The term aperture represents the diameter of the lens opening. The aperture of the camera then determines the area through which light is collected [237]. Aperture, therefore, is restrained in focal length (*f*) stops. In this understanding, the smaller aperture receives less reflected light (Figure 13.2). In other words, unclear picture results in a wide open aperture (Figure 13.2). On the contrary, the sharp dark picture results on smaller aperture because of a great depth of field.

Similarly, the antenna length of radar partly specifies the area through which it collects radar signals. The length of the antenna is also called the aperture. In general, the larger the antenna, the unique information you can obtain about a particular viewed object. With more information, it can create a better image of that object (the improved resolution). It is ridiculously expensive to place very large radar antennas in space. To overcome this problem, a SAR used the synthetic aperture antenna. This means that the short antenna with their attended wide beam can be made to behave as though they are very long. In fact, the SAR is able to transmit several hundred pulses while its parent spacecraft passes over a particular object [236].

Compared to real aperture radar, Synthetic Aperture Radar (SAR) synthetically increases the antenna's size or aperture to increase the azimuth resolution, though the same pulse compression technique is adopted for range direction. Synthetic aperture processing is a complicated data process of receiving signals and phases of moving targets with a small antenna [238]. Figure 13.3 illustrates the geometry of the real aperture radar. The strip of terrain to be imaged is from point A to point B. Point A being nearest to the nadir point is said to lie at near range and point B, being furthest, is said to lie at far range.

In addition, the distance between A and B defines the swath width. The distance between any point within the swath and the radar is called its slant range. Ground range for any point within the swath is its distance from the nadir point (point on the ground directly underneath the radar) (Figure 13.3).

13.3.1 Spatial Resolution

Spatial resolution is the keystone to determine the physical object characteristics imagined in radar sensors. In this context, it is known as the potential of radar sensor to identify the close range between two objects as separate points. For instance, if a certain radar system is capable to discriminate two closely spaced internal wave crests as separate, a lower resolution system may only imagine one internal wave crest. Internal wave is extremely clear with the fine lower resolution of 25 m as compared to 30 m and 100 m resolution, respectively.

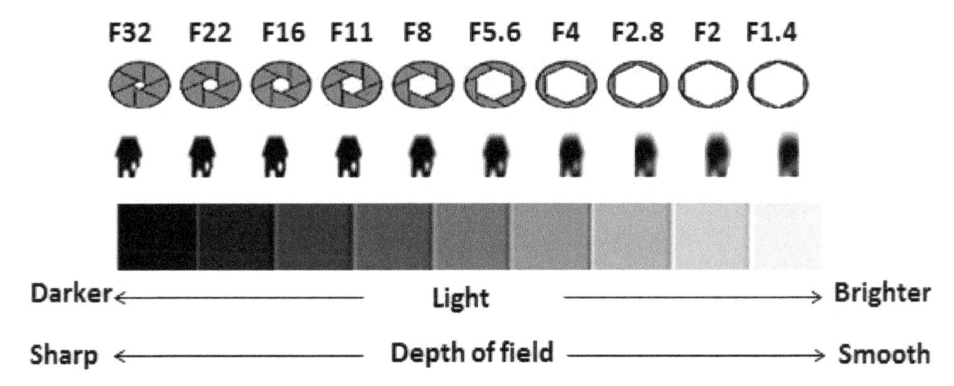

Figure 13.2 Picture quality based on aperture.

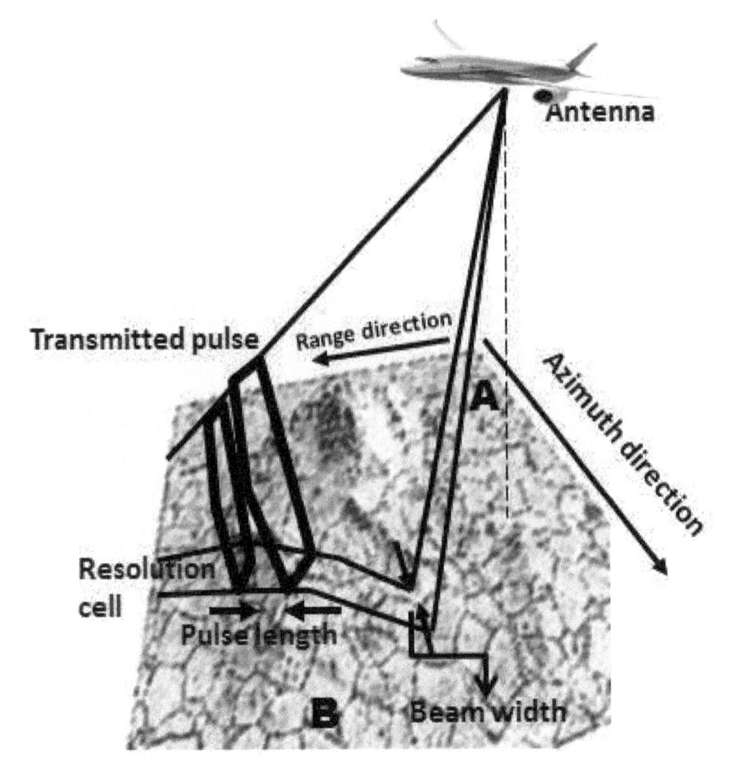

Figure 13.3 Geometry of real aperture radar.

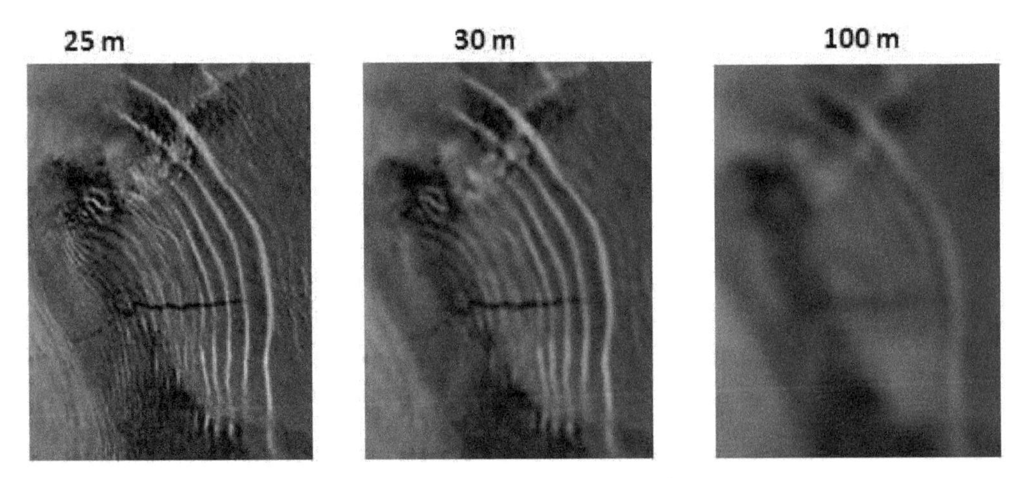

Figure 13.4 Different SAR image resolutions.

SAR spatial resolution is computed in the azimuth and range directions. Further, it is defined by the characteristics of the radar system and sensor. Resolution in the azimuth direction is, theoretically, one-half the length of the radar antenna. In the context of signal processing, azimuth resolution is independent of range. Range resolution, consequently, is governed by the frequency bandwidth of the transmitted pulse and thus by the time duration (width) of the range-focussed pulse. For instance, large band widths yield a small focussed pulse widths [238, 240].

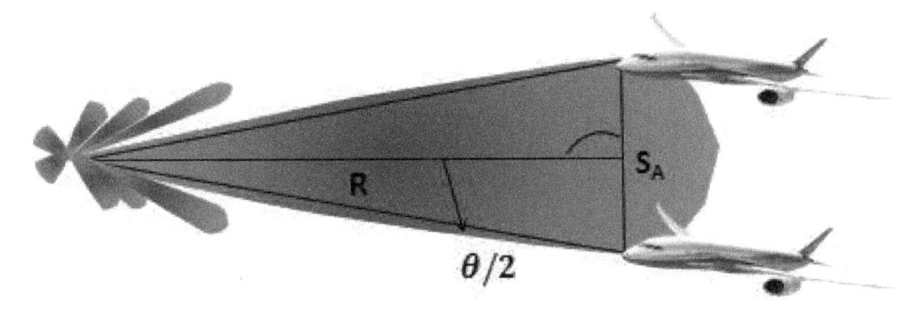

Figure 13.5 Angular resolution.

Angular resolution is the minimum angular separation at which two equal targets at the same range can be separated (Figure 13.5). The angular resolution as a distance between two objects relies on the slant-range and is formulated as [241]:

$$S_A \leq 2R.\sin\frac{\theta}{2} \qquad [\text{m}] \tag{13.1}$$

θ being antenna beam width (Theta), R is slant range aims-antenna, and S_A is the angular resolution as the distance between the two targets. The angular resolution characteristics of a radar are determined by the antenna beam width represented by the –3 dB angle θ which is defined by the half-power (–3 dB) points [238]. The half-power points of the antenna radiation pattern (i.e., –3 dB beam width) are normally specified as the limits of the antenna beam width for the purpose of defining angular resolution; two identical targets at the same distance are, therefore, resolved in angle if they are separated by more than the antenna beam width. In this regard, the smaller the beam width, θ, the higher the directivity of the radar antenna, and the better the bearing resolution [241].

13.3.2 Slant and Ground Range Resolution

As the range resolution of imaging SARs depends on signal pulse length, the actual distance resolved is the distance between the leading and the trailing edge of the pulse. These pulses can be shown in the form of signal wavefronts propagating from the SAR sensor. When these wavefront arcs are projected to intersect a "flat" Earth's surface, the resolution distance in ground range is always larger than the slant range resolution (Figure 13.6). The ground range resolution increases substantially at small incidence angles [239].

Therefore, the theoretical range resolution of a radar system is estimated via:

$$S_r = \frac{c_0.\tau}{2} \qquad [\text{m}] \tag{13.2}$$

where, c_0 is the speed of light, τ is the transmitter pulse width, and S_r is the range resolution as a distance between the two targets. In a pulse compression system, the range-resolution of the radar is given by the bandwidth of the transmitted pulse (B_{tx}), not by its pulse width [238],

$$S_r \geq \frac{c_0}{2B_{tx}} \qquad [\text{m}] \tag{13.3}$$

where S_r is the range resolution as the distance between the two targets, and B_{tx} is the bandwidth of the transmitted pulse. This allows very high resolution to be obtained with a long pulse, thus with a higher average power [237, 240].

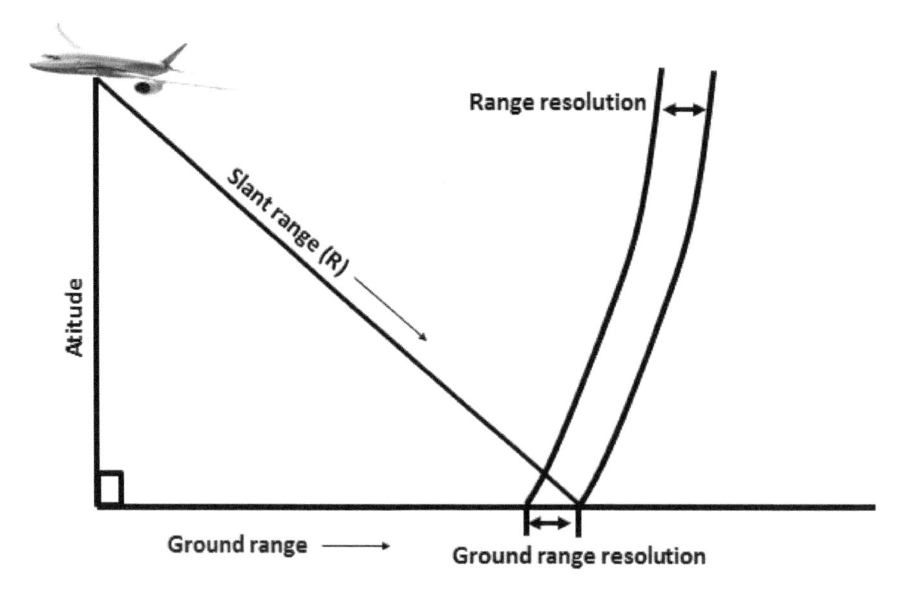

Figure 13.6 Slant and ground range resolution.

13.3.3 Resolution Cell

Consistent with Moreira (1992) [238], the range and angular resolutions lead to the resolution cell. The meaning of this cell is very clear: unless one can rely on eventual different Doppler shifts, it is impossible to distinguish two targets which are located inside the same resolution cell. The shorter the pulse with τ (or the broader the spectrum of the transmitted pulse) and the narrower the aperture angle are, the smaller the resolution cell, and the higher the interference immunity of the radar station is.

Ground-range resolution is the weakest in the near-range portion of the SAR image and the best in the far-range sector. The depression angle β brightens the near-range portion of the swath, although, small depression angles illuminate the far-range sector of the beam. The relationship between ground resolution and depression angle is casted as:

$$R_{GR} = \frac{\tau c}{2\cos\beta} \tag{13.4}$$

where R_{GR} is the ground resolution, c is the speed of light and β is the depression angle. Equation 13.4 indicates that the pulse duration, ground range and beam width are ruled by the size of the ground resolution cell. Pulse duration and ground range, consequently, prescribe the spatial resolution in the path of energy transmission which denotes the range resolution. The spatial resolution in the direction of flight is known as azimuth resolution and is determined by beamwidth.

13.3.4 Ambiguous Range

The main difficulty with range measurement and pulsed radars is how to determine the range unambiguously when a target returns a strong echo. Ambiguous range results from transmission of a sequence of pulses. In this view, pulse repetition interval (PRI) is created and is known as spacing between transmit pulses. It is also described as a pulse repetition frequency (PRF). Consequently, the delay time is created because of a spacing between transmit and return pulses. In this view, the delay time is generated and causes an ambiguity along the range direction or uncertainty. The unambiguous range R_A as a function of delay time τ_{PRI} is defined as:

$$R_{amb} = \frac{c\tau_{PRI}}{2} \tag{13.5}$$

Under the circumstance of Equation 13.5, radar can determine its range unambiguously when the object range is smaller than R_{amb}. On the contrary, an impossible range unambiguously accounts when the object range is larger than R_{amb}. Under these circumstances, the radar must have PRI to be greater the range delay frequented with the longest target ranges to circumvent range ambiguities. Alternative method to avoid the ambiguous range problem is based on multiple PRIs with waveforms. In this understanding, the waveforms assist to change the spacing between transmit pulses and detect that the target range is ambiguous. Then it can be easy to ignore the return pulse. The advanced method to resolve range delays is based on range resolve algorithm to calculate the precise target range [236, 238].

13.3.5 Range-Rate Measurement (Doppler)

Doppler frequency can measure the range rate. In fact, the Doppler frequency arises from the frequency differences by the spacing between the transmitted and received signals. Consider aircraft linear velocity of v, with different time of dt and varied range of dR as a function of plane movements from A to B (Figure 13.7). In this sense, range rate is determined from:

$$\dot{R} = \frac{dR}{dt} \tag{13.6}$$

The transmitted pulse can mathematically be written as:

$$v_T(t) = \text{rect}\left[\frac{t}{\tau_p}\right]\cos(2\pi f_c t) \tag{13.7}$$

where

$$\text{rect}\left[\frac{t}{\tau_p}\right] = \begin{cases} 1 & 0 \le t < \tau_p \\ 0 & \text{elsewhere} \end{cases}. \tag{13.8}$$

being, f_c is the *carrier frequency* of the radar, $\tau_p = 1\ \mu s$ is pulse width, and t is time of the transmitted pulse. On the contrary, the received pulse as a function of delay time or range rate v_R is defined as:

$$v_R(t) = \xi v_T(t - \tau_R) \tag{13.9}$$

where ξ is the amplitude scaling factor, and τ_R, is the range delay which is given by

$$\tau_R(t) = \frac{2R(t)}{c} \tag{13.10}$$

Then the Doppler frequency is a function of a range delay and can mathematically be given by

$$v_R(t) = \xi\text{rect}\left[\frac{t-\tau_R}{\tau_p}\right]\cos(2\pi(f_c + f_d)t + \phi_R) \tag{13.11}$$

where $\phi_R = -2\pi f_c(2R/c)$ is the phase shift owing to range delay and $f_d = -f_c(2\dot{R}/c)$ is the *Doppler frequency* of the target which can be rewritten as a function of radar wavelength λ,

$$f_d = -2\dot{R}/\lambda. \tag{13.11.1}$$

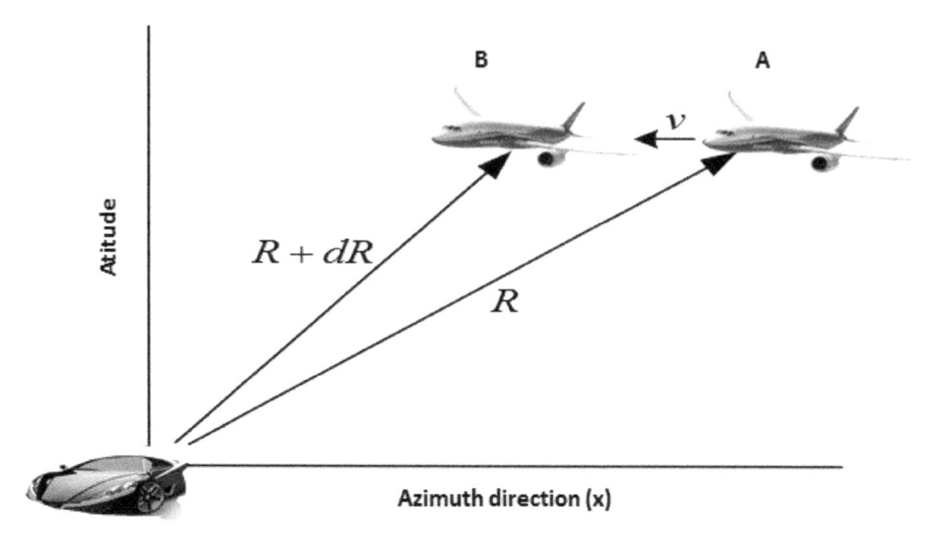

Figure 13.7 Range rate measurement by Doppler.

Equation 13.11 demonstrates that the frequency of the backscatter signal is $f_c + f_d$ in preference to simply f_c. The Doppler frequency, f_d, therefore, can be computed based on the comparison between the frequency of the transmit signal to the frequency of the received signal. Once we have f_d, we can compute the range-rate from (13.11.1).

Hypothetically, computing Doppler frequency is not as easy as implied in Equation 13.11. The difficulty is because of the comparative magnitudes of f_d and f_c. Though the quantity of Doppler frequency is difficult, it is feasible. To compute the Doppler frequency, the transmit signal must be extremely long (in the order of ms rather than μs) or the Doppler frequency must be relied on the processing of numerous signals. We will defer a detailed explanation of how Doppler frequency is measured until a later date.

13.4 Radar Range Equation

Radar range equation is a simplest mathematical description of a radar principle. Though it is one of the most effective equations, paradoxically, it is an equation that few radar analysts comprehend and plenty of radar analysts mishandle. The trouble, therefore, lies not with the equation itself but with the numerous terms that form the equation. A deep understanding of the radar range equation delivers a completely solid basis of radar principle. The radar equation is the termination of several simpler formulas.

The transmitted energy density E_T can be explained as energy per unit area and occurs in a range R. The scientific explanation of the power density at a distance of the emitted signal can mathematically be written as:

$$E_T = \frac{P_T.\tau.G_T}{4\pi R^2} \tag{13.12}$$

In Equation 13.12, $4\pi R^2$ associates the power transmitted by the radar to an isotropic sphere. In this sense, the electromagnetic energy transmits similarly in all directions. Further, G_T is the antenna and the focus of antenna signal is a function of the ratio of $\frac{G_T}{4\pi R^2}$. Consequently, P_T is called the *peak transmit power* and is the average power when the radar is transmitting a signal. P_T can be specified on the output of the transmitter with a duration time τ of or at some other point like the output of the antenna feed. It has the unit of watts.

Equation 13.12 can be developed as a function of the radar cross section (RCS) σ which is a result of the backscatter size of the target on which the radar signal is focused. Therefore, reradiated energy of an object is estimated via:

$$E_{\sigma} = \frac{P_T.\tau.G_T.\sigma}{4\pi R^2} \tag{13.13}$$

Equation 13.3 expresses the backscatter power density as a function of radar cross section (RCS). Therefore, RCS relies on the object's unique scattering characteristics. Further, the target *radar cross-section or RCS* has the units of square meters or m^2.

The energy of receiving antenna signal *S* is a function of the backscatter and is located at the same transmit antenna. This is known as monostatic radar. In this view, the term of $4\pi R^2$ in the previous equation turns into $(4\pi)^2 R^4$ with the additional parameter of A_R in the numerator. The term A_R, conversely, is the operational area of the receiving antenna and is the numerator in a ratio concerning the second $4\pi R^2$. This represents the isotropic radiation owing to a target's radar cross section. Under this circumstance, Equation 13.13 can be modified as:

$$S = \frac{P_T.\tau.G_T.\sigma.A_r}{(4\pi)^2 R^4} \tag{13.14}$$

Equation 13.14 articulates the quantity of signal which reaches on the receiving antenna. Noise, therefore, can be generated in the receiver of the radar system. This noise is known as the signal-to-noise ratio (*SNR*) and has the units of watts/watt, or w/w. Further, the radar system is also dominated by additional factors such as thermal noise temperature T_0, Boltzmann's constant *K*, and is equal to 1.38×10^{23} w/(Hz °K). The receiver bandwidth *B*, and *L* represent losses within the system itself. One last element which needs to be developed for the Equation 13.14 is a relation between antenna gain, its effective area, and signal wavelength λ.

$$A_R = \frac{G_R \lambda^2}{4\pi} \tag{13.14.1}$$

Equation 13.14.1 expresses the correlation between antenna area and gain, i.e., gain and effective aperture relation. The radar final formula results from substituting A_R and the noise constants into the previously developed formulas. In this view, the scientific explanation of radar range can mathematically be written as:

$$SNR = \frac{P_S}{P_N} = \frac{P_T G_T G_R \lambda^2 \sigma}{(4\pi)^3 R^4 k T_0 B F_n L} \tag{13.15}$$

where, F_n is the radar *noise figure* and is dimensionless, or has the units of w/w, and *L* is a term involving all losses that must be considered when using the radar range equation. *L* has the units of w/w. Equation 13.5 is the last version of the radar equation. Additional terms can be added or substituted for diverse schemes. For instance, occasionally, the transmission interval denotes the combination of countless signals that occur over the total time *t*. The main concept to consider from the radar equation is the inter-need of various diverse aspects of radar.

13.5 Radar Backscattering

A SAR backscattering cross section of a target is the equivalent area scattered equally in all directions [238]. This produces an echo at the radar equal to that from the radar. Furthermore, radar cross section is a measure of the equivalent surface area of the target that radar sees. It is an equivalent area that, if it was to intercept the power incident from the radar and scatter it uniformly in all directions, would produce a return at the radar receiver equal to that from the target [237]. Radar cross section is often alternatively referred to as effective echo area or just echo area. Therefore, radar cross section (σ) mathematically can be defined as 4 Π power unit solid angle scattered back to the receiver per power unit area incident on the target [239].

13.5.1 Characteristics of Radar Backscattering

There are three basic properties as the radar beam reflects from the surface. These properties are: (i) dielectric constant (or permittivity), (ii) the roughness (height relative to a smooth surface) and (iii) the local slope [237–240]. The interaction of electromagnetic waves with a surface is called scattering. Scattering can be divided into two categories, i.e., surface scattering and volume scattering. Moreover, surface scattering occurs at the interface between two different homogenous media such as the atmosphere and the Earth's surface. Volume scattering is the result of interaction with particles within non-homogeneous medium (Figure 13.8) [240].

An electric property of the reflected substances is termed as the complex permittivity ε_c, which significantly affects the radar backscatter. It is also known as dielectric constant which can mathematically be obtained by

$$\varepsilon_c = \varepsilon' + i\varepsilon'' \tag{13.16}$$

where ε' is the dielectric constant of the substance, ε'' is the "lossy" part of the dielectric constant, and the imaginary part, i is the square root of -1. ε' is a definite quantity of the media's reaction to an electrical domain. When an electrical domain is a response to the medium, the molecules restore to the lowest

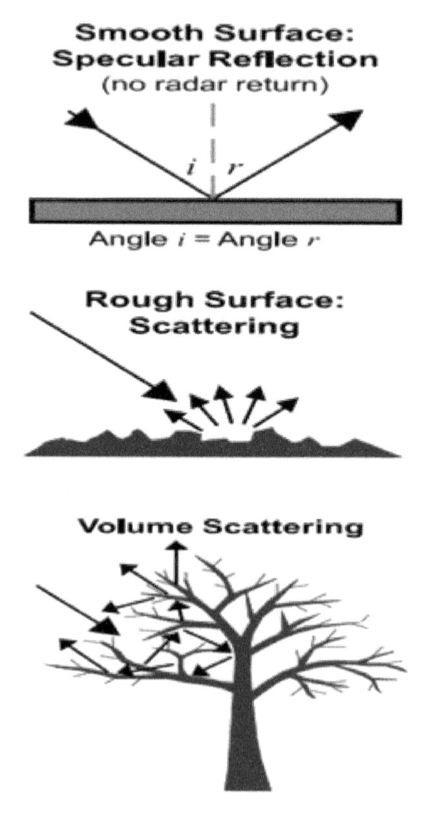

Figure 13.8 Scattering mechanisms.

power formal to fit their polarity to an electrical domain. Molecules, however, cannot absolutely associate to an electrical domain because of their crystalline structure. Consequently, the time lag between the electric domain and interaction of the molecules can be described by dielectric loss factor of the dielectric constant (ε").

A radar backscatter is proportional directly with the dielectric. A high backscatter increased due to a high material dielectric. In this view, the microwave spectra cannot penetrate the sea surface due to its high dielectric. A dry material, however, has a low dielectric, which allows the radar signal to penetrate and also produce volume reflection and form an extremely brighter backscatter at an antenna receiver.

13.5.2 Surface Scattering

Consistent with Kingsley (1999) [242], brightness in a radar image is proportional to the local amount of radar backscatter. The amount of radar backscatter is a function of surface roughness. A smooth surface would act as a mirror, which causes radiation reflection away from the radar whereas the rougher a surface is, the more power is backscattered into radar. In such case, the standard deviation of the surface height is about the same as the radar wavelength [240]. Lillesand et al. (2014) [243] stated that as a very smooth surface becomes slightly rough, the incident energy is scattered in a wide range of angles, with only a small fraction reflecting back to the SAR sensor. This is known as diffuse reflectance. Sparse, low vegetation, bare agriculture fields, and other rough surfaces will cause diffuse reflectance, and result in intermediate tones on SAR imagery as compared to a specular reflector (Figure 13.8). For instance, coarse river gravel that contains numerous stones whose dimensions are comparable to the radar wavelength, and water surfaces disturbed by strong winds are two examples of this type of reflectance. Such surfaces can appear very bright on SAR images [241].

Volume scattering is a term used to describe the multiple scattering of a radar signal within a medium, such as the vegetation canopy of a corn field or forest (Figure 13.8), or in layers of very dry soil, sand, or ice. With the volume scattering, a SAR sensor may receive backscatter from both the target surface and the interior volume of the target [243]. Further, Ahern (1995) [244] stated that an intermediate degree of volume scattering occurs within agricultural crops or other sparse or low vegetation. In these situations, there can be three important scattering mechanisms (Figure 13.9): (i) diffuse scattering from the ground (1), (ii) direct (single bounce) scattering from various vegetation components (2 and 3); and (iii) double-bounce vegetation-ground interaction (4).

13.5.3 Backscatter Coefficient

Targets scatter the energy transmitted by the radar in all directions. The energy scattered in the backward direction is what the radar records. The intensity of each pixel in a radar image is proportional to the

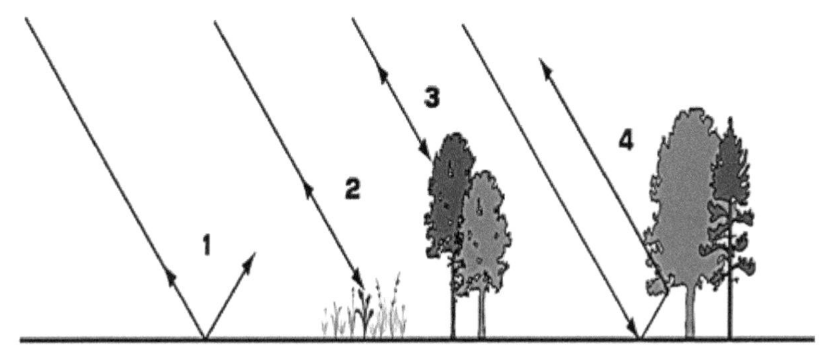

Figure 13.9 Intermediate degree of volume scattering.

ratio between the density of energy scattered and the density of energy transmitted from the targets in the Earth's land surface [237–244].

The backscatter coefficient, (σ^o) is given by:

$$\sigma^o = 10\log\sigma^o \tag{13.17}$$

The energy backscattered is related to the variable to refer as radar cross-section, and is the amount of transmitted power absorbed and reflected by the target. The backscatter coefficient σ^o is the amount of radar cross-section per unit area on the ground. σ^o is a characteristic of the scattering behavior of all targets within a pixel and because it varies over several orders of magnitude, it is expressed as a logarithm with decibel units [244].

The backscatter is measured as a complex number, which contains information about the amplitude easily converted to σ° by specific equations and the phase of the backscatter [243]. Contain speckle is an interference phenomenon produced between backscatter coming from many random targets within a pixel. The speckle represents a true electromagnetic scattering and influences the interpretation of SAR images [244]. Backscattering types in marine environments include volume scattering, subsurface (volume) scattering, surface scattering and corner reflector-like scattering.

13.5.4 Incident Angle

Incident angle θ is a major factor influencing the radar backscatter and the appearing targets in the images. The incidence angle at any point within the range is the angle between the radar beam direction (of look) and a line perpendicular (normal) to the surface, which can be inclined at any angle (varies with slope orientation in non-flat topography). The depression angle decreases outward from near to far range. Incident angle is the angle between the radar beam and a target object. The incident angle helps to determine appearing target in an image. In a flat surface, incident angle is the complement of the depression angle (Figure 13.10) [244]. A local incident angle could be determined for any pixel in the radar data. This in

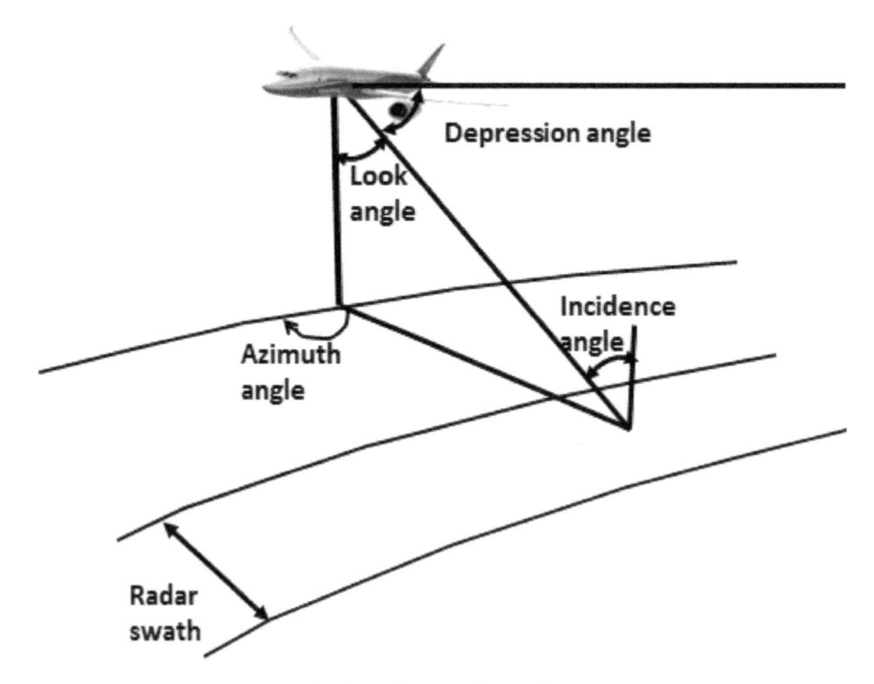

Figure 13.10 Incident and depression angles.

turn causes variations in pixel brightness. In general, reflectivity from distributed scatter decreases with increasing incident angles. Smaller incident angle results in more backscatter, although for very rough surfaces the backscatter is independent of θ [242–244].

13.5.5 Polarization

The polarization of an electromagnetic wave refers to the orientation of the electric field intensity vector. SAR usually transmits a wave such that it is horizontally polarized. Most of the received energy is still horizontally polarized, however a small portion of it may be depolarized through interaction with the terrain, creating many different components at various angles of polarization [237].

Using a filter at the antenna, all polarization can be screened out except for the desired polarization—either horizontal or vertical. The HH (transmit and receive horizontal) configuration is called the like-polarized return while the HV (transmit horizontal and receive vertical) configuration is called the cross-polarized return. The HV return requires a much higher antenna gain than the HH due to its much smaller energy. The images produced by these two returns may differ due to the differences between the scattering processes for each [239].

Depolarization, for instance, is usually explained in terms of volume scatter or multiple reflections. At short wavelengths, when the terrain can be considered very rough, like- and cross-polarized images are almost identical. At longer wavelengths, however, when the terrain is considered relatively smooth, noticeable differences stand out. This has resulted in limited geological application in identifying rock types [242].

Some of the new polarimetric SARs are capable of transmitting and receiving both horizontally and vertically, thus allowing HH, HV, VH and VV returns all at the same time. Mathematical analysis of these returns can create a geometric basis, which can be used to synthesize the image for any possible transmit/receive polarization for the entire 360° spectrum. Certain objects, especially those man-made, have been found to stand out at a specific transmit/receive polarization, thus making their detection easier in the future. Over the ocean, very few multiple reflections occur, and ocean-viewing sensors are normally HH or VV. The polarisation ratio is usually large, because vertically polarised radar reflects more strongly than horizontally polarised [243].

In practice, VV ocean backscatters are extremely appropriate for measuring ocean wind speeds than HH yields. In fact, VV ocean backscatter is larger than HH returns. The alternations between VV and HH growth with growing incident angle is high at C-band. In fact, HH is more sensitive than VV to the changes in the local incident angle produced by variations in the long-scale sea surface slope. The change in the normalized radar cross section (NRCS) of microwave backscatter from the sea surface due to the changing local incident angle is known as tilt modulation. In this view, HH has larger tilt modulation than VV which makes HH more sensitive to local incident angel. At low wind speed, nevertheless, HH is more sensitive to wind speed than VV and the discrimination between both polarizations is peaking. Therefore, HH is extremely sensitive to wave steepness and whitecaps than VV polarization.

13.5.6 Speckles

Speckle is basically a form of noise, which degrades the quality of an image and may make its visual or digital interpretation more complex. A speckle pattern, consequently, is a random intensity pattern fashioned by the mutual interference of a set of wavefronts having different phases. Under this circumstance, they add together to give a resultant wave whose amplitude, and therefore intensity, varies randomly. In this context, if each wave is modelled by a vector, then it can be seen that a number of vectors with random angles are added together. The length of the resulting vector, therefore, can be anything from zero to the sum of the individual vector lengths—a 2-dimensional random walk, sometimes known as a drunkard's walk. Further, when a surface is illuminated by a microwave spectra, according to diffraction theory, each point on an illuminated surface acts as a source of secondary spherical waves. The microwave spectra at any point in the scattered microwave field is made up of waves which have been scattered from each point on the illuminated surface. If the surface is rough enough to create path-length differences exceeding one wavelength, giving rise to phase changes greater than 2π, the amplitude, and hence the intensity, of the resultant backscatter microwave vary randomly [243].

In general, all radar images appear, to some degree, what we call radar speckle. Speckle appears as a grainy "salt and pepper" texture in an image (Figure 13.11). This is produced by random constructive and destructive interference from the multiple scattering returns (Figure 13.11) that will occur within each resolution cell [242]. Constructive interference is an increase from the mean intensity and produces bright pixels. In contrast, destructive interference is a decrease from the mean intensity and produces dark pixels (Figure 13.12).

As an example, a homogeneous target, such as a large grass-covered field, without the effects of speckle would generally result in light-toned pixel values in an image. However, reflections from the

Figure 13.11 SAR data with speckles.

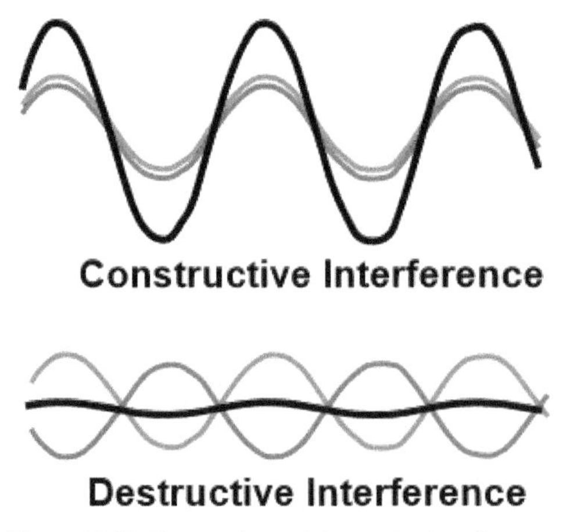

Figure 13.12 Constructive and destructive interferences.

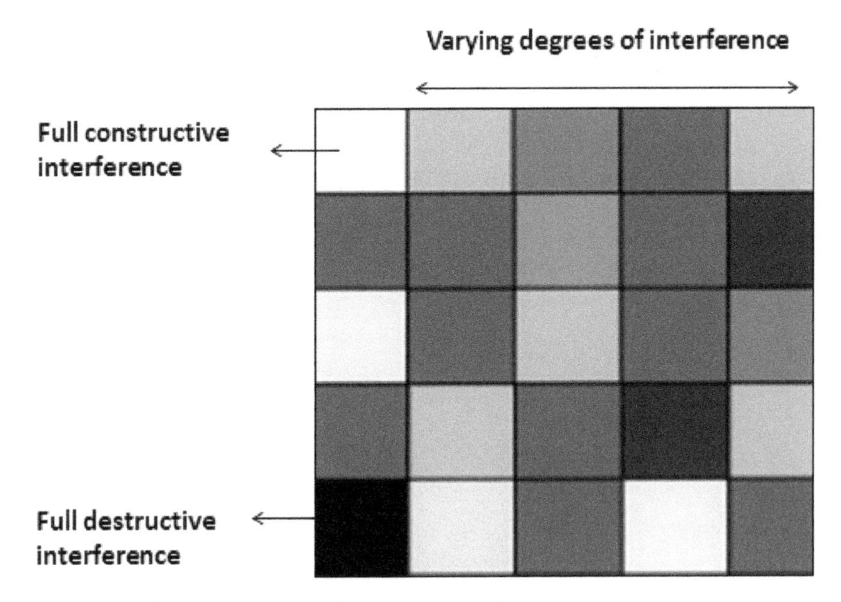

Figure 13.13 Impact of constructive and destructive interference on pixel's brightness.

individual blades of grass within each resolution cell result in some image pixels being brighter and some being darker than the average tone, such that the field appears speckled. The high speckle noise in SAR images, however, has posed great difficulties in inverting SAR images for mapping morphological features. Speckle is a result of coherent interference effects among scatterers that are randomly distributed within each resolution cell [237]. The speckle size is a function of the spatial resolution, which induces errors in the morphological feature signature detections. In order to reduce these speckle effects, appropriate filters, e.g., Lee, Gaussian, etc. [243], can be used in the preprocessing stage. The effectiveness of these speckle-reducing filters, nevertheless, is influenced by local factors and application. In fact, all speckles in SAR images are related to local changes in the Earth surface roughness.

13.6 SAR Imagine Sea Surface

According to the above prospectives, a backscatter is mainly modelled in two ways: specular reflection and Bragg scattering. Specular reflection occurs when the water surface is tilted which creates a small mirror pointing to the radar. For the perfect specular reflector, radar returns (backscatter) exist only near vertical incidence. This is due to a 90°-depression angle or the slope of the surface. Furthermore, the reflected energy is localized to small angular regions around the angle of reflection. Even for non-vertical incidence, nevertheless, backscatter can exist for a rough subsurface. This can occur if the radar is capable of penetrating deep enough. Bragg or resonant scattering is scattering from a regular surface pattern. Resonant backscattering occurs when phase differences between rays backscattered from subsurface pattern interfere constructively. The resonance condition is $\mathbf{2\lambda \sin\theta = \lambda'}$, where λ and λ' are water wavelength and the radar wavelength, respectively. $\boldsymbol{\theta}$ is the local angle of incidence (Figure 13.14). In this view, the short Bragg-scale waves form in response to wind stress, if the sea surface is rippled by a light breeze and no longer waves are present. The radar backscatter is due to the component of the wave spectrum. This resonates with a radar wavelength.

Consequently, swell waves being imaged have much longer wavelengths than the short gravity waves, which cause the Bragg resonance. In addition, a Bragg-resonance from the ocean might be considered as coming from the facets. The term facet refers to a relatively flat portion of the long wave structure with a

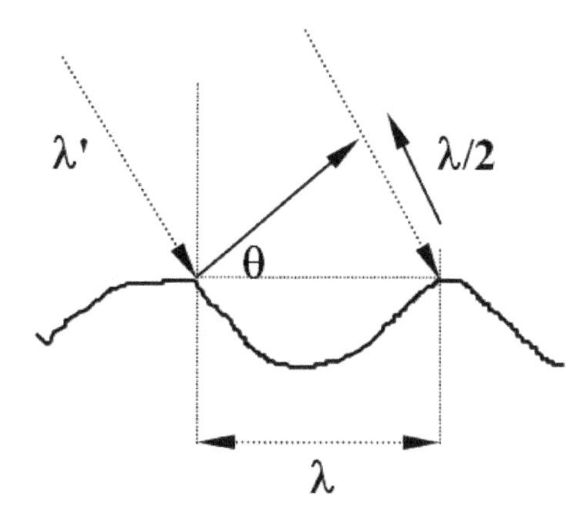

Figure 13.14 Bragg Scatter.

cover of ripples containing Bragg-resonant facets. This behaves like specular points. The beam-width and scattering gain of each facet are determined by its length in the appropriate direction.

The mathematical model of the normalized cross section as a function of the standard Bragg/composite surface scattering theory is given by:

$$\sigma_o = \sigma_c + \sigma_b \tag{13.18}$$

σ_c being the composite surface cross section and the assured wave cross section σ_b, which are expressed, respectively, by:

$$\sigma_c = \iint \sigma_B(\theta_o+\gamma, \alpha)\, P_f\, P(\gamma, \alpha \mid f)\, d\gamma d\alpha \tag{13.19}$$

$$\sigma_b = \iint \sigma_B(\theta_o+\gamma, \alpha)\, P_b P(\gamma, \alpha \mid b)\, d\gamma d\alpha \tag{13.20}$$

Both equations are based on standard Bragg scattering cross section σ_B as associated with three sorts of wave probabilities: (i) the probability of finding free waves P_f; (ii) the probability of finding bound waves (P_b); and (iii) the probability distribution of a wave sort, either free or bound ($P(\gamma, \alpha \mid x)$). Therefore, γ, α are the long wave slopes in and perpendicular to the plane of incidence and θ_o is the nominal incidence angle. Thus, on standard Bragg scattering cross section σ_B can be obtained by:

$$\sigma_B = 16\, \pi\, k_o^4 \mid F(\theta_o+\gamma, \alpha)\mid^2 \mu(2k_o \sin(\theta_o+\gamma), 0) \tag{13.21}$$

In Equation 13.21, the microwave number is k_o, the wave height variance spectrum is μ which is a function of $(2k_o sin\, (\theta_o+\gamma), 0)$. Finally, F is a function of incidence angle θ and dielectric ε.

In general, the intensity of radar cross section is a function of the incidence angle. If the radar signal strikes at right angles or at a high incidence angle to the surface, it will reflect more strongly than if it strikes at low, grazing angle. In the latter case, much of radar energy will be reflected away from the radar receiver, and will therefore appear as low or dark response on the image. Consequently, the local incidence angle is a major controlling factor in the intensity of the Bragg scattering cross section σ_B.

The radar backscatter tends to be the strongest from the slope of the wave facing towards the radar. Thus an image of backscatter modulated by tilt alone would represent a plane parallel swell-wave field as a series of parallel light and dark lines corresponding with slopes facing towards and away from the radar, i.e., displaced by 90° of phase from lines of troughs and crests.

In general, SAR ocean surface imagines count on the very high sensitivity of incident angle and radar backscatter signal which are allied with fluctuations on both the local geometry and the spectral density distribution of short gravity and gravity-capillary waves. The resonant surface waves are extremely shorter at more oblique incidence angles. In other words, the ocean backscatter returns decrease with the increase of incidence angles. Consequently, the large oblique angles view smaller amplitude of Bragg waves leading to lower backscatter. The next chapter will reveal the impact of the tsunami on sea surface waters which can be detected from SAR data.

Chapter 14

Detection of Internal Wave from Synthetic Aperture Radar Post Tsunami

The previous chapters have revealed the tsunami impacts on the coastal physical characteristics such as water masses, turbulent flows and solitary wave generations. The optical remote sensing, specially high resolution sensor, such as Quickbird delivered precise information about 2004 tsunami impacts on coastal zones. On the contrary, there were not any SAR visit data recorded on 26th December 2004. Unfortunately, there were a few SAR sensors of ENVISAT data archived post the 2004 tsunami. SAR data, however, have proved to be a promising tool for detecting sea surface phenomena. Can SAR data deliver precise clues about the tsunami inducing sea surface features, for instance, internal wave?

14.1 Internal Wave

Underneath the ocean surface, internal waves are existed and are invisible by human eyes. However, they can be detected by studying temperature or salinity changes at a given location. They exist due to the fact that the deep waters of the ocean are denser than the surface waters. If a parcel of deep (dense) water is pulled up closer to the surface, gravity might force it back downwards. Similarly, if you raised a parcel of seawater into the air; it might fall whilst free. The opposite additionally hold a reality of buoyancy. The buoyancy of surface waters, therefore, forces them to return to the surface if they may be shortly driven downwards.

In this sense, a restoring force rules the rise of depressed water, and sinking of uplifted water. In other words, restoring force implements on vertical variations of density, which triggers internal waves. In this understanding, in each case, there are two layers of fluid with a curved interface between them, as exemplified in Figure 14.1.

Internal waves are well known as gravity waves, which oscillate inside, as opposed to on the surface of, a water body. They are generated in an association with a dynamically stratified ocean. In other words, internal waves are gravity waves that oscillate inside a fluid medium, instead of on its surface. Under this circumstance, the ocean must be well stratified to generate internal waves. In this regard, the density ought to decrease continuously or discontinuously with depth because of physical characteristic fluctuations, as an instance, in temperature and/or salinity and density (Figure 14.2).

Vertical mixing takes place through the water column results in vertical velocity shear that is generated by internal waves. External forces such as tide can cause internal waves in well stratified waters. Indeed, tide displaces a water parcel which is restored by buoyancy forces. Then the restoration force may exceed the equilibrium position and forms an oscillation. In this manner, internal wave is generated. Internal tide,

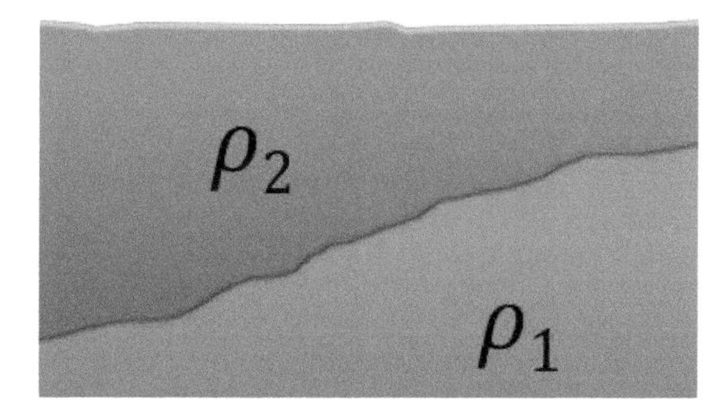

Figure 14.1 Simple explanation of internal wave generation.

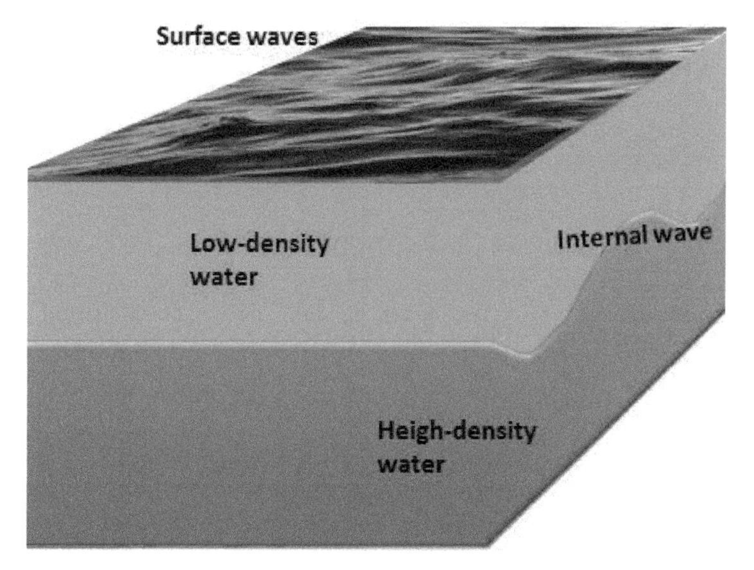

Figure 14.2 Physical conditions for generation of internal wave.

therefore, arises from the interface between the rough topography and the barotropic tide. Therefore, the internal waves are associated with drastic increases in phytoplankton and zooplankton concentrations.

It is characterized by horizontal length scales of the order 1 to 100 km and its horizontal velocity of 0.05 to 0.5 ms^{-1}. Finally, its time generation scale ranges from few minutes to days. On the shelf, therefore, the dynamic of internal wave is strongly non-hydrostatic and thus cannot be well resolved in ocean general circulation models that frequently create the hydrostatic approximation.

Similar to surface waves, internal waves change as they approach the shore. As the ratio of wave amplitude of water depth becomes such that the wave "feels the bottom," water at the base of the wave slows down due to friction with the sea floor. This causes the wave to become asymmetrical and the face of the wave to steepen, and finally the wave breaks, propagating forward as an internal bore.

14.2 Internal Wave Imaging in SAR

Internal waves can be observed from space. Satellite Synthetic Aperture Radar (SAR) can detect the radar backscatter signals from the ocean surface under all weather and night conditions, with a spatial resolution as fine as several meters. Currents induced by an internal wave form a convergence zone on one side of the internal wave and a divergence in the other. The convergence roughens the sea surface by intensifying surface waves, whereas the divergence can smooth the ocean surface by suppressing the surface waves. The rougher surface has a greater radar backscattering section, while the smoother one produces less backscatter. Therefore, in the SAR image, one can see bright and dark textures associated with internal waves beneath the ocean surface.

The internal wavelength continually decreases for the front to the rear wave, revealing the non-linearity of these internal waves. Based on their spatial shapes, these waves travel off shoreward from the plume front [164].

14.2.1 SAR Imaging of Internal Solitons

According to Da Silva et al. (2011) [164], Satellite Synthetic Aperture Radar (SAR) observations provide the best evidence for the presence of the shorter period internal solitary waves (ISWs) in the ocean. This is due to a mechanism whereby horizontally-propagating internal waves, centered on the thermocline typically some tens of meters below the surface, can generate a signature in the surface roughness field because of the modulating effect of convergence and divergence in the near-surface currents associated with the internal waves. This modulation is more effective for short period (30 minutes or shorter) ISWs because the straining of short (Bragg) surface waves (or ripples) is the strongest in these periods. It may also be possible to detect tidal period internal waves (with periods of 12.4 hours) in the presence of surface films and/or when the surface currents associated with the ITs induce alternating wind conditions relative to the surface with wind against tide exhibiting larger radar backscatter than wind with tide [165].

Da Silva et al. (2016) [166] have explained the internal wave imagine in SAR data. They demonstrated that sea surface signatures of mode 1 and mode 2 ISWs can be unambiguously identified in SAR images provided the waves propagate in relatively deep water (when the lower layer is significantly deeper than the mixed layer depth). If interfacial waves (in a two layer system) travelling along the pycnocline are waves of depression (which is often the case in deep waters), then mode 1 ISWs will be revealed in SAR images by bright bands preceding darker ones in their direction of travel. However, mode 2 ISWs will have precisely the opposite contrast in the SAR, since their radar signature consists of dark bands preceding bright bands in their direction of propagation (Figure 14.3).

In Figure 14.3, the ISWs are assumed to be moving from left to right. From top to bottom, the horizontal profiles represent the following features: SAR intensity profile along the ISWs, with bright enhanced

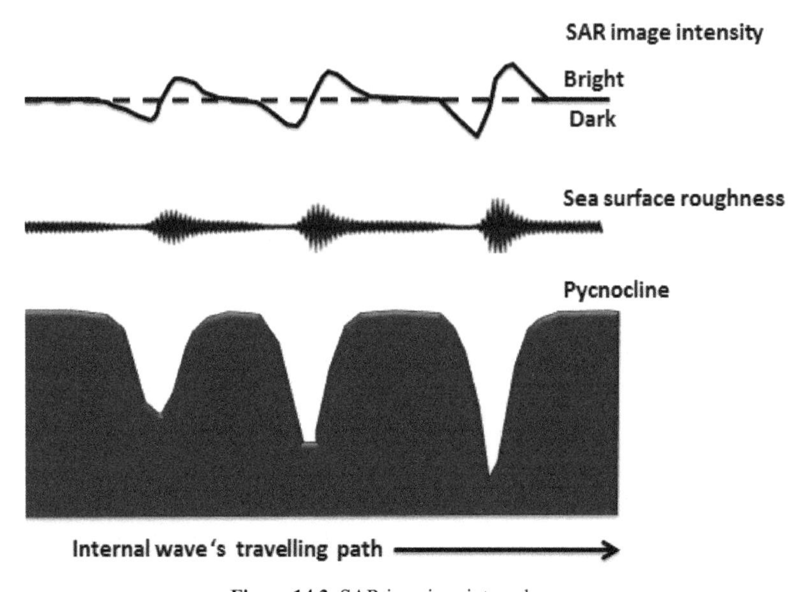

Figure 14.3 SAR imagines internal wave.

backscatter which is preceding dark reduced backscatter in the direction of internal wave propagation. Surface roughness, therefore, is indicating how rough and smooth the surface is along an ISW wave packet. Furthermore, isopycnical displacements are produced by ISW propagation.

This is because surface velocity fields induced by travelling ISWs of different modes can create different convergence and divergence patterns, which then modulate the surface roughness and thus the intensity of the radar backscatter signal. In particular, ISW is travelling along the thermocline and generates a divergence pattern on the surface, followed by a convergence pattern in the propagation direction. This dynamical variation creates the characteristic of the dark bands ahead of the bright bands in their direction of travel (Figure 14.3).

14.2.2 Mathematical Model of Internal Wave Radar Cross Section

Internal wave detection in SAR images is based on the integration of hydrodynamic interaction theory and Bragg scattering theory. In this regard, the hydrodynamic MTF defines the variations in the backscattering cross section because of the hydrodynamic modulation of the ripple energy spectral density at the mean (non-modulated) Bragg wave number by the long gravity waves. In this view, hydrodynamic interaction leads to a non-uniform distribution of the Bragg waves. The Bragg waves, therefore, ride the longest wave, and are subjected to the current stresses of the long wave orbital motion.

According to Plant (1990) [161], the radar backscatter cross section per unit area, σ_0 for the ocean surface is given by

$$\sigma_0(\theta)_{ij} = 16\pi k_0^2 \left| p_{ij}(\theta) \right|^2 \Psi(0, 2k_0 \sin\theta) . \tag{14.1}$$

where θ is incident angle, k is wave number, and the high-frequency capillary-gravity wave spectrum is Ψ. Consequently, the horizontal ($p_{HH}(\theta)$) and vertical polarization ($p_{vv}(\theta)$) states, respectively, are estimated as:

$$p_{HH}(\theta) = \frac{(\varepsilon_r - 1)\cos^2\theta}{\left[\cos\theta + (\varepsilon_r - \sin^2\theta)^{1/2}\right]^2} , \tag{14.2}$$

and,

$$p_{VV}(\theta) = \frac{(\varepsilon_r - 1)\left[\varepsilon_r(1 - \sin^2\theta) - \sin^2\theta\right]\cos^2\theta}{\left[\varepsilon_r \cos\theta + (\varepsilon_r - \sin^2\theta)^{1/2}\right]^2} , \tag{14.3}$$

where ε_r is the relative dielectric constant of seawater. In line with Zheng et al. (2004) [162], the high-frequency capillary-gravity wave spectrum Ψ is computed by:

$$\Psi = m_3^{-1} \left[m\left(\frac{W_*}{c}\right)^2 - 4\gamma^2 \varpi^{-1} - S_{\alpha\beta} \frac{\partial U_\beta}{\partial x_\alpha} \omega^{-1} \right] k^{-4} , \tag{14.4}$$

where m is a dimensionless constant (=0.04), W_* is the friction, wind speed, ϖ is the angular frequency, c is the phase speed and γ represents the viscosity. U_β represents the velocity components of the large-scale current field and $S_{\alpha\beta}$ represents the excess moments flux tensor. Subscripts α and β represent different horizontal coordinates (x or y).

$$S_{\alpha\beta} \frac{\partial U_\alpha}{\partial x_\beta} = \frac{1}{2}\left[\frac{\partial u}{\partial x}\cos^2\varphi + \left(\frac{\partial u}{\partial y} + \frac{\partial v}{\partial x}\right)\cos\varphi \sin\varphi + \frac{\partial v}{\partial y}\sin^2\varphi \right] . \tag{14.5}$$

Equation 14.5 demonstrates that the gradient current of $\frac{\partial u}{\partial y}$ and $\frac{\partial v}{\partial x}$ is related to internal wave and upper ocean parameters such as amplitude and wavelength of the internal wave, as well as water column stratification. The larger the amplitude of the internal wave, the larger the gradient current grows. Further, Equation 14.5 implies that the normalized radar cross section becomes larger in convergent flow regions when $\frac{\partial U_\alpha}{\partial x_\beta} < 0$. In contrast, if $\frac{\partial U_\alpha}{\partial x_\beta} > 0$, the divergent flow is created and then the normalized radar cross section becomes smaller.

Thus, the radar backscatter cross section per unit area has a form of

$$\sigma_0(\theta)_{ij} = 16\pi k_0^{\ 2} m_3^{\ -1} \left| g_{ij}(\theta) \right|^2 \left[m\left(\frac{W_*}{c}\right)^2 - 4\gamma^2 \varpi^{-1} + \frac{1}{2}\frac{\eta_0 c_0}{h} a(1+\frac{1}{a^2 b^2})\cos^2\phi \frac{\sinh 2a\xi}{(\cosh^2 a\xi + \frac{1}{a^2b^2}\sinh^2 a\xi)^2} \right] k^{-4} \quad (14.6)$$

The soliton induced radar backscatter cross section, $\sigma_{0IS}(\theta)_{ij}$ is the last term in the right-hand side of the Equation (14.6), which is written as

$$\sigma_{0IS}(\theta)_{ij} = 8\pi k_0^{\ 2} m_3^{\ -1} \left| p(\theta) \right|^2 k^{-4} \frac{\eta_0 c_0}{h} a(1+\frac{1}{a^2 b^2})\cos^2\phi \left[\frac{\sinh 2a\xi}{(\cosh^2 a\xi + \frac{1}{a^2b^2}\sinh^2 a\xi)^2} \right] \quad (14.7)$$

where η_0 is the vertical displacement of the interface between two layers of water, c_0 is soliton phase speed, h is water depth, and ϕ is propagation direction. The nonlinear least square algorithm must be applied to obtain the parameters a, b, and sea surface elevation ξ. Consequently, parameter b is calculated based on the water depth layer h and vertical displacement η_0 as

$$b = \frac{4h_1^{\ 2}}{3\eta_0} \quad (14.7.1)$$

while a is at wavenumber like parameter satisfying the relationship

$$ab \tan(aH) = 1, \quad (14.7.2)$$

As stated by Pan and Jay [163], the normalized form of the backscatter cross section caused by an internal wave is casted as:

$$\hat{\sigma}_{0IS}(\theta)_{ij} = \frac{1}{M} \frac{\sinh 2a\xi}{(\cosh^2 a\xi + \frac{1}{a^2b^2}\sinh^2 a\xi)^2}, \quad (14.8)$$

where M is the maximum rate of the function $\frac{\sinh 2a\xi}{(\cosh^2 a\xi + \frac{1}{a^2b^2}\sinh^2 a\xi)^2}$.

Equation 14.8 implies that $\xi = x\text{-}ct$, where x is the horizontal distance of the internal wave propagation, c is the phase speed and t is time. ξ influences the radar backscatter which depends on radar wavelength, incidence angle, and relaxation rate. The relaxation rate is quite variable, relying on, among other parameters, for instance, wind speed and direction. In addition, the theories require wind speeds to be above the threshold for Bragg wave generation (approximately 2–3 m s^{-1}) and below approximately 10 m s^{-1}, at which point the wind-generated roughness and the internal wave-generated roughness patterns can no longer be distinguished from each other.

14.3 Algorithms

14.3.1 Wavelet Transform for Internal Wave Detection

Wavelet transform tool is mainly used for analysing time-varying signals. This technique generates spectral decomposition through the scale model. In remote sensing images investigation, the two-dimensional wavelet transform (2-DWT) serves up as an exceedingly effective band-pass filter. In this regard, 2-DWT operates to discrete procedures with altered scales. In fact, the 2-DWT produces a time scale view of the signal which means stretching or compressing of the signal. Under this circumstance, it can deliver precise evidence on the feature description in remote sensing data. These functions provide the wavelet transform a valuable algorithm for extorting physical properties precisely in remote sensing data.

The continuous wavelet transform of a signal $f(t)$ is then defined as

$$CWT_f^{\psi}(s,\tau) = \int_{-\infty}^{\infty} f(t)\psi_{s,\tau}(t)dt \tag{14.9}$$

Two dimensional wavelet function having an oscillation in x direction only, can be written as follows:

$$\Psi_{s,\tau}(t) = \frac{1}{\sqrt{s}}\Psi\left(\frac{t-\tau}{s}\right) \tag{14.10}$$

where $\Psi(t)$ is the transforming function, and is called the mother wavelet. The two new variables, s and τ, are the scale and translation of the daughter wavelet. The term $\sqrt{s}$ normalizes the energy for different scales, whereas the other terms define the width and the translation of the wavelet. The Continuous Wavelet Transform (CWT) is defined as follows. The asterisk denotes a complex conjugate function.

$$\gamma(s, \tau) = \int f(t)\Psi^{*}_{s,\tau}(t)dt \tag{14.11}$$

There are, however, two conditions that the wavelet has to fulfill: (i) the admissibility condition; and (ii) the regularity condition. Thus the admissibility condition is achieved by

$$\int_{-\infty}^{\infty} \frac{\left|\hat{\Psi}(\omega)\right|^2}{\omega} d\omega < \infty \tag{14.12}$$

where ω is angular frequency. Equation 14.12 must achieve the following rules: (i) The transformation is invertible. In this regard, all square integrable functions that satisfy the admissibility condition can be used to analyse and reconstruct any signal. Therefore, (ii) the function must have a value of zero at zero frequency. Therefore, the wavelet itself has an average value of zero. This means that the wavelet describes an oscillatory signal, where the positive and negative values cancel each other. This gives the 'wave' in wavelet.

The regularity condition requires that the 2-DWT has to be locally smooth and concentrated in both the time and frequency domains. Under this circumstance, vanishing moments can be introduced as a new approach in dealing with regularity condition. If the Fourier transform of the wavelet is M times differentiable, the wavelet has M vanishing moments.

$$\int x^n \Psi(x)dx = 0 \quad n \in [0,M] \tag{14.13}$$

where n is role of shift parameter. For data analysis, the basic wavelets frequently used are the Gaussian modulated sine and cosine wave packet (the Morlet wavelet) and the second derivative of a Gaussian (the Mexican-hat wavelet). In this study, the analysing wavelet is defined as the Laplacian of a Gaussian as:

$$\omega(t) = \frac{1}{\sqrt{2\pi}\sigma} e^{\frac{-t^2}{2\sigma^2}} \tag{14.14}$$

The mother wavelet can be given by

$$\psi(t) = \frac{1}{\sqrt{2\pi}\sigma^3} (e^{\frac{-t^2}{2\sigma^2}} . (\frac{t^2}{\sigma^2} - 1)) \tag{14.15}$$

The Morlet wavelet is defined as

$$\omega(t) = e^{iat} . e^{-\frac{t^2}{2\sigma}} \tag{14.16}$$

where a is the modulation parameter and σ is the scaling parameter that affects the width of the window. Each high-pass filter creates an exhaustive version of the imaginative signal and the low-pass a smoothed form. Three bands are involved in Morlet wavelet (i) lowpass band $S_j f$; (ii) horizontal highpass band $S_j^H f$; and (iii) vertical highpass band $S_j^V f$. In this context, the edge detection of internal wave based on both $S_j^H f$; and vertical highpass band $S_j^V f$ is formulated as:

$$S_j f = \sqrt{\left|S_j^H f\right|^2 + \left|S_j^V f\right|^2} \tag{14.17}$$

Then the gradient angle ϑ_j of the internal wave edge domain is determined from:

$$\vartheta_j f = \tan^{-1}\left(\frac{S_j^H}{S_j^V}\right) \tag{14.18}$$

Equation 14.17 also describes 1-D local maxima in the direction of the gradient which is considered as edge pixels of internal wave in SAR data. In this context, the local maxima is a function of internal wave boundary morphology in a SAR image.

The algorithm is considered as a conventional edge detector where edges are sought on an individual pixel level, which is often undesirable when processing images contaminated heavily with speckles such as SAR. Varying scaling factor σ would accentuate features with different spatial scales [170]. The parameter scale in the wavelet analysis is similar to the scale used in maps. As in the case of maps, high scales correspond to a non-detailed global view (of the signal), and low scales correspond to a detailed view. Similarly, in terms of frequency, low frequencies (high scales) correspond to a global information of a signal (that usually spans the entire signal), whereas high frequencies (low scales) correspond to a detailed information of a hidden pattern in the signal (that usually lasts a relatively short time). In practical applications, low scales (high frequencies) do not last for the entire duration of the signal but they usually appear from time to time as short bursts, or spikes. High scales (low frequencies) usually last for the entire duration of the signal [167–169].

14.3.2 Particle Swarm Optimization (PSO)

Succeeding Kennedy and Eberhart (1997) [273] and Marghany (2015) [274], Particle Swarm Optimization (PSO) is a population-based random searching process. It is assumed that there are N "particles", i.e., physical properties of internal wave and its surrounding environments: bright and dark backscatter variations, linearity, depressions, and direction of propagation in ASAR data. These internal wave features invasive contacts randomly seen in a "solution space" [275]. Thus, as the internal wave edges are determined, the optimization problem can be solved for data clustering, there is always a criteria (for example, the squared error function) for every single particle at their position in the solution space [273]. The N particles will keep moving and calculating the criteria in every position, the remaining of which is named as fitness in PSO, given that the criteria reaches satisfied threshold. Therefore, each internal wave feature (particles) maintains its coordinates in the solution space of ASAR which are combined with the finest fitness that has been accomplished by requested physical features, i.e., particles.

Two procedures are involved in implementation of PSO: (i) Initialization; and (ii) velocity update. The first step involves generation of random population, which is a need for the potential solutions. In this view, particles are referred to as random population. Lastly, the generated particles fly from side to side of SAR data as a function of updating their velocity. Under this circumstance, the particles can realize their past flight and those of their acquaintance.

Following Kennedy and Eberhart (1997) [273], each agent moves the particle with a direction and velocity $v_{m,n,l}$,

$$p_{m,n,l} = p_{m,n,l} + v_{m,n,l}, \tag{14.19}$$

where $p_{m,n,l}$ represents particle and $v_{m,n,l}$ is the velocity of the 3-D particle in the i, j, k agents, respectively.

$$v_{m,n,l} = v_{m,n,l} + c_1 r_1 (lbest_{m,n,l} - p_{m,n,l}) + c_2 r_2 (gbest_{m,n,l} - p_{m,n,l}) \tag{14.20}$$

where $lbest_{m,n,l}$ is the local best particle, $gbest_{m,n,l}$ is the global best particle, r_1 and r_2 are random variables and c_1 and c_2 are the swarm system variables. After each iteration, the global best g_{best} particle and the agent local best l_{best} particle are evaluated based on the maximum fitness functions of all particles in the solution space. Then equation 14.20 can be expressed as follows [277]:

$$\begin{aligned} v_{m,n,l} &= w \cdot v_{m,n,l}(t-1) + c_1 \cdot r_1 (p_{m,n,l}(t-1) \\ &\quad -\psi_{m,n,l}(t-1)) + c_2 \cdot r_2 (p_{m,n,l}(t-1) \\ &\quad -\psi_{m,n,l}(t-1)) \end{aligned} \tag{14.21}$$

$$\psi_{m,n,l} = \psi_{m,n,l}(t-1) + v_{m,n,l}(t) \tag{14.22}$$

where $\psi_{m,n,l}$ is the position of the particle for wavelet transform, and $v_{m,n,l}$ is the current velocity of the particles in $m \times n \times l$. The velocity is regulated by a set of rules that influence the dynamics of the swarm. Further, there are several parameters must be considered such as initial population, representation of position and velocity strategies, fitness function identification and the limitation (Dorigo et al. 2008). These parameters are for PSO performances. Following Ibrahim et al. (2010) [278], the initial swarm particles proposed PSO is initialized to contain 3000 points of particles for $\psi_{m,n,l}$ and velocity $v_{m,n,l}$. The points had been randomly selected in the azimuth and range directions wavelet transform of ASAR data.

14.4 Tsunami Derived Internal Wave in SAR Data

14.4.1 SAR Data Acquisition

In this study, Advanced Synthetic Aperture Radar (ASAR) is a ground range projected detected image in zero-Doppler SAR coordinates, with a 12.5 m pixel spacing and nominal resolution at a range and an azimuth of 30 m × 30 m. It has four overlapping looks in Doppler covering a total bandwidth of 1000 Hz, with each look covering a 300 Hz bandwidth. Sidelobes reduction is applied to achieve a nominal PSLR of less than –21 dB. ASAR data are radiometrically calibrated, and are numerically scaled such that a Beta0 value of 0 dB corresponds to a product digital number (DN) value of 682.3. ENVISAT's mission was ended on 9 May 2012.

In this study, two images have been selected to investigate the impact of tsunami on the sea surface pattern changes. Prior to the tsunami, ASAR data were acquired on 29th April, while the post-tsunami ASAR data was acquired on 31st December 2004 (Figure 14.4). Therefore, the two images were acquired in different polarization: the one prior to the tsunami was in VV polarization while the one acquired post the tsunami was in HH polarization (Table 14.1).

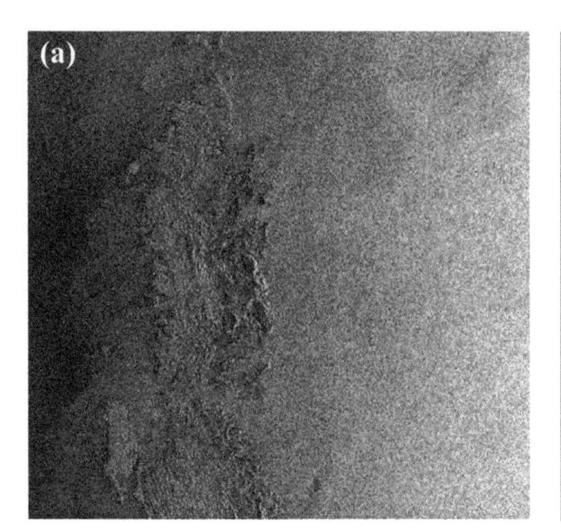

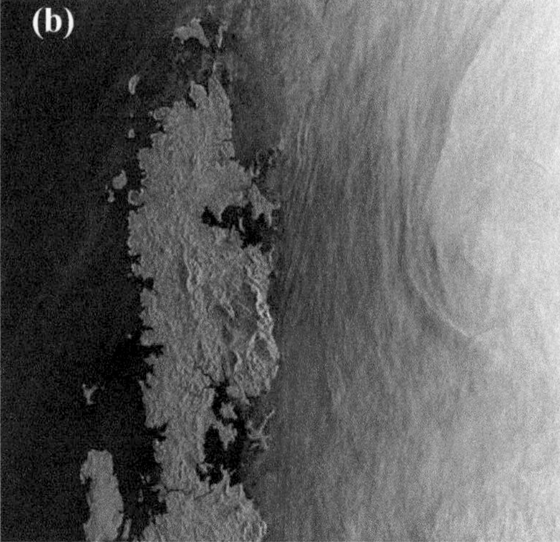

Figure 14.4 ENVISAT data were acquired (a) pre-tsunami and (b) post-tsunami.

Table 14.1 ENVISAT data characteristics.

ENVISAT	Pre-Tsunami	Post Tsunami
Date	29/4/2004	30/12/2004
Polarization	VV	HH
Product	ASAR-IMP	ASAR-IMP
Incident Angle (°)	15–45	15–45
Band	C	C
Swath (km)	58–110	58–110
Resolution (m)	30–150	30–150

14.4.2 Andaman and Nicobar Islands

Figure 14.5 shows the bathymetry variations of the Andaman and Nicobar Islands. The water depth along Andaman and Nicobar Islands is ranged between 200 m to 3000 m. The Andaman and Nicobar Islands on the western side of the Andaman Sea are volcanic in origin. The sills between the islands, as well as a number of underwater volcanic seamounts, are all potential sources of internal waves. The result is an area rich in internal wave excitations and complex soliton-soliton interaction.

It is well known that an exceptionally large selection-like internal waves have contended with the Andaman Sea of the Indian Ocean. The internal waves propagate beneath the Andaman Sea surface in the water layers whose density fluctuate as a function of the water depth. Further, there is a large packet of internal wave groups that have been found in the western verge of the Strait of Malacca between Phuket, Thailand, and the northern coast of Sumatra.

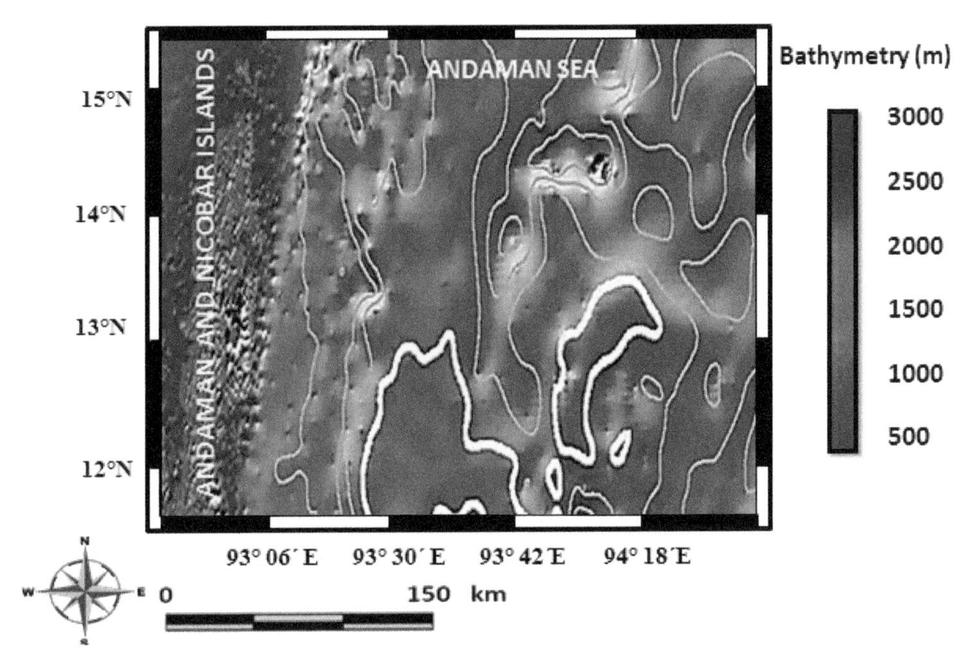

Figure 14.5 Bathymetry along Andaman and Nicobar Islands.

14.4.3 Backscatter Distribution in ASAR Data

Figure 14.6 shows the ENVISAT SAR radar image (Orbit 148) that was acquired on 31 December 2004, five days after the tsunami hit Asia. It shows the Indian Andaman Islands and the Ritches Archipelago. The normalized radar cross section is ranged between –24 to –4 dB. The lowest normalized radar cross section of –28 dB is described as the low window zone shelter along the Andaman and Nicobar Islands. However, the highest backscatter of –4 dB describes the occurrence of whirlpool in the east of the Andaman Sea. This whirlpool is located between latitude of 14° N to 15° N and longitude of 94° E and 96° E. The whirlpool has a radius of 1.9 km and located above water depth gradient of 1000 m. It was rotated in an anticlockwise direction (Figure 14.7).

It is interesting to observe internal wave occurrence post-tsunami rather than pre-tsunami. In this regard, Figure 14.8 indicates the prevalence of the internal wave along the 72 km of coastal water of two Indian Andaman Islands and the Ritches Archipelago. The internal wave has an approximate maximum wavelength of 1000 m and is shifting westward to Andaman and Nicobar Islands. It moves into an irregular parallel group and phase speed vectors, analogous to propagation inside a waveguide due to the irregular bathymetry pattern gradient of about 1000 m. These waves are internal waves, and they run through the lowest layers of ocean water, in no way swelling the surface.

It is clearly noticed that these internal waves have higher backscatter of –4 dB than the surrounding sea environment. The internal wave crests dominate by higher backscatter while the troughs dominate by lower backscatter of –10 dB. Therefore, the internal wave edges appear as discontinuities in the backscatter. Under this circumstance, it is barely to detect the internal wave edges have discontinuity intensities in SAR data because of the multiplicative speckle noise. In fact, the signal backscattered from the sea surface might be comparable to one from the land.

In addition, the internal waves modulate the overlying surface waves and the sea surface manifestations of internal waves are readily observed in synthetic aperture radar (SAR) images. They appear as alternating bright and dark bands corresponding to regions of increased surface roughness with maximum normalized radar cross section of –4 dB alternating with smooth regions with the lowest one of –10 dB (Figure 14.8).

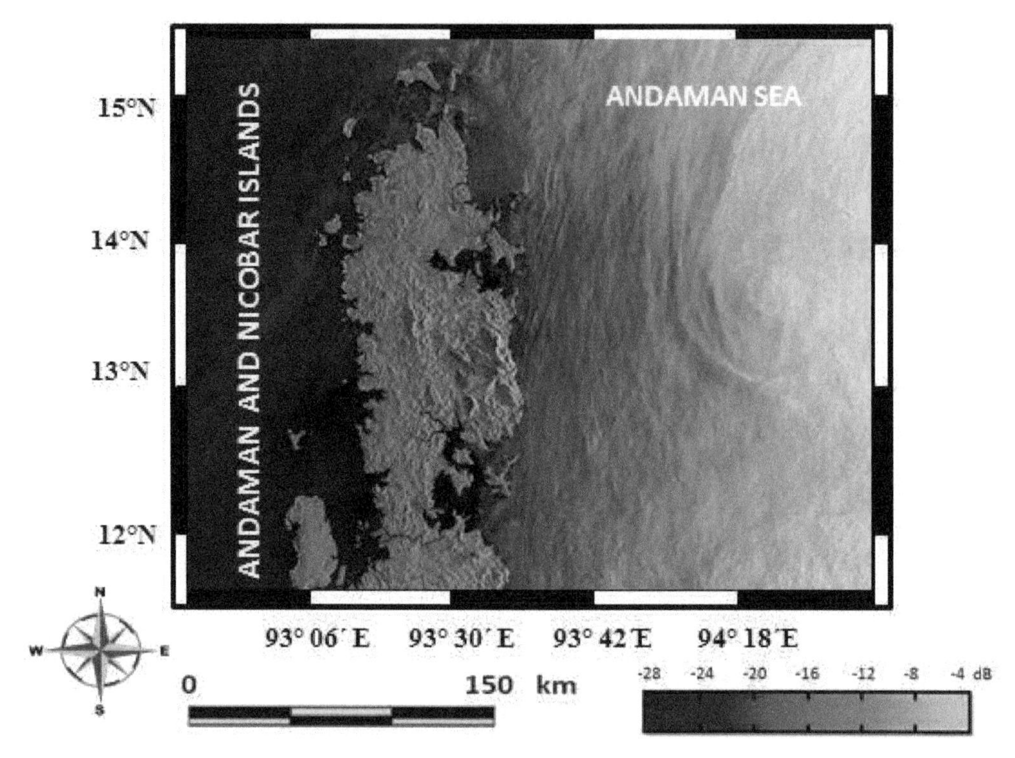

Figure 14.6 ENVISAT of Andaman and Nicobar Islands.

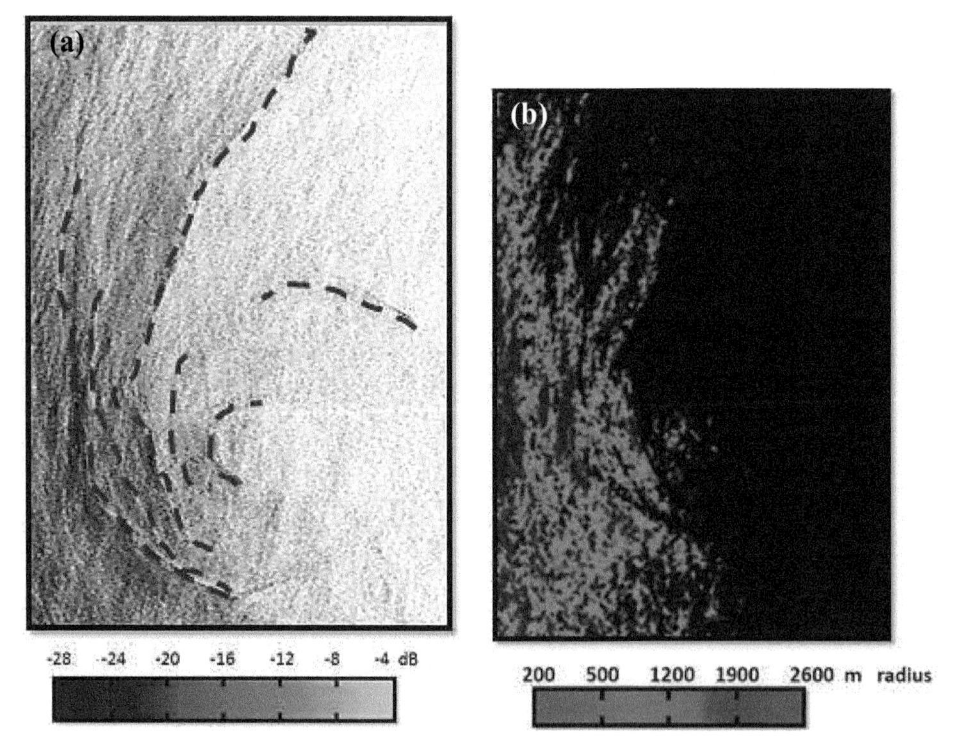

Figure 14.7 Whirlpool characteristics (a) Backscatter and (b) radius.

Figure 14.8 Backscatter of internal wave.

It is worth mentioning that the Bragg scattering is the main concept of the scattering mechanism for ocean surface. In this view, the backscatter variations are mainly a function of incidence angle and wavelength as well as the sea surface conditions base of wind speed. For ASAR, wavelength is 5.3 cm which indicates that the Bragg scattering is approximately equivalent to its wavelength. Further, current shear, wind speed, long gravity waves associated with whirlpool and internal waves created modulation in the short waves. Under these circumstances, the ASAR data show high variability of backscatter as a function of surface roughness. The large incidence angle of ASAR that ranges from 15° to 45° allows for extreme discrimination between water/land boundary as a function of a large radar backscatter contrast (Figure 14.8).

14.5 Automatic Detection of Internal Waves

14.5.1 Wavelet Transformation for Automatic Detection of Internal Wave

The automatic detection of internal wave has been performed by using a 2-DWT algorithm. However, the 2-DWT is not able to distinguish between internal wave and whirlpool features. It may be the case that both features have a similar shape, especially along the edges. In addition, 2-DWT also detected the edges of look-alikes, and current shear boundary. In fact, look-alikes and current shear have occasionally the quasilinear periodic lines which are similar to signature of internal waves in SAR data (Figure 14.9).

This may explain the difficulties that are raised up for automatic detection of internal waves in SAR data. In fact, there are many features associated with internal wave look-alikes: upwelling, eddies, oil slicks, current shear zones, wind sheltering by land, natural film, and threshold wind speed zones. These are considered as secondary information in SAR data in addition to impact on the sea surface roughness which do not allow 2-DWT to accurately detect internal wave in ASAR data (Figure 14.9).

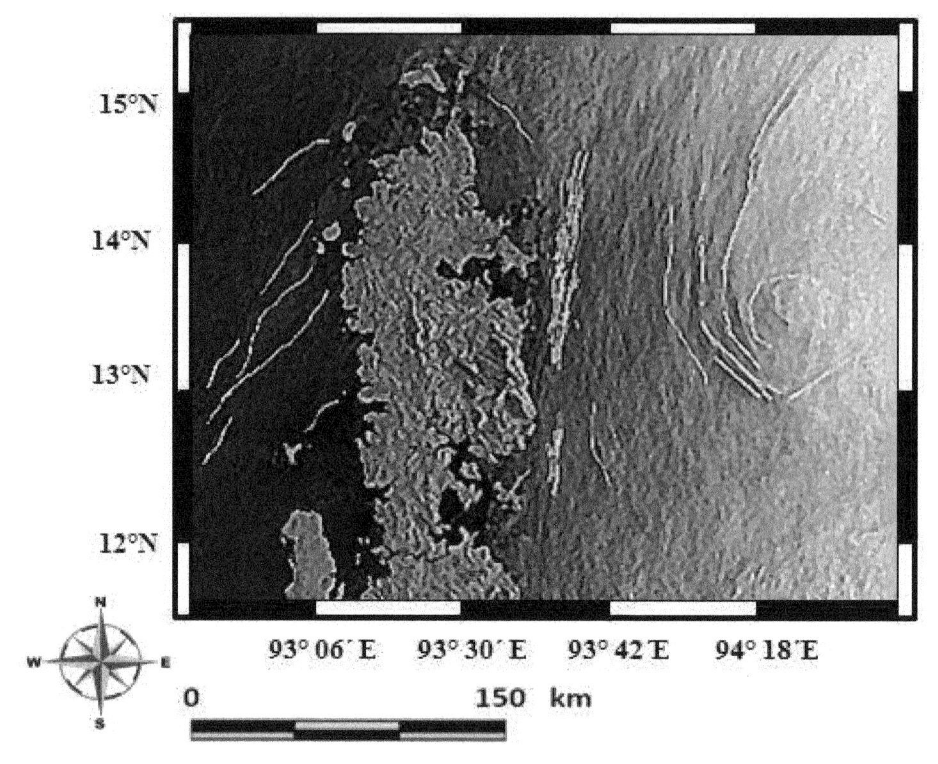

Figure 14.9 2-DWT results of internal wave detection in ASAR data.

14.5.2 PSO for Automatic Detection of Internal Wave

The performance of PSO algorithm has been achieved through different numbers of particles, fitness, and clamping velocity. In this investigation, the *gbest*-to *lbest* PSO is implemented for ASAR data. In this algorithm, *lbest* begins with a zero-radius neighbourhood, then *gbest* is achieved as neighbourhood radius is linearly increased to avoid being trapped in local optima. In this view, the algorithm converges into the best solution by using *gbest* approach. Figure 14.10 shows different swarm sizes ranging between 200 to 800 with the number of iterations of 60. It is worth noting that the internal wave morphology structures are well organized in 50 to 60 iterations with a swarm size of 800. In fact, as long as the number of particle swarm is increasing, the internal wave morphology structures are well detected. This is proved by the best fitness function of 95 and RMSE of ±5.

Figure 14.11 shows the best performance for automatic detection of internal wave with RMSE of ±0.65. This occurred within clamping of velocity updates and swarm particle sizes of 1000. In this iteration of 80 and fitness function of 93, the internal waves are distinguished precisely from the surrounding sea features and land too.

Generally, an increase in the number of swarm particles enlarges diversity, constraining the weights of preliminary circumstances and decreasing the probability of being blocked in local minima. Furthermore, accurate detection of internal wave is a function of less compressing of swarm's velocity. Indeed, this allows the swarm to be extra accelerated in the exploratory domain.

In fact, the PSO circumvents a decreasing resolution by making a weighted combination of running average with the neighbours surrounding pixels. This reduces the noise in the features' edge areas without losing edge sharpness. The implementation of PSO assists to determine optimal growth regions across the continuing and discontinuing internal wave edges. In fact, in PSO system, particles are impartial of

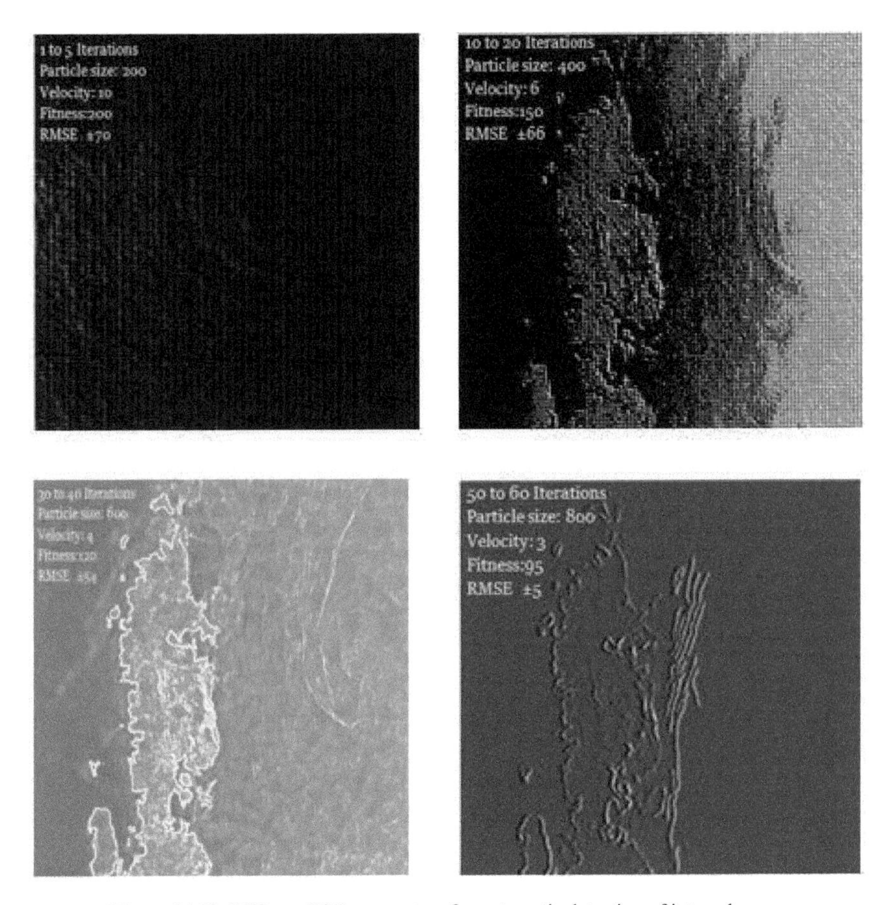

Figure 14.10 Different PSO parameters for automatic detection of internal wave.

Figure 14.11 Best PSO parameters for accurate detection of internal wave.

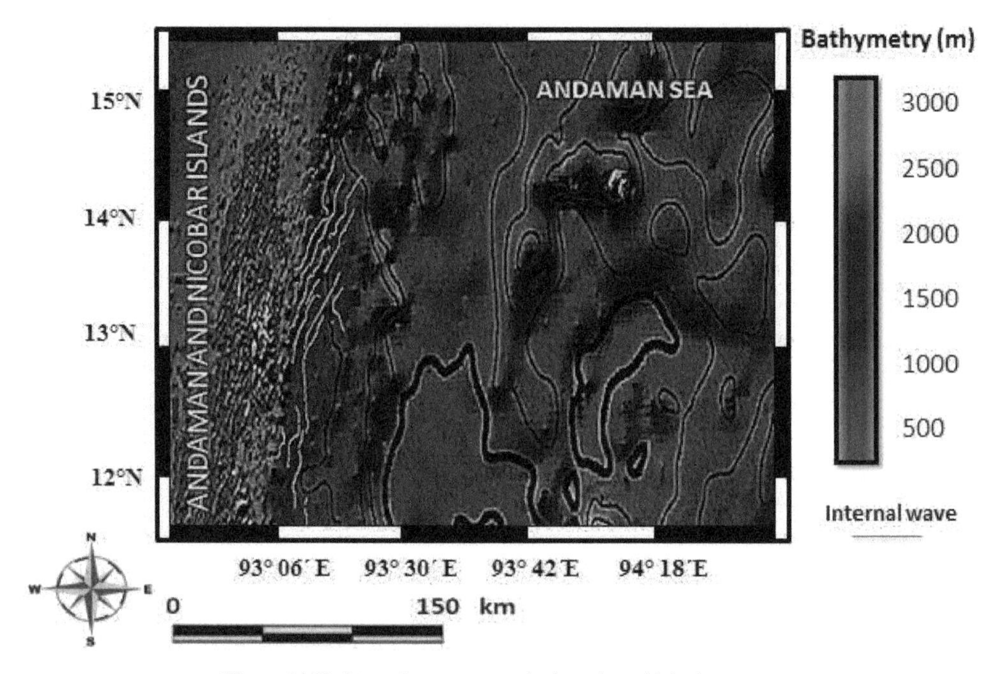

Figure 14.12 Internal wave automatic detection with bathymetry.

every other swarms and their movements are ruled with the aid of a set of rules. With this concern, PSO synchronized sequences sides of the internal wave edge. The PSO algorithm commences by creating random locations for the particles within an initialization pixel of the internal waves in ASAR image. Particles in the PSO algorithm can also be modified to zero or two minor random values to avoid particles from withdrawal of the search space of internal wave pixels during the first iterations. Throughout the core loop of the algorithm, the velocities and the locations of the particles are iteratively rationalized until an ending condition is encountered. In addition, PSO algorithm constructed the discontinuity in quality order. This is appropriate in the high intensity line or curve of fixed length, and locally low curvature boundary is known to exist between edge elements and high noise levels in ASAR data.

The automatic detection of 10 packets of internal wave has been performed by using a PSO algorithm. It is interesting that the PSO algorithm provides automatic detection of the full length and packets of internal waves (Figure 14.12). Moreover, these internal waves have approximately 10 packets. These internal waves occurred with ocean bathymetry gradient of 1000 m. In fact, the PSO analysis picks out a number of curves that are the most prominent boundaries in the image of this scale [170].

14.6 Internal Wave Variations with Physical Water Properties

Figure 14.13 shows the spatial propagation of internal wave with high salinity of 38 psu (Figure 14.13a), cold water of 25.8°C (Figure 14.13b) and high concentrations of Chl-a of 5.0 mg/m^3 (Figure 14.13c). Internal waves have occurred because the Andamen Sea is layered. Deep water is cold, dense, and salty, while shallower water is warmer, lighter, and fresher. The differences in density and salinity cause the various layers of the ocean to behave like different fluids. When the tsunami drag the sea over a shallow barrier such as a ridge in the sea floor, it creates waves in the lower, denser layer of water. These waves are internal waves, which can be tens of kilometers long and can last several hours [163].

Moreover, thermoclines are regularly related to chlorophyll maximum layers (Figure 14.13c). Internal waves, consequently, constitute oscillations of those thermoclines and subsequently have the capability

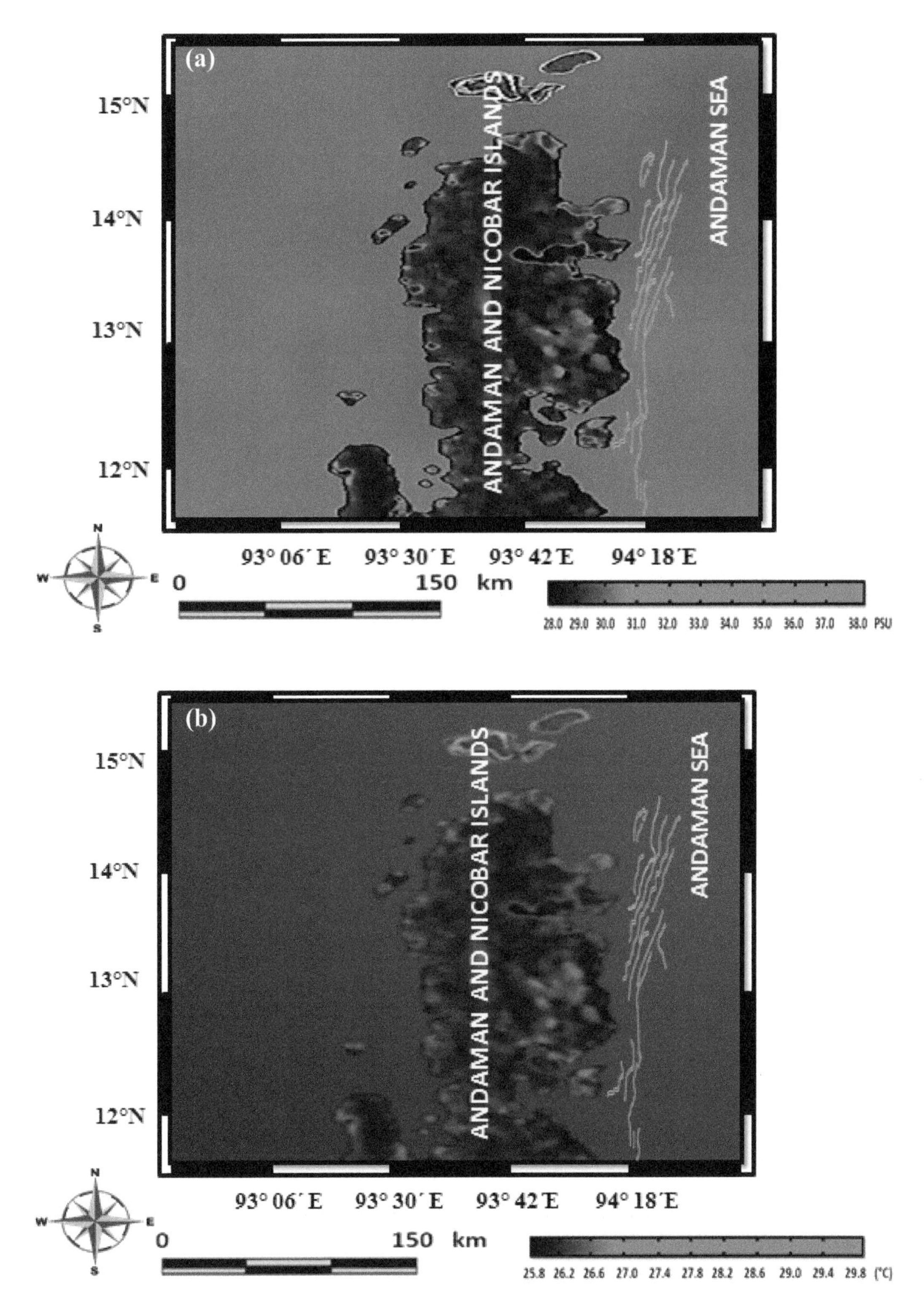

Figure 14.13 contd. ...

...Figure 14.13 contd.

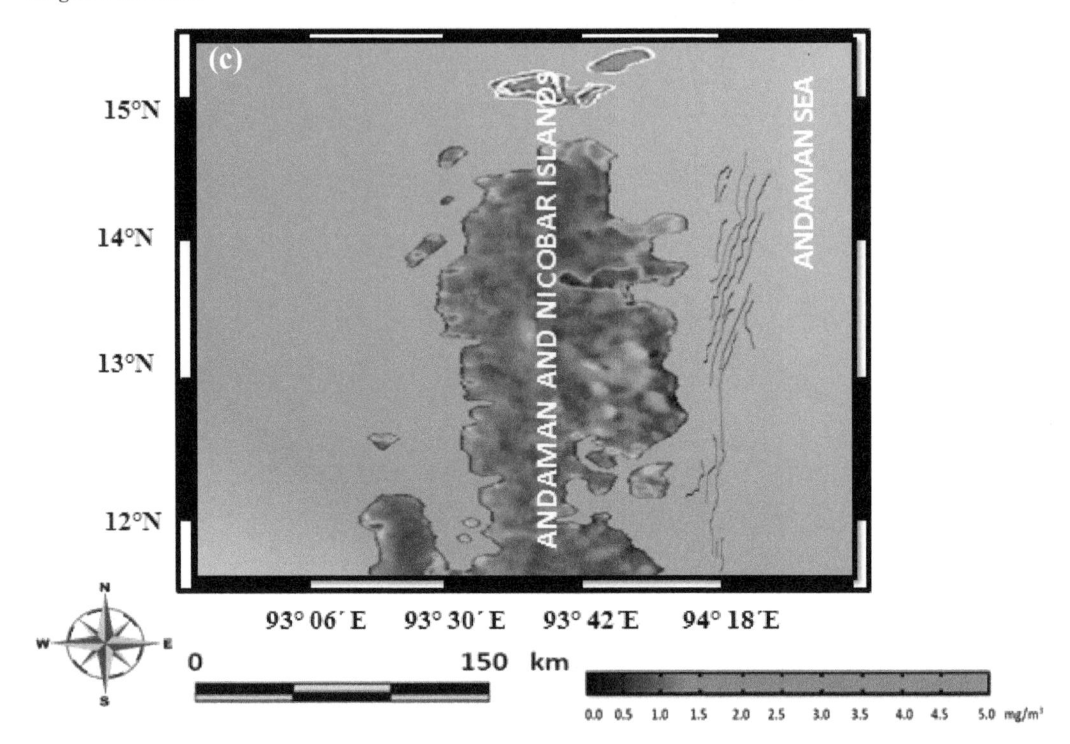

Figure 14.13 Internal wave with spatial variations of (a) salinity; (b) temperature; and (c) Chl-a on December 28, 2004.

to switch these phytoplankton rich waters downward, coupling benthic and pelagic systems. Coastal zones exaggerated by those occasions reveal the better increase rates of suspension feeding ascidians and bryozoans, probably because of the periodic inflow of excessive phytoplankton concentrations. Periodic melancholy of the thermocline and related downwelling may additionally play an essential function within the vertical delivery of planktonic larvae.

The ENVISAT also showed the occurrence of slick (Figure 14.14) which was associated with internal wave event. This slick is because of floating of phytoplankton and chlorophyll due to upwelling occurrence. This is proof of upwelling evidence which has been discussed previously.

It can be said that strong earthquakes have an adequate reserve of energy for essential transformation of the ocean stratification structure. Tenths of a percent of the energy of an earthquake are sufficient for formation on the ocean surface of a temperature anomaly with a characteristic horizontal dimension, measured by hundreds of kilometers and with a temperature deviation of the order of 1°C. Note, that a comparable amount of energy (less than 1% of the earthquake energy) is spent on the formation of tsunami waves. The formation of a temperature anomaly of the ocean surface is most probable in the case of a shallow thermocline and for seismic events, characterized by a persistent process at the source or by a large number of aftershocks. The most striking manifestation of the effect is to be expected in the case of realization of the turbulence generation mechanism with a scale exceeding 10 m [159, 160].

Local variations of the vertical temperature distribution should serve as a source of internal waves even in those cases when temperature variations are insignificant. Internal waves are ocean waves that propagate underwater along the thermocline. They can have an extremely large amplitude (~ 100 meters) and can readily be seen in optical and radar images from space.

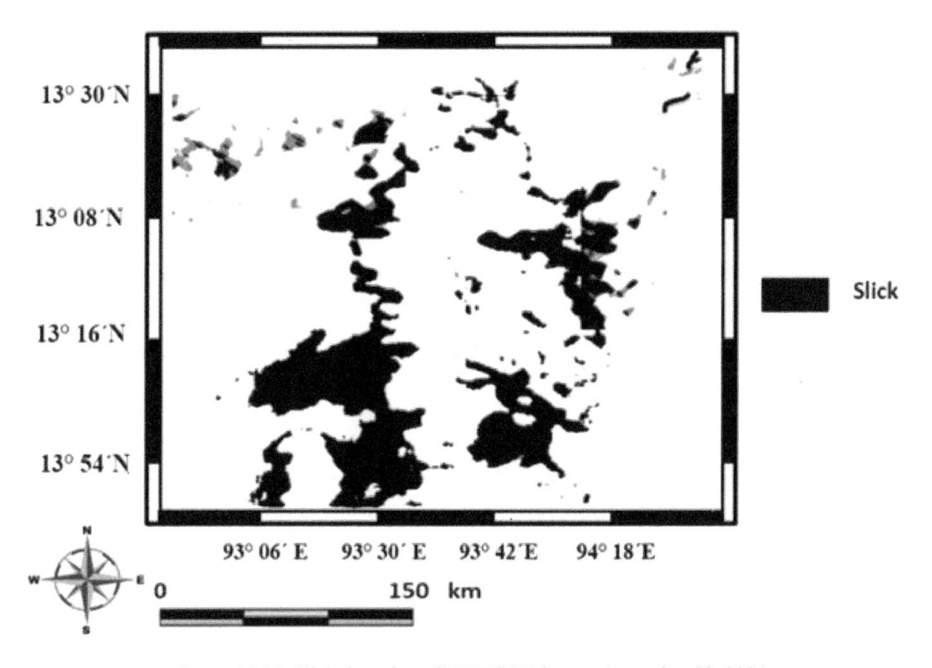

Figure 14.14 Slick detection of ENVISAT data on December 28, 2004.

14.7 Tsunami Deriving Internal Wave from Optical Satellite Data

Figure 14.15 shows the MODIS satellite data which were acquired on 26 December 2004, at 05:15 GMT. MODIS data has captured the tsunami impact on the east coastal water of Sri Lanka with 2 h and 30 minutes post the tsunami and caused high concentration of sediments along the east coast which can be seen in the bright color of Figure 14.15.

Figure 14.16 shows the internal waves have been extracted from combination of band 1 (620–670 nm), band 4 (545–565 nm), and band 3 (459–479 nm). These bands correspond to red, green, and blue, respectively. The MODIS instrument is capable of detecting significant disturbances in the sea by changes in the reflectivity of the water column [139]. In addition, the vector layer of internal wave indicates two sorts: (i) linear wave features and (ii) curvature waves. The linear waves are located in the south over a water depth of 3000 m and has a wavelength of 50 km. These linear wave patterns could be generated by reflection. The curvature waves are located in north with maximum wavelength of 100 km. The curvature waves propagated eastward. It might be generated by the existence of canyons and sea mounts [139].

Figure 14.17 exposes the Multi-angle Imagine SpectroRadiometer (MISR) satellite data which were acquired on December 26, 2004 at 05:15. MISR captured approximately of 30–40 km from the southwestern of Sri Lanka coastal waters. It is interesting to find that MISR data is able to detect the ripple-like wave pattern. Figure 14.18 proves that these internal waves are generated because of the undersea boundary of the continental shelf. The wavelength of these internal waves is about 1000 m. In north, these internal waves propagated parallel to coastal waters in a linear pattern while in the south, the internal wave with two packets tended to curve away from the coastal water. These internal waves could be generated by reflection from the continental land mass.

Internal solitary waves are generated by the nonlinear deformation of long waves like internal tides or tsunamis. When the wave approached the continental slope of Sri Lanka, parts of it were transmitted, reflected, and scattered and possibly generated internal waves. Internal solitary waves occur, particularly,

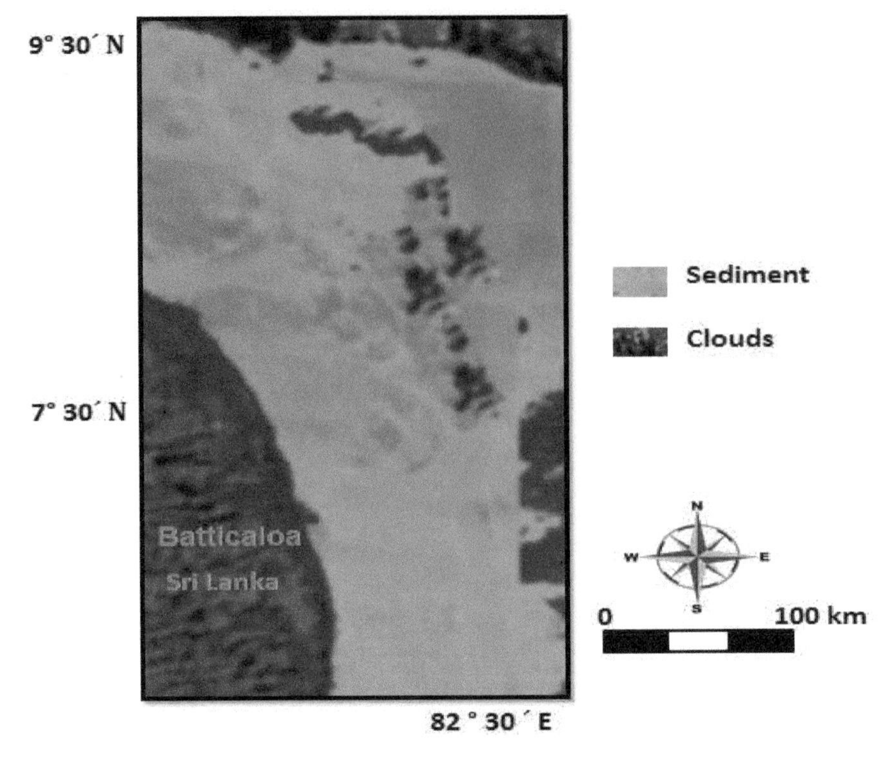

Figure 14.15 MODIS data along Batticaloa, Sri Lanka on December 26, 2004.

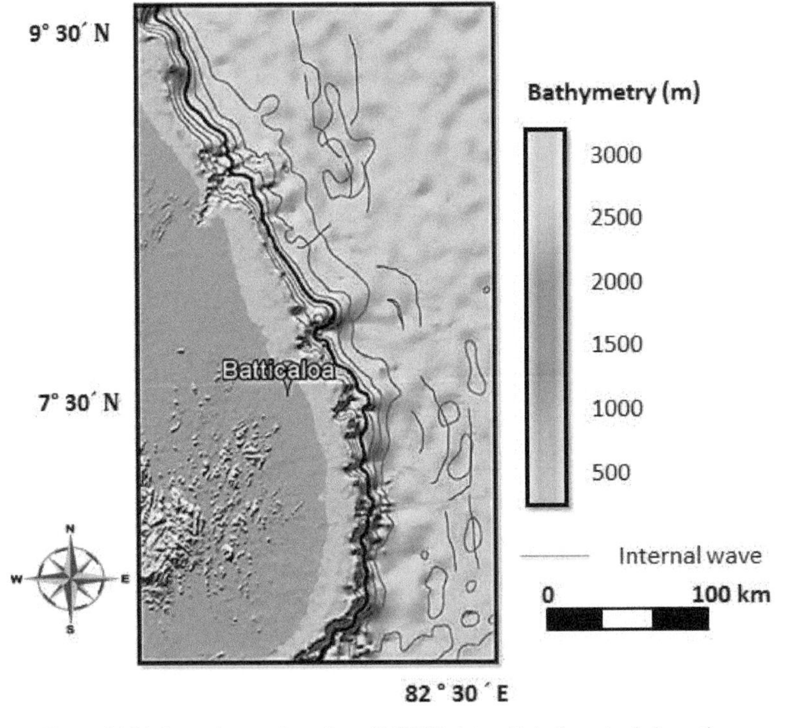

Figure 14.16 Internal wave detection of MODIS data with bathymetry information.

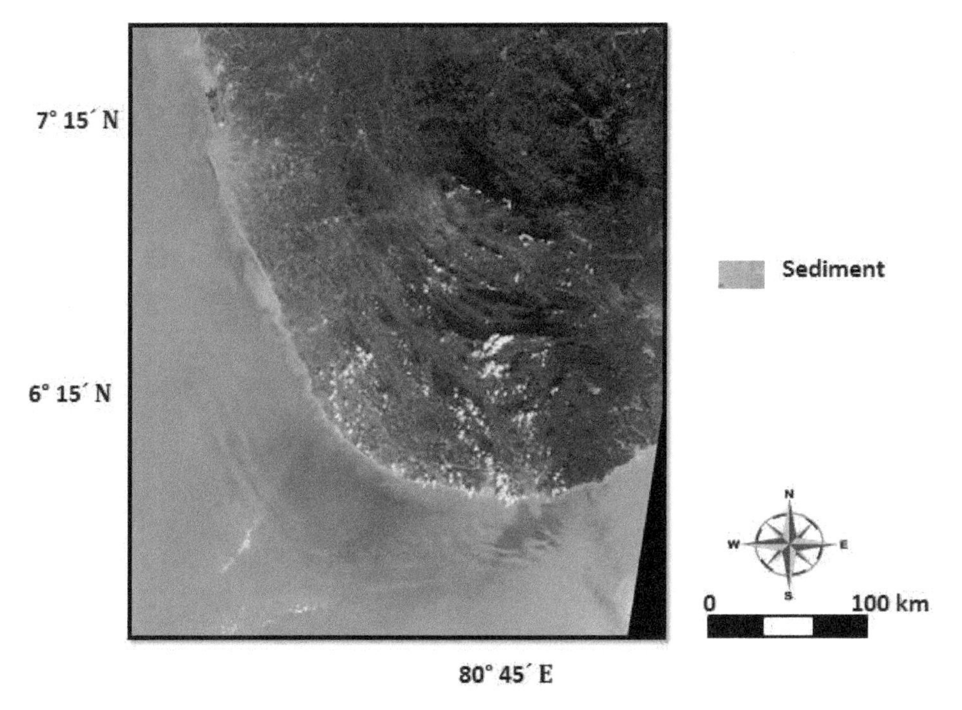

Figure 14.17 Multi-angle Imagine SpectroRadiometer (MISR) satellite data which is acquired on December 26, 2004.

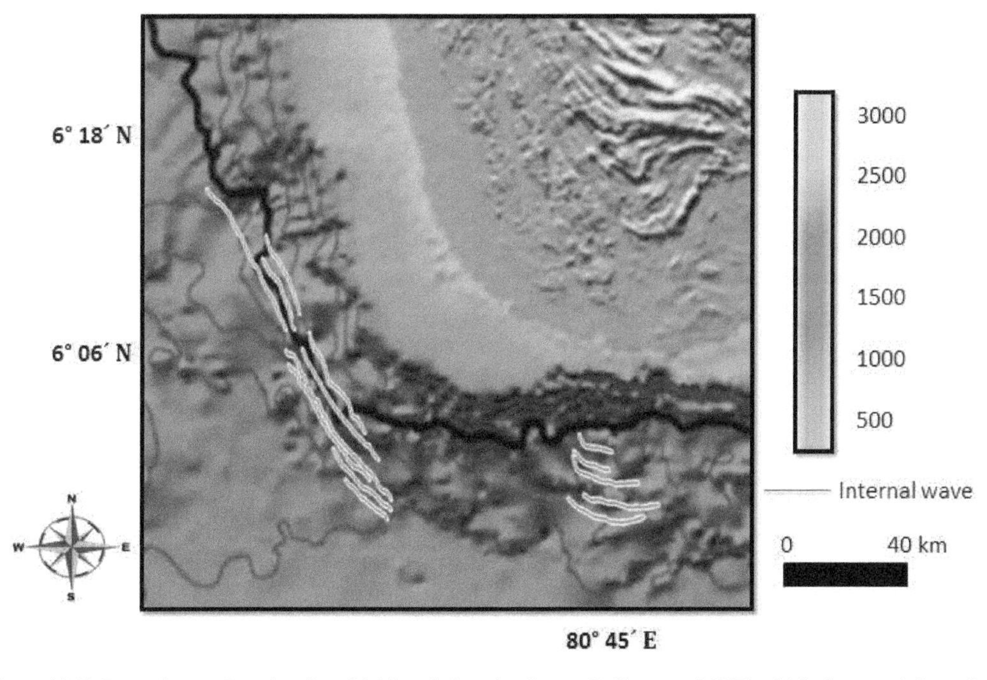

Figure 14.18 Internal wave detection from Multi-angle Imagine SpectroRadiometer (MISR) with bathymetry information.

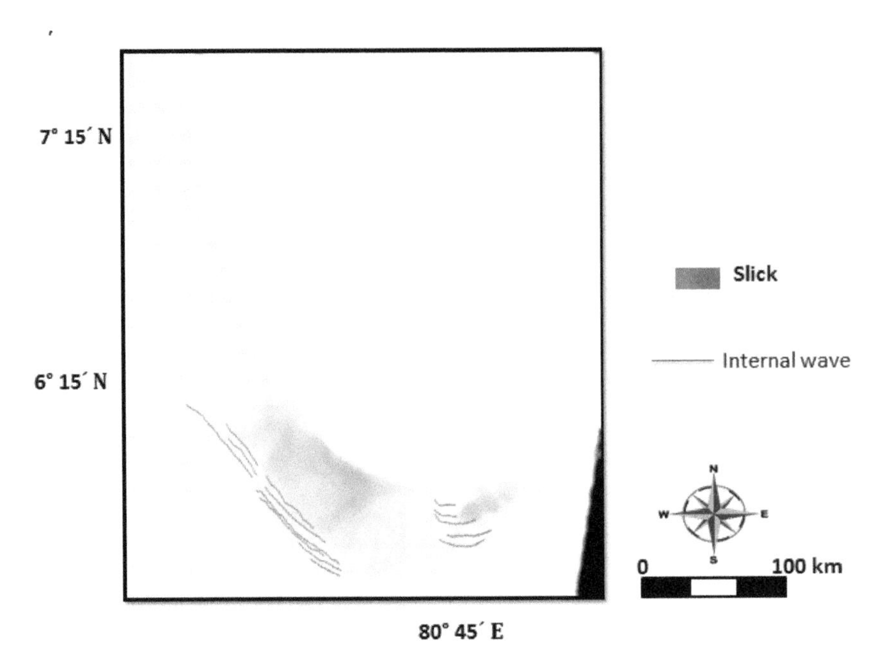

Figure 14.19 Slick associates with internal waves from Multi-angle Imagine SpectroRadiometer (MISR).

near regions of variable bathymetry, such as shelf edges, seamounts, sills, and submarine canyons, where the bathymetry forces the pycnocline to oscillate with the frequency of the tidal wave or tsunami. Such conditions apply to the coast of Sri Lanka [139, 164].

Internal waves happen because the ocean is layered. Deep water is cold, dense, and salty, while shallower water is warmer, lighter, and fresher. The differences in density and salinity cause the various layers of the ocean to behave like different fluids. When tides drag the ocean over a shallow barrier such as a ridge on the ocean floor, it creates waves in the lower, denser layer of water. These waves, internal waves, can be tens of kilometers long and can last several hours [166].

Ultimately, Figure 14.19 suggests that the intense packets of internal waves are related to natural slicks. While internal waves of higher magnitudes will frequently smash after crossing over the shelf break, smaller trains will continue throughout the shelf unbroken. At low wind speeds, those internal waves are evidenced by means of the formation of extensive surface slicks, orientated parallel to the bottom topography (Figure 14.19), which develop shoreward with the internal waves. Waters above an internal wave converge and sink in its trough and upwell and diverge over its crest. The convergence zones related to internal wave troughs frequently acquire oils and flotsam that occasionally develop shoreward with the slicks (Figure 14.19) [165].

Finally, the PSO algorithm was implemented in optical and ASAR images for automatic detection of internal waves. In this context, PSO shows a great promise for internal wave detections in different remote sensing sensors. It can be said that the internal wave and whirlpool features are associated with the 2004 tsunami. This chapter has proved the occurrence of upwelling which has been discussed in Chapter 9 as associated with the appearance of natural slick along the coastal waters of Andaman and Nicobar Islands and Sri Lanka.

Chapter 15

Altimeter Satellite Data Observed Tsunami Spreading

Altimeter satellite data were the first devices that observed and monitored the tsunami wave propagation. Two hours after the 2004 earthquake, altimeter data started to record the echo signal from the tsunami wave spreading. The sensors first recorded the wave travelling from Aceh coastal water towards the Bay of Bengal and Sri Lanka. The altimeter sensors continued to observe the wave spreading and dissipation widely across the Indian Ocean.

This chapter critically evaluates the existing altimeter sensors which monitored the tsunami wave dissipation to bridge the gap found between various remote sensing data recorded as a part of tsunami scenario. In this understanding, the wide range of tsunami wave dynamic spreading is required to forecast and predict any tsunami which can occur in the future in the Indian Ocean. It is concluded that different comprehensive approaches are necessary to develop a forecasting tool for assessing and monitoring tsunami dynamic dissipation.

15.1 Microwave Altimeter

An altimeter is referred to as an altitude meter. In this view, it is a device to compute the object's height above a stable point. In this regard, the estimation of altitude is known as an altimeter. The bathymetry, consequently, is associated with altimeter which is the computation of depth beneath the sea surface. The dimension is usually measured from the altimeter platform, i.e., satellite or aircraft and the Earth's surface.

15.2 Principles of Altimeter

Like a synthetic aperture radar (SAR), altimeter emits a radar signal and then records the backscattered signal from the objects. Unlike SAR, the altimeter emits and receives radar waveform perpendicular to object. This mechanism allows to estimate the object height from the inverted backscatter signal. In this context, the ocean wave height can be inverted more easily than SAR as a function of perpendicular backscatter signal's amplitude [102]. The main signal bands used with altimeter are E band, k_a band, and S-band. Advanced sea-level retrieving parameters are easily made by S band. In this regard, the reliable and precise of ocean wave height is delivered by altimeter than SAR sensors.

15.2.1 Sort of Radar Altimeter

There are two main components of radar altimeters: (i) frequency modulated continuous wave (FMCW) and (ii) pulse altimeters which are a function of used radar signals. Two sorts of FMCW altimeters are mainly implemented (i) broad-beamwidth types and narrow-beamwidth. Both FMCW altimeters are a function of antenna beamwidth. In contrast, the pulse altimeters are well known as short-pulse altimeters or pulse-compressions which are function of intrapulse modulation. Beside, an altimeter is also operating in optical bands, for instance, laser altimeters.

15.2.2 The Geoid

The geoid is termed as the shape of the ocean surface which it would have if it were enveloped with ocean surface at comparative relaxation to the Earth rotating. In this view, geoid is the reference used to ensure the precise height measurements by altimeters (Figure 15.1). Therefore, mass concentration pulls the geoid away from a perfect sphere shape which is mainly affected by the Earth's rotation. In this context, the geoid is also considered as the total of the Earth rotation impact and gravity [120].

Geoid heights, moreover, are obtained relative to a reference ellipsoid. Indeed, the reference ellipsoid is ultimately a suitability. In other words, using a reference ellipsoid is reducing implantation of massive numbers, which allows for precise computations.

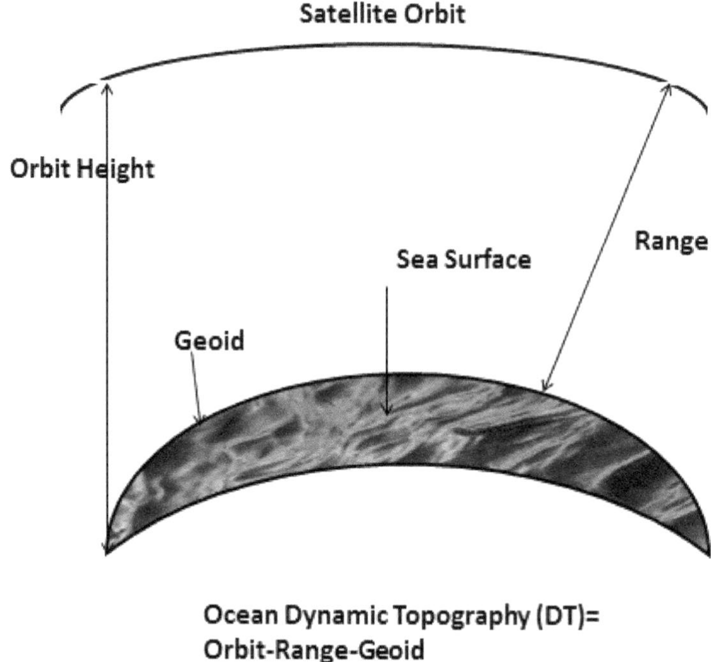

Figure 15.1 Principle of altimeter measurements.

15.2.3 Reference Ellipsoid

It is the best fitting ellipsoid to the geoid. An ellipsoid is essentially a sphere with a bulge at the equator. To first order, this accounts for over 90% of the geoid (Figure 15.2). Consequently, sea surface peak measurements from the center of the Earth are in the order of approximately 6000 km. With the aid of putting off a reference surface, the heights corresponding to the ellipsoid are about 100 m.

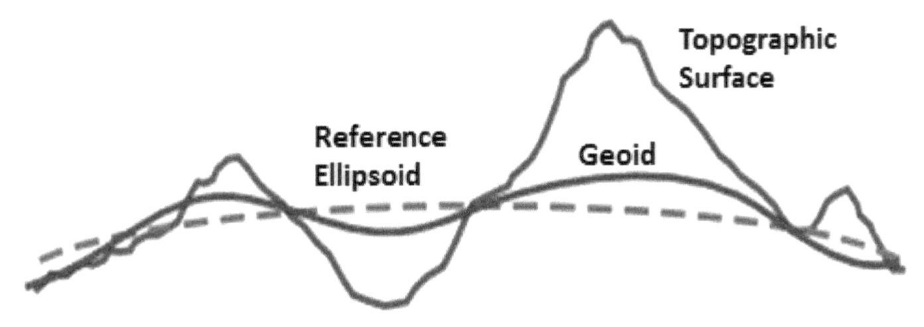

Figure 15.2 Surface reference ellipsoid.

In reality, any reference surface can be used. A sphere could be paintings; nevertheless, sea surface peak differences can be as huge as 20 km. In this regard, an ellipsoid can create less accurate information due to the fact that the geoid isn't widely recognized regionally. In this context, altimeters are commonly flown in orbits which have a precisely repeating ground track, each 9.9156 days. By subtracting sea-surface height from one traverse of the ground track from height measured on a later traverse, changes in topography can be observed without knowing the geoid. Modifications in topography, therefore, can be revealed without identifying the geoid.

Generally, the geoid is steady in time, and the deduction eliminates the geoid, revealing modifications because of exchanging currents, along with mesoscale inconsistency, supposing tides being eliminated from the facts. Mesoscale changeability contains eddies with diameters more or less than 20 and 500 km. The tremendous accuracy and precision of Topex/Poseidon's altimetric device approve the measurements of the oceanic topography over ocean basins with an accuracy of ±5 cm.

15.2.4 Range and Azimuth Resolutions

Like SAR, range *R* measurement is included in altimeter devices. In time t, altimeter devices transmit more than 1700 pulses per second as signals to the Earth's surface and then receive the backscatter signal in waveform. This mechanism is a function of a range *R*. Altimeter range resolution is mathematically estimated by:

$$R_r = \tau\ c/2\cos(\gamma) \tag{15.1}$$

where τ is pulse length (τ), γ and *c* is the speed of light which converts pulse length from units of time to distance. Far field targets are seen at high resolution than near-field targets! The travel time is longer, so a shorter distance can be determined at a specified pulse length. When the travel time is short, either the targets must be spaced out or the pulse must be short. In this regard, Range resolution grows with decreasing the pulse length.

On the contrary, the nearer range resolution is higher than the far range resolution in the azimuth direction. To be settled, targets must be more apart than the beam width. The function of the wavelength of pulse (λ), slant range distance (S), and antenna length (D) is given by:

$$R_a = (0.7)(S)(\lambda)/D \tag{15.2}$$

where S = Altitude/cos(90-γ) and γ is depression angle.

Nevertheless, as electromagnetic waves propagate through the atmosphere, they may be decelerated by using water vapor or ionisation. Once those phenomena had been corrected, the final range can be expected with high accuracy. The ultimate aim is to compute the surface height. This requires independent measurements of the satellite's orbital trajectory, i.e., genuine latitude, longitude and latitude coordinates.

15.2.5 Satellite Altitude

The important orbital parameters for satellite altimeter missions are altitude, inclination and duration period of the transmitted signal. The altitude of a satellite altimeter is the satellite's distance with respect to an arbitrary reference (e.g., the reference ellipsoid, a difficult approximation of the earth's surface). The altitude of altimeter relies upon some of the constraints (e.g., inclination, atmospheric drag, gravity forces performing on the satellite, area of the sector to be mapped, and so on). The period or 'repeat orbit' is the time wished for the satellite altimeter bypass over the equal function on the ground, uniformly sampling the earth's surface. Inclination delivers the highest range at which the satellite can perform the measurements.

15.2.6 Surface Height

The simple mathematical formula to identify the surface height (H) is given by:

$$\text{(Corrected) height} = \text{altitude} - \text{(Corrected) variety} \tag{15.3}$$

In case of the sea surface height, there are several circumstances that must be considered. These include (i) ocean surface height, ocean circulation, and other physical parameters, for instance, wind speed, eddies and seasonal variations. Ocean surface height is determined without referring to other physical parameters which are associated with the sea surface, for instance, tide, wind speed, and currents. In fact, geoid is governed by the sea surface due to the impact of gravity distribution over the world. Under this circumstance, geoid fluctuates due to changes in the water masses and densities. In other words a hill at the geoid is noticeable as the seafloor has a denser rock zone at the seafloor which would distort sea level by tens of metres. Furthermore, dynamic topography, which is known as ocean circulation, is a function of the Earth's rotation. It is the derived impact of about 1 m. By removing the geoid from sea surface height, the dynamic topography is then computed. In practice, mean sea level is deducted to yield the variable component (sea degree anomalies) of the ocean signal.

15.2.7 Pulse-limited Altimetry

Consider a radar pulse emanating from a radar beacon propagating downwards and interacting with a flat ocean surface. Figure 15.3 shows an illustration of the vertical cross-section and top-down view of the radar pulse.

Implementing the Pythagorean theory, the leading edge r_p of the pulse is casted as:

$$r_p = \sqrt{Hct_p} \tag{15.4}$$

where H is satellite Height, c is the speed of light and t_p is pulse time.

Equation 15.4 demonstrates that the pulse time fluctuation of the backscatter signal off the ocean or land surfaces are identified as (i) the period before the pulse arrives, (ii) the period after the pulse arrives and before the tail of pulse has been received by the antenna, and after the tail pulse has been received by the antenna.

In order to determine the power signal of the delay-Doppler radar as a function of time, we'll need to assume that the footprint of the pulsed radar is small enough to be considered two rectangles of width W. This can be expressed by:

$$P(W) = \begin{cases} 0 & \mathrm{t} < \mathrm{t}_0 \\ 2W_r(t) & \mathrm{t}_0 < \mathrm{t} < \mathrm{t}_0 + t_p \\ 2W_r[r(t) - r(t - t_p)] & \mathrm{t} > \mathrm{t}_0 + t_p \end{cases} \tag{15.5}$$

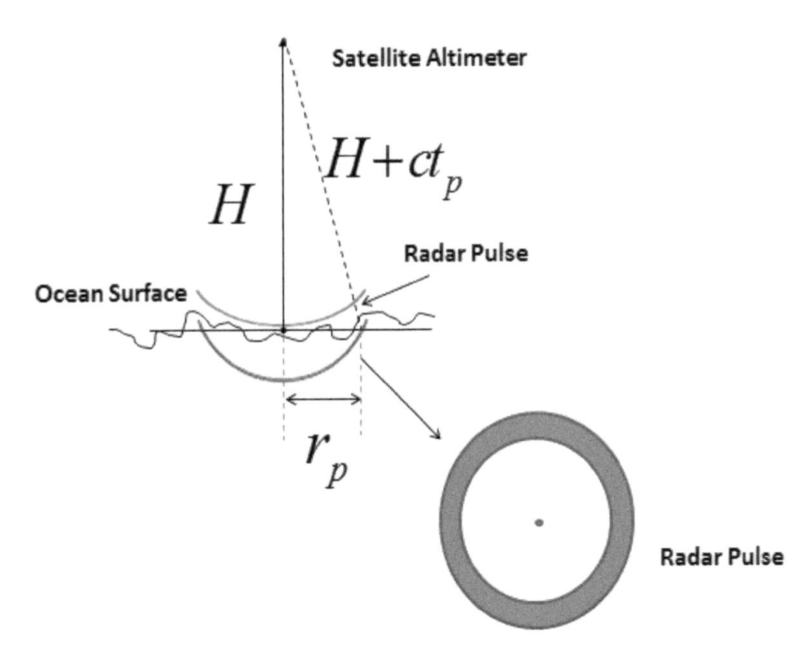

Figure 15.3 Radar pulse.

The altimeter radiates a pulse towards the Earth's surface. The time, which intervenes from the transmission of a pulse to the reception of its backscatter of the Earth's surface, is proportional to the satellite's altitude. Some theoretical details of the principle of radar are applied to altimeter which can assist a great understanding of the different behaviors and characteristics of the pulse in function of irregularities on the surface encountered. In this regard, the magnitude and shape of the echoes (or waveforms) also contain information about the characteristics of the surface which caused the backscatter. The greatest results are acquired over the ocean, which is spatially homogeneous, and has a surface which conforms with known statistics. In contrast, land surfaces which are not homogeneous and contain discontinuities or significant slopes, make precise analysis further challenging. Even in the best case (the ocean), the pulse should last no longer than 70 picoseconds to achieve an accuracy of a few centimetres. Technically, this means that the emission power should be greater than 200 kW, and that the altimeter would have to switch every few nanoseconds.

These problems are solved by the full deramp technique, making it possible to use only 5 W for emission. The range resolution of the altimeter is about half a metre (3.125 ns) but the range measurement performance over the ocean is about one order of magnitude greater than this. This is achieved by fitting the shape of the sampled echo waveform to a model function which represents the form of the echo.

15.2.8 Frequencies used and their Impacts

Numerous altered frequencies are exploited for radar altimeters. The choice be determined by regulations, mission objectives and constraints, technical possibilities—and impracticalities, for each frequency band has its advantages and disadvantages.

ku band with 13.6 GHz is the utmost regularly-operated frequency for Topex/Poseidon, Jason-1, Envisat, ERS, etc. It is sensitivity to atmospheric alarms and disconcertion by effect of ionospheric electrons (Figure 15.4).

Further, C band with 5.3 GHz is known to be more sensitive than Ku to ionospheric perturbation, and less sensitive to the effects of atmospheric liquid water. Its main function is to enable correction of

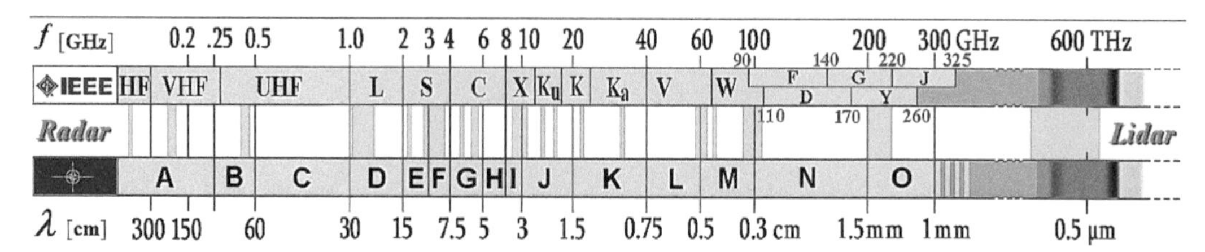

Figure 15.4 Electromagnetic waves for altimeter bands.

the ionospheric delay in combination with the Ku-band measurements. To obtain the best results, an auxiliary band like this must also be as far as possible from the main one [10]. Consequently, S band with 3.2 GHz is also used in combination with the Ku-band measurements for the same reasons as the C band. Finally, signal frequencies in the Ka band with 35 GHz enable better observation of ice, rain, coastal zones, land masses (forests, etc.) and wave heights. Due to international regulations governing the use of electromagnetic wave bandwidth, a larger bandwidth is available than for other frequencies, thus enabling higher resolution, especially near the coast. It is also better reflected in the ice. Nevertheless, attenuation owing to water or water vapor in the troposphere is high, meaning that no measurements are produced when the rain rate is higher than 1.5 mm/h [110].

15.3 Altimetric Measurements over the Ocean

The echoes or reflected waves received by altimeter are a function of the sea surface situations which creates variations in echo powers over a time. In this regard, the echo amplitude increases since the leading pulse strikes the flat surface. Nonetheless, if the surface is extremely rough, it causes the reflected waves to increase gradually. Ocean wave height estimation is a function of the reflected waves since the slope of the curve representing its amplitude over time is proportional to wave height (Figure 15.5).

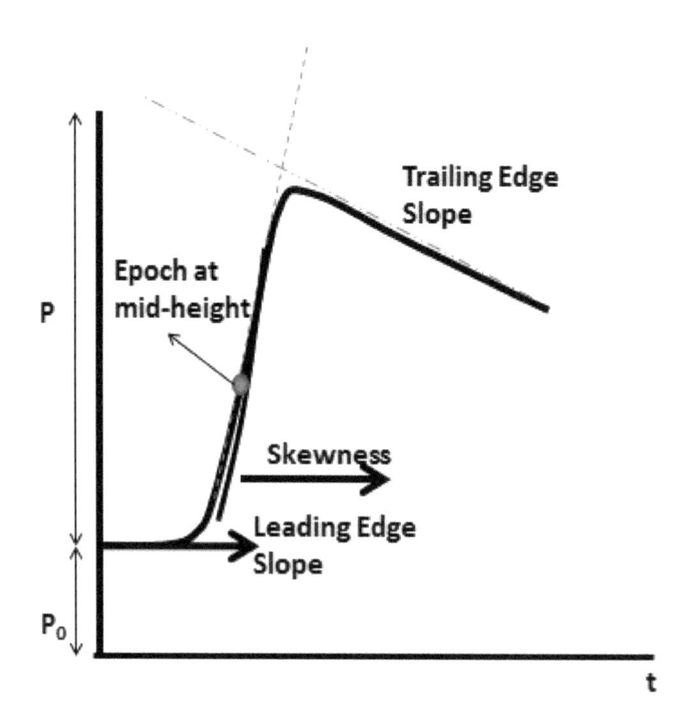

Figure 15.5 Concept of ocean surface measurements by Altimeter.

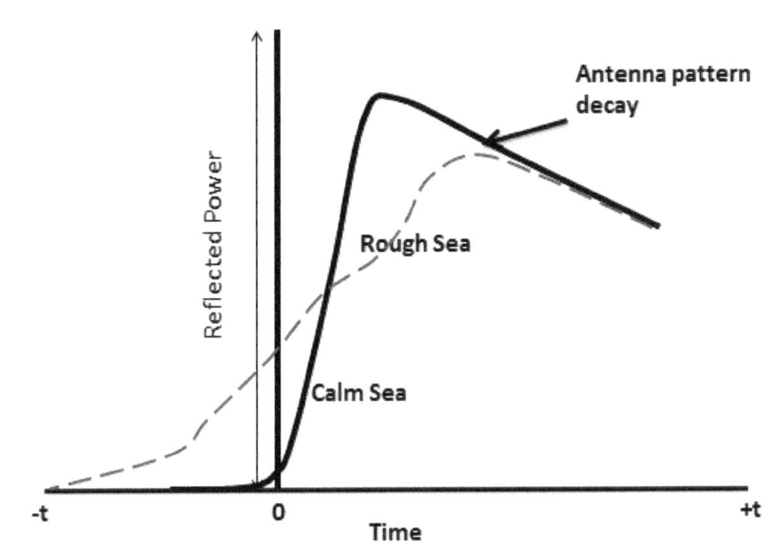

Figure 15.6 Reflected pulses from different ocean surface conditions.

Figure 15.5 is a keystone to retrieve six parameters from the shape of reflected waves [104]. Epoch at mid-height delivers the delay time of the reflected pulse which is a function of range. The amplitude of radar signal power reveals the backscatter coefficient and sigma0. Therefore, significant wave height can be estimated from the leading edge slope. The rate of leading curvature can be acquired by skewness. The mispointing of radar antenna is associated with trailing edge slope. In other words, it means any deviation from radar altimeter nadir of the radar pointing. Furthermore, the leading edge slope must occur above the thermal noise P_0 [103].

The altimeter irradiates a circle of the beam over ocean or land surfaces with a 3 to 5 km wide diameter which depends on the sea condition, the wave height or the ridged land. A rough sea surface or land delivers a wider footprint of approximately 10 km (Figure 15.6). In contrast, the calm sea surface or flat land provides a narrow footprint of about 2 km [105].

15.4 Altimeter Sensors for 2004 Tsunami

The Sumatra-Andaman tsunami is the first for which detailed concurrent measurements of the sea surface height (SSH) and radar backscattering strength at nadir in deep water are available. These measurements were made with microwave radars on board the Jason-1, Topex/Poseidon, ENVISAT, and Geosat Follow-on (GFO) altimetric satellites (Figure 15.7) [106].

The satellite altimetry data for the December 2004 tsunami provided the opportunity to compare and validate tsunami propagation models with a clear and unique measurement of the wave in the open ocean, before the effect of coastal processes contaminates the distinct tsunami signal that is generally observed in the high seas. The details of the coastal wave interaction, for the most part, do not account for in the propagation models that simulate open ocean wave dynamics. Therefore, coastal measurements are difficult to use for validation and comparison of propagation model data. Despite the fact that no deep ocean tsunami sensor (DART system) was available in the Indian Ocean at the time of this event, offshore tsunami data were recorded by at least four altimeters-equipped satellites flying over the Bay of Bengal in the hours following the earthquake. All four satellites provided unambiguous deep ocean measurements of the wave, to be used in propagation models. Waves generated by the 2004 Sumatran–Andaman tsunami reached offshore amplitudes in excess of 50 cm as recorded by several of the satellites. For the first time, the magnitude of the offshore tsunami amplitude was sufficiently large to distinctly emerge from the background noise of the altimeter data generated by tides, storm systems and swell [109].

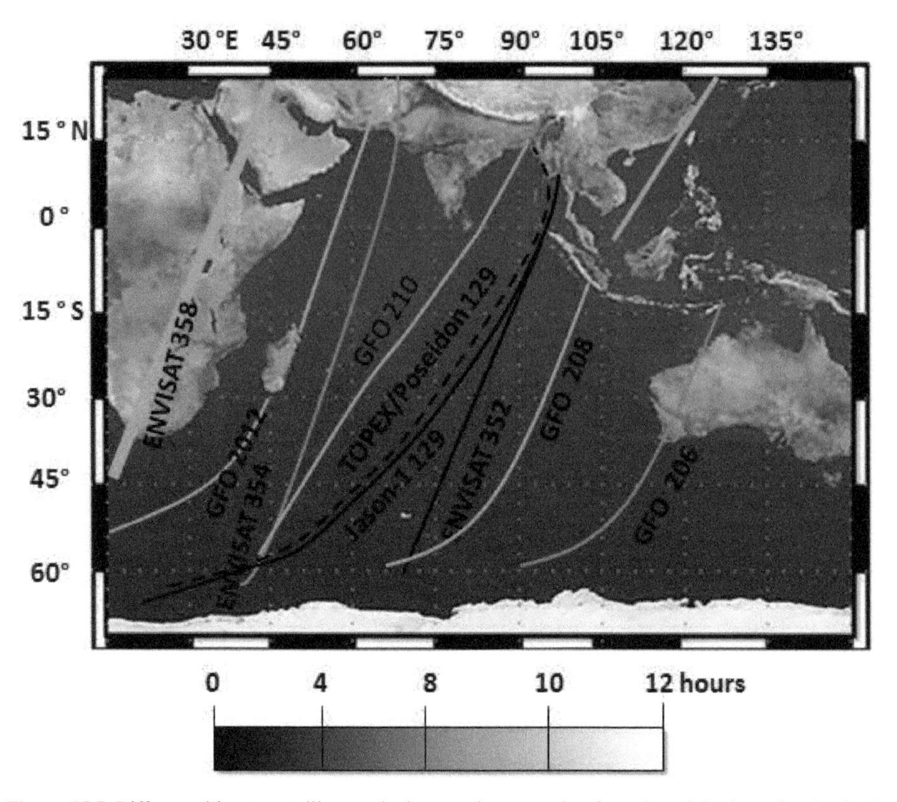

Figure 15.7 Different altimeter satellites tracked tsunami propagation from the origin time of main shock.

15.4.1 Retrieving Tsunami Wave Height and Propagation from Altimeter Data

Three methods have been used to study physical properties of tsunami wave propagation: (i) multisatellite time-spatial interpolation; (ii) investigation of tsunami wave height; and (iii) estimation of Background Level [111]. Multisatellite time-spatial interpolation is performed to define the reference heights, which must be estimated at sea level anomaly (SLA), "the SLA under the assumption of no tsunami." The reference heights are defined by the weighted mean as [111]

$$SLA_{ref}(\phi,\vartheta,t)=\sum(e^{(r_i^2/R^2-t_i^2/T^2)}\bullet SLA_{obs,i})/\sum w_i \tag{15.6}$$

where r_i is the distance between the location of the ith datum and the tsunami searching point; ϕ, ϑ, and t are the latitude, longitude, and date of a tsunami searching point; t_i is the time difference of observations between a tsunami observation point (t) and the ith datum; and R and T are scale parameters [111]. Then the tsunami height ($h_{tsunami}$) at any sampling point is derived from

$$h_{tsunami}(\phi,\vartheta,t)=SLA_{obs}(\phi,\vartheta,t)-SLA_{ref}(\phi,\vartheta,t) \tag{15.7}$$

where SLA_{obs} is an anomaly in an observed sea surface height, and SLA_{ref} is the reference height defined by Equations 15.4. Finally, the background level at any sampling point (ϕ and ϑ) is defined as the root mean square (RMS) of the residual error calculated by

$$r_k(\phi,\vartheta,t)=SLA_{obs}(\phi,\vartheta,t+kc)-SLA_{obs}(\phi,\vartheta,t+(k-1)c) \tag{15.8}$$

where k = –5, –4, –3, –2, –1, 1, 2, 3, 4, 5, and c is the recurrence cycle, for instance Jason-1 has 9.9156 cycles, day, TOPEX/POSEIDON is 9.9156 cycles, day, GFO is 17.0506 cycles, day and ENVISAT is 35.0000 cycles, day [111].

15.4.2 Jason-1

Jason-1 was the first altimeter device which provided the earliest observations of the Sumatra-Andaman tsunami and has the most extensive data records. Jason-1 encountered the tsunami 1 h 53 min post the earthquake [106]. Figure 15.8 shows the detected tsunami propagation two hours post the 2004 earthquake. The maximum wave height was 0.5 m and was travelling towards Sri Lanka and India. On the contrary, the offshore coastal water of Aceh was dominated by tsunami wave height of less than 0.5 m [107].

Figure 15.9 shows another simulation of tsunami wave propagation across Indian Ocean using Jason-1 satellite data. The simulation of tsunami wavefronts in the Indian Ocean was about 10 hours after the Sumatra earthquake of 26 December, 2004 together with the altimeter sea-level profile taken along Track 129 of the Jason-1 satellite during Cycle 109. The wavefront was computed from the leading edge of the wavefront point on the satellite track [108]. It is interesting to note that the tsunami approached coastal water of Madagascar within less than 8 hours and coastal water of Australia within 2 hours.

The maximum wave crest of about 80 cm was associated with the second peak. The spatial scale of wave coverage was roughly 500–800 km. In this regard, the first wave was maximum in all records near the source area, including the Maldives. Consistent with Kulikov (2006) [108], the leading edge of the wave components with 10-km wavelength were significantly delayed in comparison with the main wavefront components with about 1000 km wavelengths. It is also worth noting that oscillations for all frequencies (and wavenumbers) were abruptly dampened northward of 8° N (delineated by the vertical dashed line

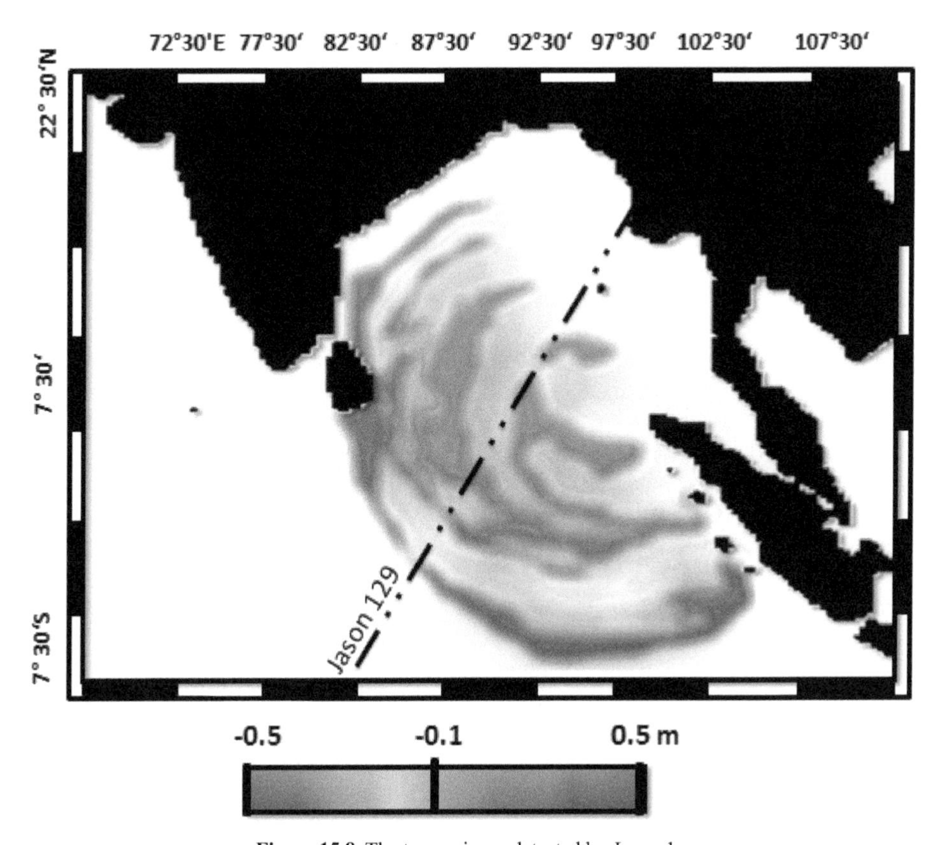

Figure 15.8 The tsunami was detected by Jason-1.

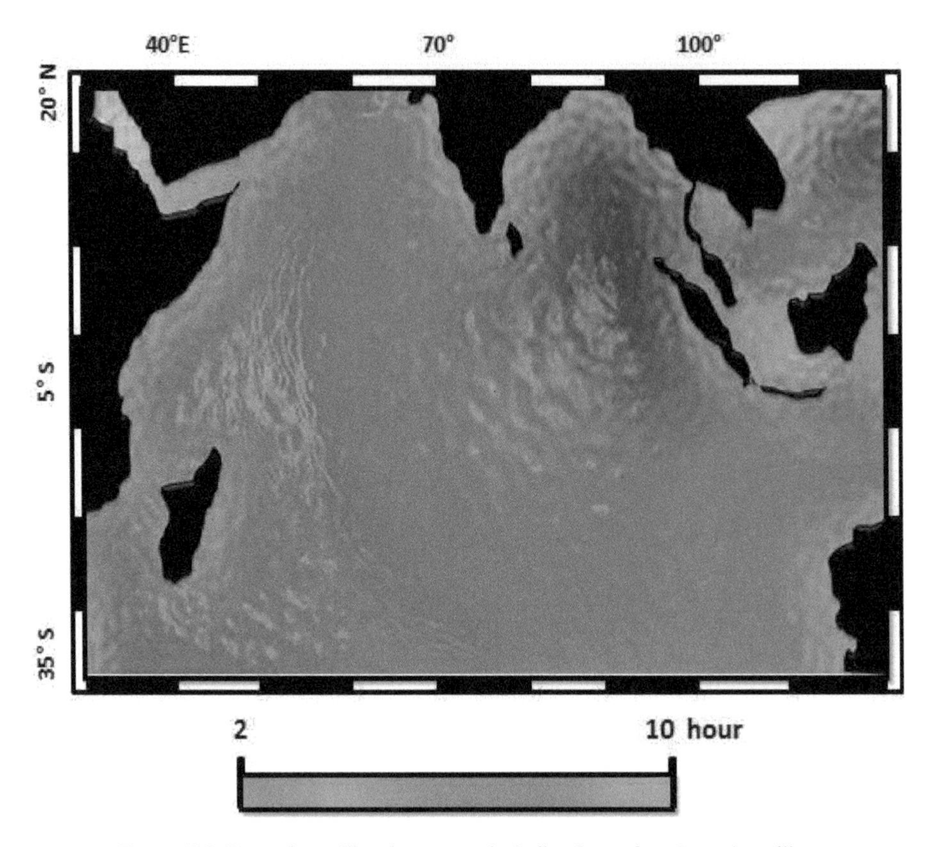

Figure 15.9 Tsunami travelling time across the Indian Ocean from Jason-1 satellite.

in the figure). The latter zone consists of short tsunami waves with wavelengths $L = T(gH)^{1/2} \approx 1400$ km, where $T = 2$ hours is the time of wave propagation, g is gravity, and $H \approx 4000$ m is the ocean depth. Based on this calculation, it is possible to state that the "boundary of the high-frequency wave generation zone" was located at between 7–8° N.

Further, Kulikov [108] reported that most of the energy was associated with a wave period of about 50–60 min. Although the satellite trace does not coincide with the wave ray trace, it was assumed that this period (50–60 min) is the fundamental period of the Sumatra tsunami waves. In addition, the leading edge of the extracted tsunami signal (the first forked maximum) is probably associated with the initial tsunami signal in the source area. Such a complicated structure for the tsunami "signature" is probably related to a complex rupture process that had a start-stop character.

Due to the fact that the ground velocity for the Jason-1 satellite is 5.8 km/s, extremely faster than tsunami transmission, the information may be thought of as a footprint of tsunami amplitudes alongside the Jason-1 track. Jason-1 measured the peak of the tsunami wave wherein its track line crossed these arcs of the Indian Ocean. To the east, the local tsunami continues to cause major wave action along the Sumatran coast [116]. The high potential of Jason to track the tsunami due to the five instruments are attached in the satellite. The height of the tsunami was measured by Poseidon 2 using C-band and Ku band. The pulse delay was corrected by Jason Microwave Radiometer (JMR) which is a function of water vapor estimation. Beside, orbit determination within 10 cm or less precisely is determined by Doppler Orbitography and Radiopositioning (DORIS) which can also be used to correct ionospheric effects, specially for Poseidon 2. Therefore, the Global Positioning System (GPS) was able to determine the accurate orbit. Lastly, the altimeter collected information about the tsunami was calibrated and verified by ground station using Laser Retroreflector array [107].

15.4.3 Topex-Poseidon Satellite

Topex-Poseidon satellite had two on-board altimeters involving the same antenna. However, Topex only operated at different time. Consequently, Topex had an average of 9 in 10 cycles [112]. It operated with two bands which is C-band with 5.3 GHz and Ku band with a frequency of 13.6 GHz. Both bands were used to measure the height above the sea surface. However, Poseidon was operated by Ku band [113].

Figure 15.10 shows that the tsunami wave propagation was simulated from TOPEX/Poseidon post two hours after the earthquake shock. The maximum wave height is approximately 0.8. The maximum height occurred close to coastal waters (Figure 15.10). The tsunami profile along TOPEX/Poseidon track 129 cycles 452 indicates the flat peak of the front wave [117].

In fact, the altimeters, the TOPEX Microwave Radiometer was operated at 18, 21, and 37 GHz, which was used to correct atmospheric wet path delays. Knowing the satellite's precise position to within 2 centimetres (less than 1 inch) in altitude was a key component in making accurate ocean height measurements possible [114].

Smith et al. (2005) [117] compared sea level retrieved from TOPEX/Poseidon with Method of Splitting Tsunamis model (MOST). They found that TOPEX/Poseidon data agreed with MOST model. However, the TOPEX/Poseidon provided discontinued record (Figure 15.11). In fact, the probability of a satellite altimeter observing a tsunami is low because it requires that the satellite overflies the tsunami wave almost immediately after it originates, due to the tsunami's great propagation speed (about 800 km/h in an ocean 5,000 m deep). Tsunami signals in the open ocean are also quite weak.

Lastly, the accurate determination of the ocean height is made by first characterizing the precise height of the spacecraft above the center of the Earth. This is achieved through a technique called "precise

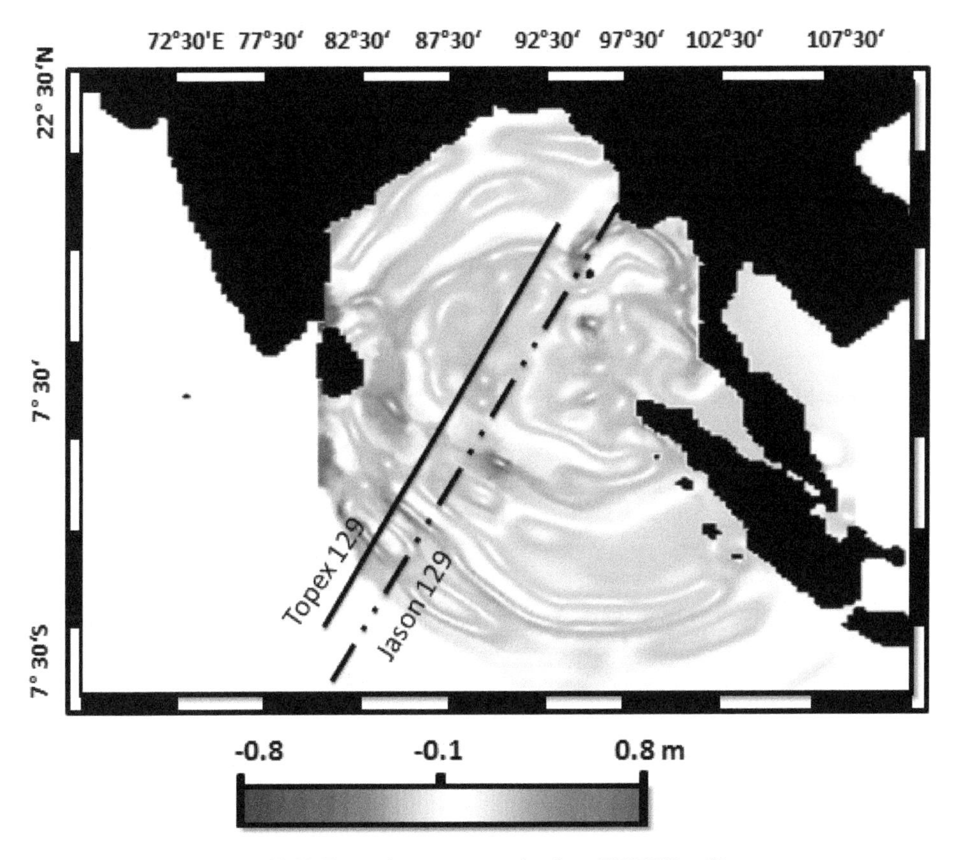

Figure 15.10 Tsunami wave propagation from TOPEX/Poseidon.

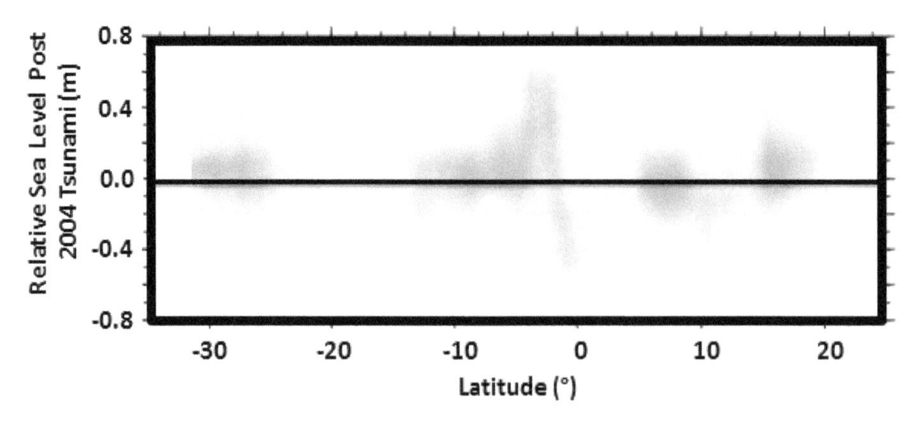

Figure 15.11 Discontinued sea level records from TOPEX/Poseidon satellite data.

orbit determination," of which satellite-tracking information is the most important ingredient. The second component of the ocean height measurement is the range from TOPEX/Poseidon to the ocean surface. TOPEX/Poseidon carried two radar altimeters for providing this information. To take a measurement, an onboard altimeter bounces microwave pulses off the ocean surface and measures the time it takes the pulses to return to the spacecraft [115].

15.4.4 ENVISAT

ENVISAT (Environmental Satellite) is the follow-on to ERS-1 and ERS-2. Devoted to environmental studies, and climate change in particular, its mission is to observe Earth's atmosphere and surface. Built by ESA, the European Space Agency, ENVISAT is carrying ten complementary instruments for observing parameters ranging from the marine geoid to high-resolution gaseous emissions. Among these instruments are a radar altimeter, and the DORIS orthography and precise location system [110].

ENVISAT's orbital period is 35 days, like ERS-2 and some of the ERS-1 phases. As it is integrated in new international climate study programmes such as GOOS and GODAE. ENVISAT thus forms part of the coming operational era in oceanography, offering near-real-time data access.

The main objective of the microwave radiometer (MWR) is to measure the integrated atmospheric water vapor column and cloud liquid water content, which are used as correction terms for the radar altimeter signal. Once the water content is known, we can determine the correction to be applied for radar signal path delays for the altimeter. In addition, MWR measurement data are useful for determining surface emissivity and soil moisture over land, for surface energy budget investigations to support atmospheric studies, and for ice characterization [110].

The DORIS system uses a ground network of orbitography beacons spread around the globe, which send signals at two frequencies to a receiver on the satellite. The relative motion of the satellite generates a shift in the signal's frequency (called the Doppler shift) that is measured to derive the satellite's velocity. These data are then assimilated in orbit determination models to keep permanent track of the satellite's precise position (to within three centimetres) on its orbit [110].

Figure 15.12 shows the tsunami wave propagation which was retrieved from ENVISAT data. This was captured within 3 hours, post the tsunami occurrence. It is clear that the tsunami propagated widely within 3 hours. The wave height increased to approximately 0.35 m as compared to initial time from Ache to Sri Lanka.

Figure 15.13 shows the excellent continuous record of sea level data which was delivered by ENVISAT as compared to TOPEX/Poseidon. In fact, ENVISAT is considered more advanced than TOPEX/Poseidon. With this concern, ENVISAT contains the Radar Altimeter 2 (RA-2) as an instrument for determining the two-way delay of the radar echo from the Earth's surface to a very high precision: less than a nanosecond. It also measures the power and shape of the reflected radar pulses. The RA-2 is derived from the ERS-1

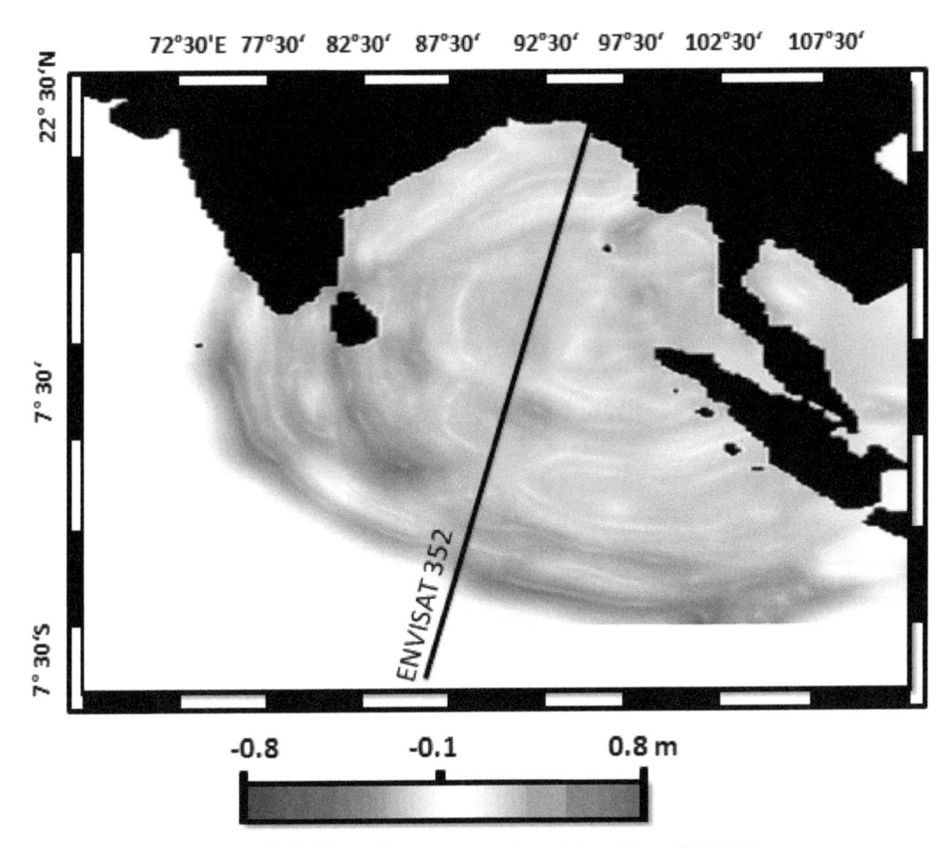

Figure 15.12 Tsunami wave propagation retrieved from ENVISAT.

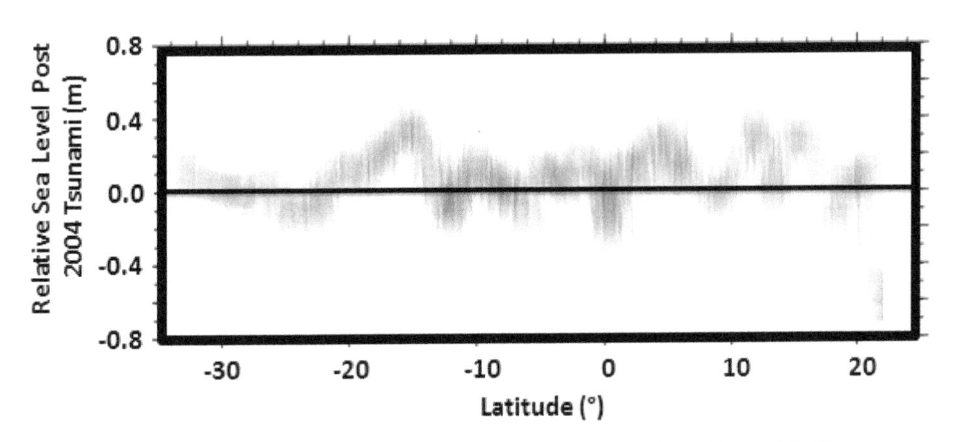

Figure 15.13 Continuous record of sea level data which was delivered by ENVISAT.

and 2 radar altimeters, providing improved measurement performance and new capabilities. Operating over oceans, its measurements are used to determine the ocean's topography, thus supporting research into ocean circulation, bathymetry and marine geoid characteristics. Measurement of the radar echo's power and shape enables wind speed and significant wave height at sea to be determined, thus supporting weather and sea state forecasting [110]. This is proved by excellent correlation between ENVISAT record and MOST model (Figure 15.13).

15.4.5 Geosat Follow-on

Geosat Follow-On was launched in February 1998, and retired in November 2008. Its mission is to provide real-time ocean topography data to the US Navy. Scientific and commercial users have access to these data through NOAA (National Oceanic and Atmospheric Administration). Its primary payload is a radar altimeter. GFO follows the 17-day repetitive orbit of Geosat.

According to Gower (2007) [118], the Geosat Follow On (GFO) altimetry satellite was not well placed to observe this event. On pass 210, GFO moved towards the equator over the Bay of Bengal, about 500 km to the west of pass 129 with an equator crossing time of 09:48 UT, nearly 9 h after the earthquake. The front had moved down to the southern Indian Ocean by this time and was greatly reduced in amplitude. It is also located in an area of much greater eddy activity. No identification on the front has been made on GFO data. The slow-moving, shorter waves also appear to have dissipated. Figure 15.14 shows the wide propagation of tsunami wave across the Indian Ocean within 7 hours and 10 minutes from the shock.

8 hours and 50 mins post the earthquake, the wave dispersed over the wide space of the Indian ocean and changed into quite small waves in most zones and ranged between 0.05 to 0.1 m. Nonetheless, in the Bay of Bengal, the wave maintained its height of 0.3 m. On the contrary, wave dissipated across the Indian Ocean due to its wide spreading with a height of 0.2 m (Figure 15.15). Nevertheless, in the latitude of 4°, the sea level was dominated by ebb sea level of –0.8 m (Figure 15.16).

Corrected data show that the Sumatra tsunami was detected by three of the four altimeters, presently giving sea-surface height information. Each detected the spreading wave front as it moved south-westward into the Indian Ocean at about 750 km/h (shallow-water wave speed at 4500-m water depth), causing an initial rise in sea level. They also appear to have detected the front moving to the north, into the shallower

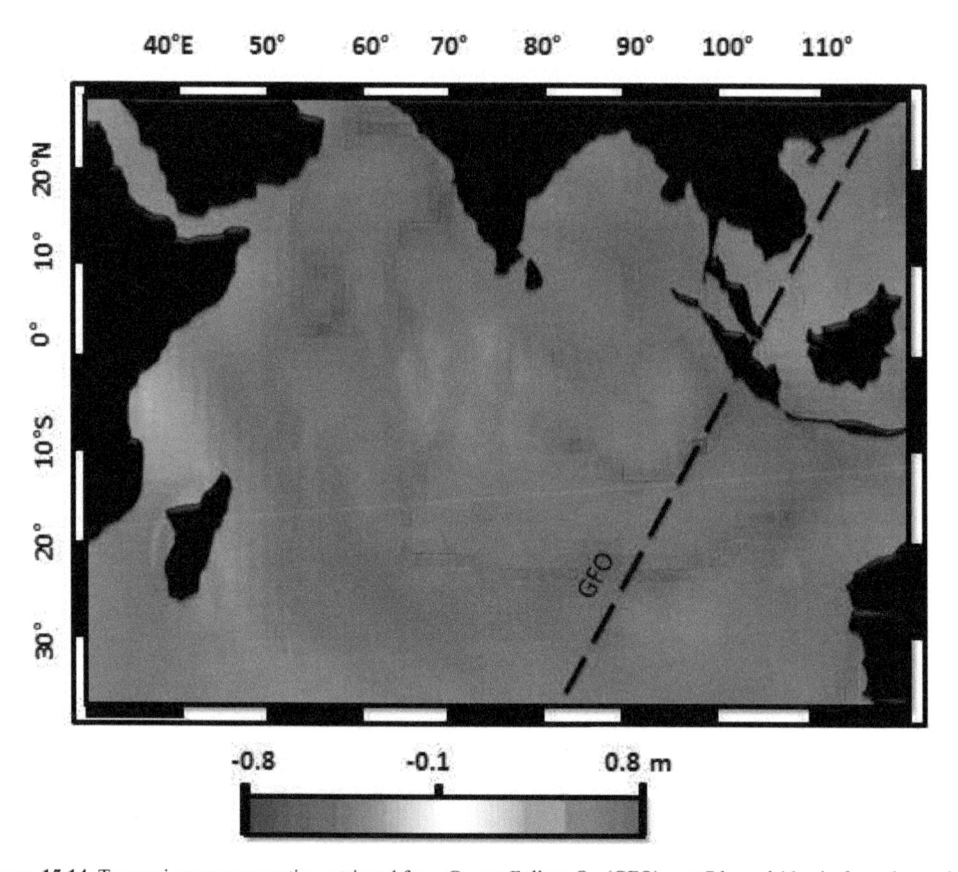

Figure 15.14 Tsunami wave propagation retrieved from Geosat Follow On (GFO) past 7 hr and 10 min from the quake.

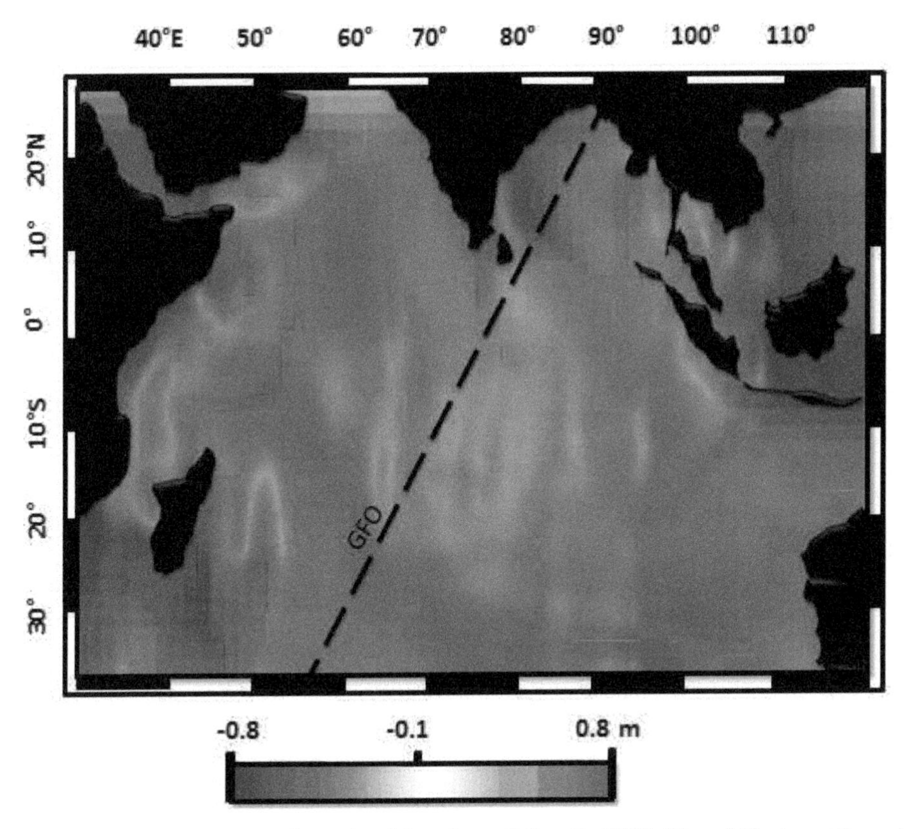

Figure 15.15 Tsunami wave propagation retrieved from Geosat Follow On (GFO), 8 hr and 50 min past the quake.

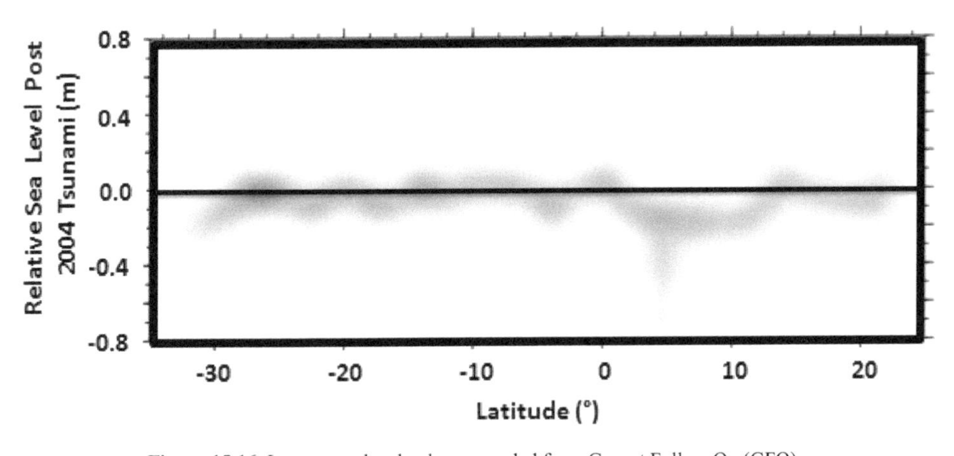

Figure 15.16 Lowest sea level values recorded from Geosat Follow On (GFO).

water of the Bay of Bengal. This signal was a smaller amplitude and appeared as an initial drop in sea level. The altimeters also detected the disturbed region closer to the epicentre that expanded with the slower velocities of higher-frequency waves [118].

Consistent with Gower (2007) [118], the Jason-1 observation provides a clear and complete enough record to show the main pattern of the tsunami wave. The model record uses a tsunami wavelength of

580 km and an amplitude of 0.6 m. The phase is adjusted to fit the leading edge of the tsunami. The model provides a good fit to the observed signal over the latitude range from 5° S to about 2° N, covering just over a full cycle of the wave. The satellite is closest to the epicentre at 6° N, after which it should again intersect about a cycle and a quarter of the spreading tsunami. The model does not take into account the shallow water to the north. It is clear that the tsunami wave amplitude is very less at latitudes north of 6° N in the Bay of Bengal. At these northern latitudes, Jason-1 shows a linear gradient of height increasing northwards at about 0.04 per degree of latitude.

It would be tempting to suggest that this 100-km component could have started at the same time and location as the main wave but had been slowed by dispersion. However, the observed delay in arrival of the 100-km wavelength component is almost an order of magnitude too large, suggesting a generation zone further east and closer to the position of maximum negative ocean bottom displacement computed by Chen (2005) [119].

However, Topex/Poseidon data show that the 100-km wavelength component is much less evident than in Jason data, but should now be appearing at the trailing edge of the first maximum. Additional confirmation of the model is given by the short segment of data in the latitude range 2.4–2.8° N [118].

The improbability of Jason-1 was being 'in the right place at the right time' for detection of this tsunami propagation across the Indian Ocean. This suggests that the probability is about 1/20 for ascending passes alone, and hence about 1/10 if a descending pass would have equally been acceptable [118]. Topex/Poseidon is locked in orbit to Jason-1 and so does not affect a computation of the overall 'improbability'. ENVISAT is in an independent orbit, with a pattern similar to that shown in Figure 6.8. The probability of all three satellites observing the front is therefore only about 1%. The fact that all three satellites appear to have observed the front a second time reduces the overall probability to significantly below 1% [118].

Chapter 16

Schrödinger Theory for Future Tsunami Forecasting in Malacca Straits, Indian Ocean, Red Sea and Nile River

16.1 Quantum for Wave Propagation

Quantum mechanics is a keystone to understand the propagation of tsunami waves. In this sense, the tsunami has several characteristics, for instance, energy and momentum, besides wave-particle duality. For instance, wave-particle duality is the perception of quantum mechanics that each particle or quantic entity can be partially expressed in terms not only of particles but also of waves. It expresses the incapacity of the classical theories of "particle" or "wave" to absolutely designate the behavior of wave phenomena in quantum-scale.

The uncertainty in tsunami propagation, dissipation and run-up can also be discussed through uncertainty theory. Tsunami forecasting requires a comprehensive theory to solve its propagation non-linearity from deep water to shallow water in addition to its run-up. In this regard, the quantum modern concept is formulated in numerous advanced mathematical formalisms. In one among them, a mathematical characteristic, the wave function, is a probability amplitude, momentum, and other bodily residences of a particle. The critical question is, can the quantum field theory predict and explain the future tsunami in the Indian Ocean? Further, how does quantum mathematics provide an explanation for the existence of the tsunami? Particularly, the nonlinear Schrödinger equation might assist the quantum theory to deliver a full understanding of tsunami transformation across the ocean. In fact, the nonlinear schrödinger equation depicts the evolution of the envelope of modulated water wave groups. Indeed, the rate of the nonlinearity parameter relies upon the relative water depth. In other words, if the deep water is extremely larger than the wavelength, the nonlinearity parameter is negative and envelope solitons may additionally rise up. In contrast, if shallow water depth is 4.6 times smaller than the wavelength, the nonlinearity parameter is first-rate and wave groups with envelope solitons do not exist. In other words, the shallow water surface-elevation solitons or waves of translation do exist. However, they are not ruled by the nonlinear schrödinger equation.

Exactly well-timed detection and simulation of tsunami propagation is an essential phase of competent tsunami early warning system. Observing and monitoring systems and forecasting of long ocean waves must be in place in the Asian seas and predominantly positioned in view of environmental monitoring rather than for tsunami observations. Nonetheless, they are believed to be valuable for tsunami recording, along with the present and forthcoming oceanographic events, such as monitoring, forecasting, simulation and modelling.

Satellite technology provides sea-level data because of numerous altimeters accessibility. Nonetheless, altimeters data are presently imperfect to pathways which are excessively vague from one another (low space resolution) and have an unsatisfactory long recurrence time (low time resolution), and additional sea-level resolves resulting from handling of satellite raw data which are currently very far from being accessible in real time.

Tsunami modelling is accepted to be an important means to elucidate the monitoring and detection of a tsunami, to devise observational strategies for detection network design and to assess tsunami hazards, receptiveness and hazard. In the context of tsunami early warning systems, tsunami modelling and simulations can be expanded to deliver tsunami valuation and real-time tsunami forecast in terms of arrival times, expected coastal amplitude and effects. Examples of products that can be provided through tsunami simulations are: numerical sea-level records, flow velocity and direction, maximum and minimum sea level height maps, propagation maps, travel time maps, inundation maps, forces on structures, etc.

The existing tsunami models deal with a tsunami as a deep wave propagation due to a submarine earthquake without including other forces such as volcanic eruption, pyroclastic flow, caldera collapsing and asteroid/comet impacts. They further treat the propagation of the waves in the ocean as well as the interaction of the tsunami waves with the coastal morphology and coastal flooding.

Tsunami propagation is usually computed through models based on shallow-water and Boussinesq approximations. To cover very large basins and to study the tsunami propagation along the entire path across the ocean from the source to the coast, techniques of domain nesting and domain decomposition are used.

Simulation of tsunami attack on the coast has to be performed by means of models that account for non-linear terms, for energy dissipation through bottom friction and coastal roughness, for wave breaking, for bore formation, for computation of water flows and currents, for run-down and run-up heights, for water penetration and retreat distances, for effects on the coastal structures, etc. In most cases, not all of these requirements are met by the same model, and simplifying approximations are used, that would need verification on quantitative empirical data.

Tsunami propagation calculations are dependent on the input ocean depth. More accurate bathymetry data sets are needed, as well as access to data set has to be ensured. Shallow-water and Boussinesq approximation models seem to be adequate for tsunamis induced by earthquakes, but could be inappropriate for other sources such as landslides, where sea vertical fluid displacements and velocities may play a role. Techniques such as domain decomposition and nesting should be further developed for accurate computation of tsunami propagation from the source region to the coastal areas.

Consistent with Chubarov and Fedotova (2003) [231], the shallow water model is not applicable for steep slopes, as the vertical acceleration of the particles is significant. Noble approximation can be achieved by substituting the shore slope by an impenetrable vertical wall with the corresponding modification of the boundary conditions. Again the maximal run-up height can be estimated from the equations of the shallow water theory.

In general, to model water waves, numerical methods can be classified into four approaches. These approaches are grid based methods (e.g., Finite Volume Method), methods combining a grid with particles (e.g., Marker-and-Cell method), a method combining a grid with a surface finder which use a fixed grid to solve the PDE's governing fluid motion (e.g., Volume-of-Fluid method), and particle methods without grid (e.g., Smoothed Particles Hydrodynamics method).

16.2 Schrödinger Equation for Tsunami Propagation

The scientific explanation of the Schrödinger equation can mathematically be written as

$$i\partial_t \psi = -0.5\partial_x^2 \psi - \kappa |\psi|^2 \psi \tag{16.1}$$

where κ is nonlinearity parameter, ψ is the complex field and function of position x and time t, and i is complex number. Equation 16.1 is delivered from the mathematical expression of the Hamiltonian as

$$H = \int dx \left[0.5 \left| \partial_x \psi \right|^2 + \frac{\kappa}{2} \left| \psi \right|^4 \right] \tag{16.2}$$

Consequently, the amplitude and phase of the water waves due to the vertical displacement are related to the complex field ψ (Equation 16.1). When the sudden earthquake shock perturbs the upper sea surface, the vertical displacement η(x, y, t) of each point of the surface is generated. In this regard, z = η(x, y, t) is considered as the boundary condition of the sea surface. Further, the waves must have a small amplitude and long wavelength. In other words, this causes a slow surface carrier modulation to the sea surface elevation and can be mathematically identified by:

$$\eta = \mathrm{Re}[\zeta(X, T)e^{(i(\omega_0 t - k_0 x))}] \tag{16.3}$$

where $\zeta(X, T)$ is the modulation amplitude of the wavetrain, k_0 is a wave-number, and ω_0 is the wave frequency of the carrier Stokes wave, respectively. The mathematical expressions of X and T are

$$X = \zeta k_0 x \tag{16.3.1}$$

$$T = \zeta k_0 t \tag{16.3.2}$$

This displacement travels from initial source of shock across the water surface with the potential velocity ϕ of

$$\phi = \omega\zeta \frac{\cosh((k(z+h))}{k\sinh(kh)}\sin(kx - \omega t) \tag{16.4}$$

The wavenumber k_0 of the carrier wave and about the envelope amplitude ζ can be extended into the Taylor series expansion as:

$$\omega = \omega_0 + K\frac{\delta\omega}{\delta k} + 0.5K^2\frac{\delta^2\omega}{\delta^2 k} \tag{16.5}$$

In Equation 16.5, the last term, which is a function of ζ, is neglected. In fact, $\zeta = \zeta_0 = 0$. Under the circumstance, the group velocity of the envelope is twice smaller than the phase velocity of the carrier wave. Then, the nonlinear Schrödinger equation for the evolution of the amplitude of the envelope of the wavetrain is casted as:

$$i\left(\frac{\partial\zeta}{\partial T} + \frac{\omega_0}{2k_0}\frac{\partial\zeta}{\partial X}\right) - \frac{\omega_0}{8k_0^2}\frac{\partial^2\zeta}{\partial^2 X} - 0.5\omega_0 k_0^2|\zeta|\zeta = 0 \tag{16.6}$$

Equation 16.6 can be formulated as a function of $k_0\zeta$ and expressed as:

$$i\frac{\sqrt{2}k_0^2\zeta}{\partial\tau} + \frac{\partial^2(\sqrt{2}k_0^2\zeta)}{\partial(X - v_g T)} + 2\left|\sqrt{2}k_0^2\zeta\right|^2\sqrt{2}k_0^2\zeta = 0 \tag{16.7}$$

Where $\tau = -\dfrac{\omega_0}{8k_0^2}\zeta k_0 t$ (16.7.1)

Taking this period to infinity, the Peregrine breather is then achieved. This solution is doubly-localized, in space and time, and earthquake pulsates only once, which makes the description of an earthquake caused tsunami extremely ideal. In this context, Equation 16.7 can be written in term of polynomials which is based on doubly-localized Akhmediev-Peregrine solutions as:

$$M(\zeta,t) = \sqrt{2}k_0^2\zeta e^{\left(2i\left|\sqrt{2}k_0^2\zeta\right|^2 - \frac{\omega_0}{8k_0^2}\zeta k_0 t\right)}\left[(-1)^j + \frac{4 + i(16\left|\sqrt{2}k_0^2\zeta\right|^2 - \frac{\omega_0}{8k_0^2}\zeta k_0 t}{1 + 4\left|\sqrt{2}k_0^2\zeta\right|^2(X - v_g T)^2 + 16\left|\sqrt{2}k_0^2\zeta\right|^4(\frac{\omega_0}{8k_0^2}\zeta k_0 t)^2}\right] \tag{16.8}$$

where, $M(\zeta, t)$ is a modulated wave surface due to submarine earthquake which is solved in 1st order polynomials. This is considered as an initial source of tsunami generation due to the fault displacement. Under this circumstance, the altimeter data have recorded the reflected echo of $M(\zeta, t)$. In this understanding, the altimeter equation for computing sea surface displacement cannot be based on classical Equation 15.4 which was presented in Chapter 15. Then, the sea surface modulation because of the submarine earthquake can mathematically be written as

$$M_{obs}(\phi,\vartheta,t) = M(\zeta,t) + [\sum (e^{(r_i^2/R^2 - t_i^2/T^2)} \bullet SLA_{ref,i}) / \sum w_i] \tag{16.9}$$

where $SLA_{obs,\,i}$ is the observed sea level anomaly (SLA), r_i is the distance between the location of the *ith* datum and the tsunami searching point, ϕ, ϑ, and t are the latitude, longitude, and date of a tsunami searching point, t_i is the time difference of observations between a tsunami observation point (t) and the ith datum, and R and T are scale parameters [111]. Sea surface height data are revised from all altimeter errors and from a precise mean sea surface to eliminate the geoid signal. Further, sea surface height data are not exclaimed on a steady along-track grid and that the along-track sampling is not precisely the same for the different missions, for instance, it differs between 5.8 and 7.5 km [232].

The estimated M_{obs} (ϕ, ϑ, t) is directly proportional to the earthquake slip S and the fraction of slip α. In this view, this relationship can be expressed as:

$$M_{obs}(\phi, \vartheta, t) = \alpha \Delta S \tag{16.10}$$

where ΔS is the earthquake slip and α is the fraction of slip that transforms into tsunami-making uplift. M_{obs} (ϕ, ϑ, t) is the initial tsunami source amplitude observed from altimeter Jason-1 satellite. This factor depends upon the style of the fault

$$\alpha = (1 - \phi/180^{o}) \sin\phi \sin|\varphi| \tag{16.10.1}$$

where ϕ and φ and are the dip $0 < \phi < 90^{o}$ and rake angles $-90^{o} < \varphi < 90^{o}$ of the faulting in degrees. The most efficient mechanism for tsunami generation have ϕ near 45° and φ at plus or minus 90°. Note that Equation 16.10.1 makes no attempt to model source radiation patterns, so in this sense it represents the worst case.

16.3 Numerical Model of Tsunami Travelling

Following Suleimani et al. (2003) [227], the numerical model used in this study is based on the vertically integrated nonlinear shallow water equations of motion and continuity with friction and Coriolis force. Written in a spherical coordinate system, they are [19, 20]:

$$\frac{\partial U}{\partial t} + \frac{U}{R\cos\phi}\frac{\partial U}{\partial \lambda} + \frac{V}{R}\frac{\partial U}{\partial \phi} - fV = -\frac{g}{R\cos\phi}\frac{\partial M_{obs}}{\partial \lambda} - \frac{oUW}{D} \tag{16.11}$$

$$\frac{\partial V}{\partial t} + \frac{U}{R\cos\phi}\frac{\partial V}{\partial \lambda} + \frac{V}{R}\frac{\partial V}{\partial \phi} + fU = -\frac{g}{R}\frac{\partial M_{obs}}{\partial \phi} - \frac{oVW}{D} \tag{16.12}$$

$$\frac{\partial M_{obs}}{\partial t} = \frac{\partial S}{\partial t} - \frac{1}{R\cos\phi}\left[\frac{\partial (DU)}{\partial \lambda} + \cos\phi \frac{\partial (DV)}{\partial \phi}\right], \tag{16.13}$$

where λ is longitude, ϕ is latitude, t is time, U and V are horizontal velocity components along longitude and latitude, $W = \sqrt{U^2 + V^2}$, M_{obs} is variation of sea level from equilibrium, S is the bottom displacement, g is the gravity acceleration, R is radius of the Earth, f is the Coriolis parameter, $D = (H + M_{obs} - S)$ is the total water depth, and o is the bottom friction coefficient.

Various approaches to deriving a numerical solution of the above system of equations were outlined in Imamura (1996) [221] and Titov and Synolakis (1998) [222]. In this study, we apply a space-staggered grid, which requires either sea level or velocity as a boundary condition. The first order scheme is applied in time and the second order scheme is applied in space. Integration is performed along the north-south and west-east directions separately in a way that is described in Kowalik and Murty (1993) [223]. To apply this procedure, Equations 16.11 and 16.12 are split in time into two subsets. First, these equations are solved along the longitudinal direction,

$$\frac{U^{m+1} - U^m}{T} + \left(\frac{U}{R\cos\varphi}\frac{\partial U}{\partial \lambda}\right)^m + \left(\frac{V}{R}\frac{\partial U}{\partial \varphi}\right)^m - (fV)^m =$$

$$= -\frac{g}{R\cos\phi}\left(\frac{\partial M_{obs}}{\partial \lambda}\right)^m - \left(\frac{oUW}{D}\right)^m \quad (16.14)$$

$$\frac{1}{2}\frac{M_{obs}{}^{*} - M_{obs}{}^{m}}{0.5T} = \left(\frac{\partial S}{\partial t}\right)^m - \frac{1}{R\cos\phi}\frac{\partial\left(D^m U^{m+1}\right)}{\partial \lambda}, \quad (16.15)$$

and next along the latitudinal direction

$$\frac{V^{m+1} - V^m}{T} + \left(\frac{U}{R\cos\varphi}\frac{\partial V}{\partial \lambda}\right)^m + \left(\frac{V}{R}\frac{\partial V}{\partial \varphi}\right)^m + (fU)^{m+1} =$$

$$= -\frac{g}{R}\left(\frac{\partial M_{obs}}{\partial \phi}\right)^m - \left(\frac{oVW}{D}\right)^m \quad (16.16)$$

$$\frac{1}{2}\frac{M_{obs}{}^{m+1} - M_{obs}{}^{*}}{0.5T} = -\frac{1}{R}\frac{\partial\left(D^m V^{m+1}\right)}{\partial \phi}. \quad (16.17)$$

The calculation of sea level starts from time step m, and the intermediate value of sea level $M_{obs}{}^{*}$ is obtained after integration along the first direction. Afterwards, this value is carried over the other direction to derive the sea level of the $(m+1)$ time step.

In order to propagate the wave from a source to various coastal locations, we use embedded grids, placing a coarse grid in a deep water region and coupling it with finer grids in shallow water areas. We use an interactive grid splicing; therefore, the equations are solved on all grids at each time step, and the values along the grid boundaries are interpolated at the end of every time step [224]. The radiation condition is applied at the open (ocean) boundaries [225]. At the water-land boundary, the moving boundary condition is used in those grids that cover areas selected for inundation mapping [226]. At all other land boundaries, the velocity component normal to the coastline is assumed to be zero.

16.3.1 Finite Difference Model Method

Let u be the velocity of the tsunami particle propagations as a function of M_{obs}, and water depth d which yields

$$u = M_{obs}\sqrt{\frac{g}{d}} \tag{16.18}$$

The leapfrog method is then used to discritize the u. A central finite difference scheme yields

$$\frac{M_{obs\,j}^{\,i+1} - M_{obs\,j}^{\,i-1}}{2\Delta t} + \bar{M}_{obs}\frac{u_{j+1}^{i} - u_{j-1}^{i}}{2\Delta x} = 0 \quad \text{Mass Conservation} \tag{16.19}$$

$$\frac{u_j^{i+1} - u_j^{i-1}}{2\Delta t} + g\frac{M_{obs\,j+1}^{\,i} - M_{j-1}^{i}}{2\Delta x} = 0 \quad \text{Momentum Conservation} \tag{16.20}$$

where $\bar{M}_{obs}$ is average of mean surface. The mathematical solutions of $\bar{M}_{obs}$ and u, respectively, are given by:

$$M_{obs\,j}^{\,i+1} = h_j^{i-1} - \bar{M}_{obs}\frac{\Delta t}{\Delta x}\left(u_{j+1}^{i} - u_{j-1}^{i}\right) \quad \text{Mass Conservation} \tag{16.21}$$

$$u_j^{i+1} = u_j^{i-1} - g\frac{\Delta t}{\Delta x}\left(M_{j+1}^{i} - M_{j-1}^{i}\right) \quad \text{Momentum Conservation} \tag{16.22}$$

The initial conditions at spatial variation times are required for Equations 16.21 and 16.22, respectively. Beside, the boundary conditions are required. In fact, the independent values are specified for left and right boundaries. Consequently, a forward finite difference scheme is implemented to avoid the identifications of initial conditions at two different times. The forward finite difference scheme for the Equations 16.21 and 16.22, respectively, are given by:

$$M_{obs\,j}^{\,i+1} = M_{obs\,j}^{\,i} - \bar{M}_{obs}\frac{\Delta t}{2\Delta x}\left(u_{j+1}^{i} - u_{j-1}^{i}\right) \quad \text{Mass Conservation} \tag{16.23}$$

$$u_j^{i+1} = u_j^{i} - g\frac{\Delta t}{2\Delta x}\left(M_{obs\,j+1}^{\,i+1} - M_{obs\,j-1}^{\,i+1}\right) \quad \text{Momentum Conservation} \tag{16.24}$$

Consistent with Chubarov and Fedotova (2003) [231], a finite-difference scheme in a fixed line segment represents the simplest method for a run-up modelling. The numerical domain includes "underwater" and "dry" points. The method is defined by the grid resolution, time step and the minimal depth, d_0, such that a layer of water of depth less than d_0 is considered dry. The total depth and velocity over the dry regions are set to zero. Calculations in the entire domain, including water and the shore, were performed using several schemes as the MacCormack scheme on a non-staggered fixed grid and the second-order scheme with central differences on a staggered fixed grid. The value of d_0 can be chosen by gradually decreasing this parameter in a series of computations until the shoreline becomes stable. In practice, this value corresponds to the characteristic size of the irregularities on the bottom. For a monotone scheme and a gentle bottom slope, it can be set to zero.

16.3.2 Grid Generation

In the practical application of 2D models, the grid generation is a decisive step. Often it decides upon the success or failure of a model, both with respect to the quality of the results as well as the time consumption for the user. Triangles can be represented by the appearance of acute-angled triangles with one angle is larger than 90° (Figure 16.1). Digital terrain models on the basis of TINs (Triangulated Irregular Networks), which are the basis for the spatial discretization in 2-D, often contain such triangles. That means the triangulated DTM is usually not directly utilisable as discretization grid.

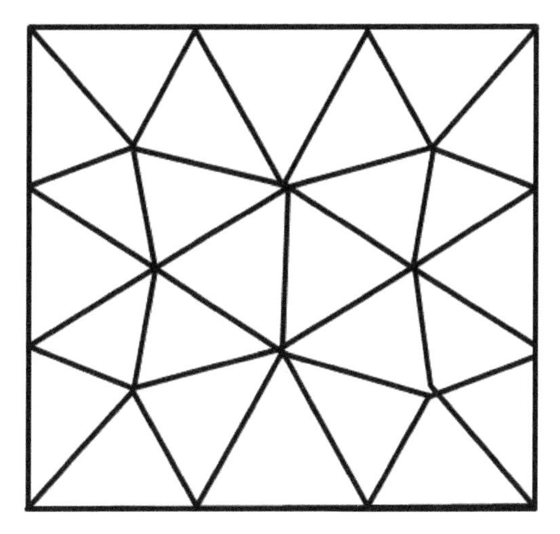

Figure 16.1 Acute-angled triangles.

16.4 Different Study Cases

Two study cases are used to examine the above proposed model. These cases include Malacca Straits and the Gulf of Aden. Therefore, the Malacca Straits is examined based on the submarine earthquake due to Sunda plate tectonic displacement. On the contrary, the expected tsunami along the Gulf of Aden is investigated as a function of the African Horn Rift displacement. In this view, it is assumed that the Grand Ethiopian Renaissance Dam (GERD) can cause an earthquake which is associated with the African Horn Rift.

16.4.1 Malacca Straits

Consistent with Patel et al. (2010) [230], the bathymetric grid is derived from the General Bathymetric Chart of the Ocean (GEBCO) 30 second database and updated (Figure 16.2). Besides, Shuttle Radar Topography Mission (SRTM) data are used to precisely map the land heights (topography).

In practice, it is important to predict the tsunami wave propagation and its timely arrival and geographical locations that could be struck by a tsunami. In doing so, the epcientre is chosen far north of the Boxing day one (Table 16.1). The selected fault degree is 20° and the trench downleft is –5 m and uplift is 5 m which is caused by 9 Richter. It is assumed that the location of the earthquake is further north of the Aceh and located along the Sunda Arc (Figure 16.3).

The 2-D intervals are divided into 200 × 400 in x and y Cartesian, respectively. A finite element mesh based on acute-angled triangles is used which consist of 160000 elements and 80000 nodes (Figure 16.4). This is considered suitable for shallow water area as the Malacca Straits (< 50 m) and for deep water, such as Sundae Arc and Andaman Sea (5000 m).

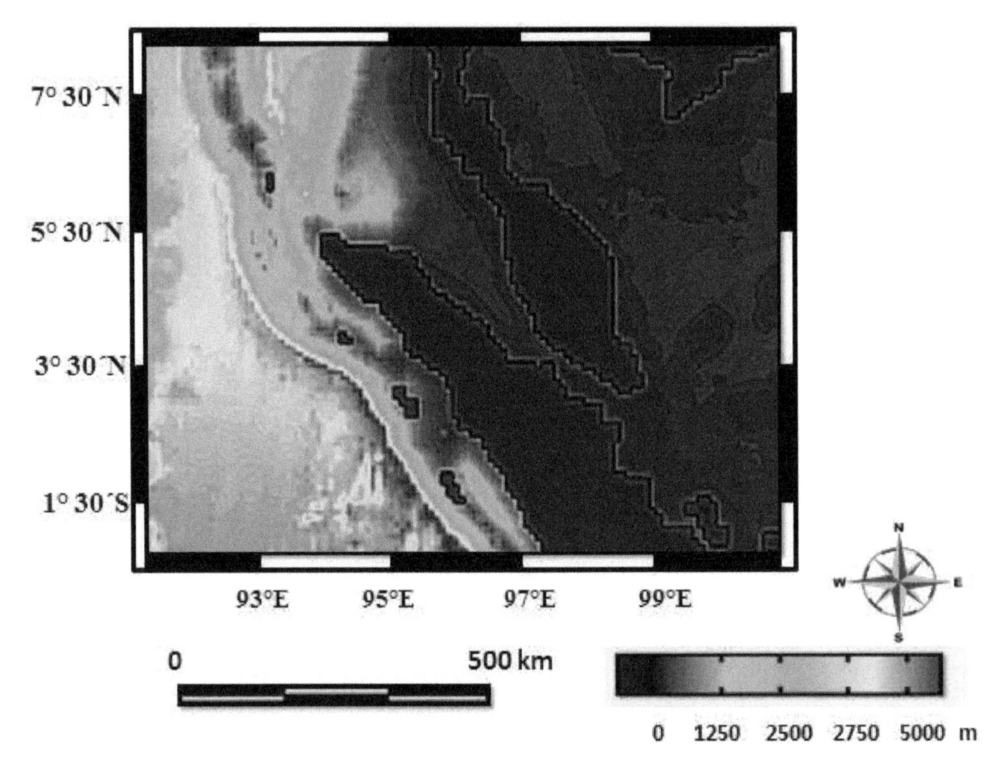

Figure 16.2 Initial Malacca Straits bathymetry.

TABLE **16.1** Fault parameters for possible tsunami occurrence.

Segment	Values
Epicenter longitude (°E)	93
Epicenter latitude (°N)	7.41
Fault Length (km)	400
Fault Width (km)	140
Strike Angle (°)	338
Dip Angle (°)	90
Displacement (m)	12
Focal Depth (km)	25
Time of Occurrence (s)	900

The assumed velocity of tsunami propagation is 35 km/hr. This tsunami is assumed to be propagated from further north of Sunda (7° 41' N) towards the Malacca Straits. The next section shows the simulation results of the tsunami proportion into the Malacca Straits. The most important parameters which would be discussed are the arrival time and variations of the tsunami wave height, tsunami breaking height and run-up.

It is assumed that the next earthquake should occur with a magnitude of 9.0 Richter further north east part of the Sumatra-Java subduction zone. In fact, this zone contains the accumulated stress due the relative plate motions which could be released through a rupture along the plate interface between the Australia-India plates and the Sunda micro-plate. Recent studies in the region have revealed that this segment of the plate boundary has been locked for quite some time since the three largest earthquakes that occurred in

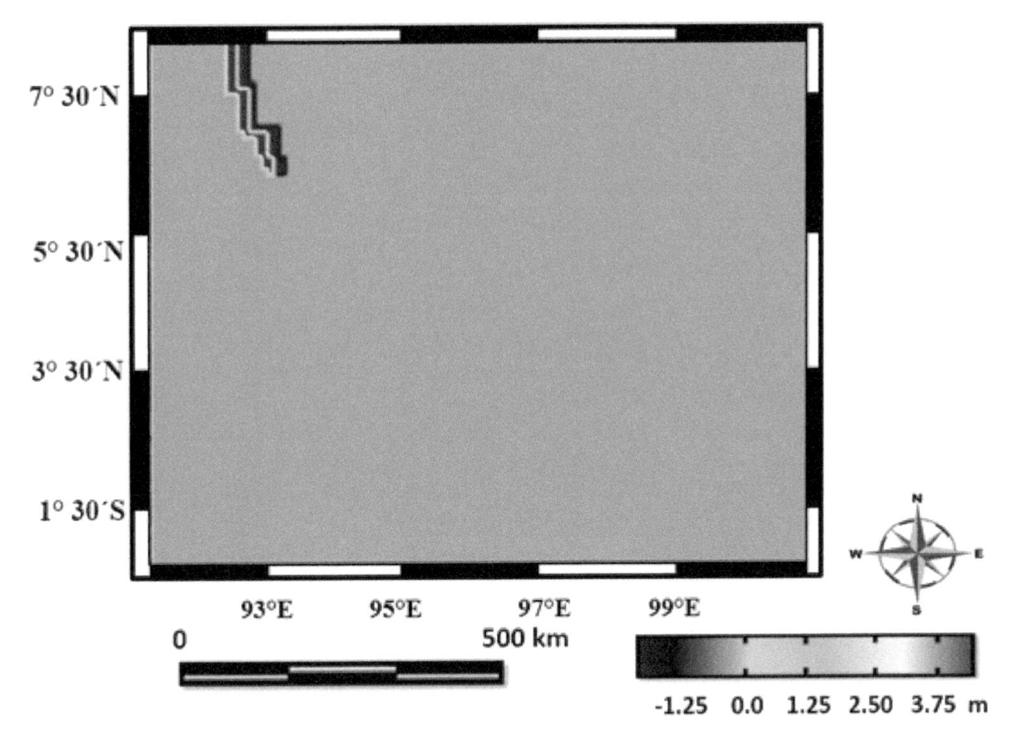

Figure 16.3 Proposed location of an earthquake along Sunda Arc.

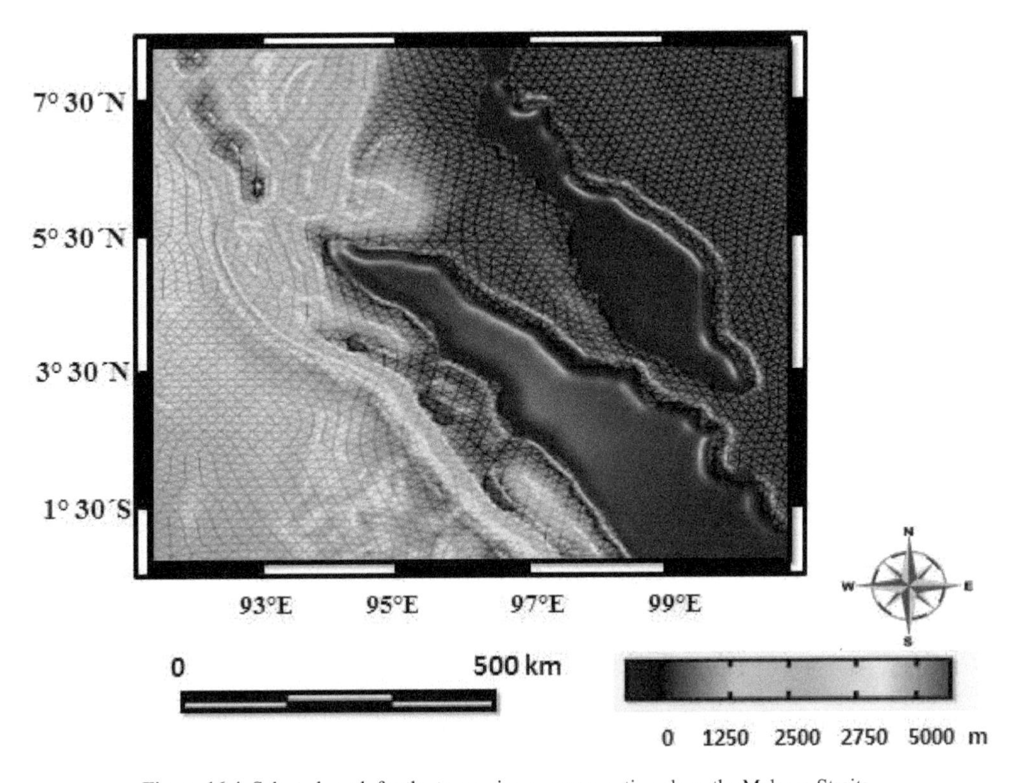

Figure 16.4 Selected mesh for the tsunami wave propagation along the Malacca Straits.

the same general area towards both north and south of the 2004 earthquake in 1833, 1861 and 1881. The seismic slip at greater depths is postulated to be the result of pressure and/or temperature induced steady-state brittle sliding along the plate interface, possibly favored by the fluids released from the subducting slab. The mega-thrust earthquake on Dec 26, 2004 was the latest manifestation of the repeated earthquake cycles along the convergent plate boundary.

The tsunami wave propagations have been simulated previously from altimeter data of TOPEX-1 and Jason-1, respectively [233]. It is clear that tsunami wave propagated around Aceh and into the Malacca Straits with maximum wave heights of 0.8 m. These results are restricted to coastal water of Penang, Malaysia and did not involve the details, or clues about the tsunami wave propagation in shallow water of Johor, Malaysia and Singapore. The Malaysian-Singapore coastal waters are dominated by water depth of 5 m.

Conversely, Figure 16.5 shows the initial tsunami wave propagation with maximum water level of +4.5 m and minimum value of –4.5 m at the source. This explains explain how the tsunami can propagate across the Andaman sea and how it can build up into the Malacca Straits. It is clear that the tsunami wave propagated in curvature geometry near the epicenter with a crest height of 4.5 m and trough height of –4.5 m with a speed of 35 km/hr. Where the ocean is over 6,000 meters (3.7 miles) deep, unnoticed tsunami waves can travel at the speed of a commercial jet plane, over 800 km per hour (500 miles per hour).

After 30 minutes, the tsunami wave propagated from the assumed epicenter, then curves along Indonesia and Penang, Malaysia and then start to flow into the Malacca Straits. The initial tsunami wave could be propagated from the source towards the southern part of Andaman with maximum height of 4 m. It is interesting to predict that tsunami can be propagated towards Thailand and the Malacca Straits with a trough wave value of –2 m. As the tsunami wave propagates, it will be curved along the Andaman sea, and Thailand coastal waters due to the impact of the bathymetry (Figures 16.2 and 16.4).

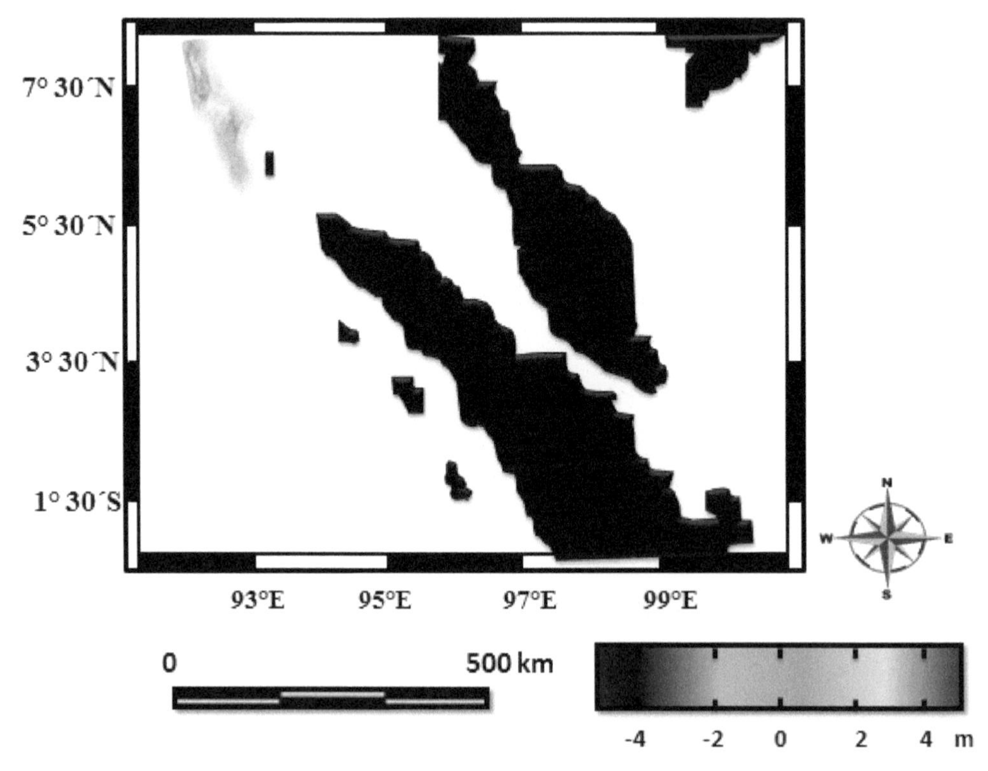

Figure 16.5 Initial tsunami propagation.

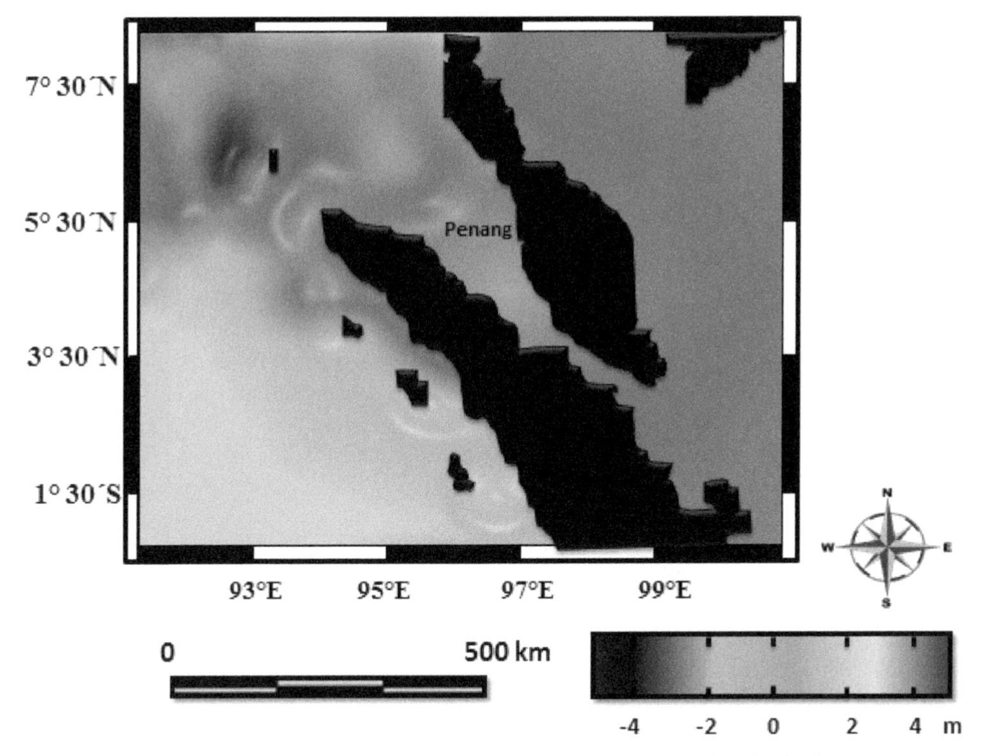

Figure 16.6 Tsunami wave propagation with 30 minutes after earthquake shock.

The arrival time in Penang and Kedah coastal waters can take approximately about 60 minutes, which is faster than 2004 tsunami. This occurs when the epicenter of the earthquake along the Sunda Arc is located further to the north of 5.22° N. The estimated wave height can strike both coastal waters of Myanmar, Thailand and Aceh in approximately 8 and 9 m, respectively. This wave height decreases gradually to 3.5 m along Penang coastal waters (Figure 16.7). The successive breaking of tsunami wave within 6 m can be due to the shallow water bathymetry of less than 30 m.

The tsunami would continue propagation with maximum wave heights of 8.6 m over 30 m water depths in the center of the Malacca Straits. Hence, the tsunami propagation could generate eddy turbulent flow in the Andaman Sea and the northern part of the Malacca Straits due to the fast speed of the tsunami and breaking of the group of the successive tsunami wave crests and troughs (Figure 16.7). The maximum wave height which could be predicted in the Andaman Sea is 8 m within 150 minutes (Figure 16.8). In two hours, the successive tsunami wave height of 8.7 appears in the centre of the Malacca Straits between Malaysia and Indonesia (Figure 16.8).

Singapore, Johor, Pulau Bulan, and the Batam Islands will strike with 9 m of tsunami height in 180 minutes. This will be occurring because of the zone of the Singapore Strait and Batam Island. Further, the shallow water depth which is above 10 m is the dominant feature of the Singapore Strait (Figure 16.9). It is worth noticing that in 3 hours, the tsunami turns into a stream turbulent flow as the Malacca Straits becomes narrow in the south.

Combined with the speed at which the tsunamis travel of up to 700 km/h over the Malacca Straits, slowing to 100 km/h as they approach the Singapore coastal zone, they can break with heights of 24 meters (Figure 16.10), which can cause extensive damages. The Penang and Kedah will be the first coastal waters in Malaysia to be struck by the tsunami. In fact, they are located close to the south of the Andaman Sea. The tsunami wave could break with wave height of 21 m along the coastal waters of Penang and then a tsunami wave would continue its propagation to the Singapore Strait with wave height of 8 m (Figure 16.10a).

Figure 16.10b shows the tsunami wave breaking with more than 20 m along the Johor, Singapore and Pekanbau. In fact the narrow Singapore Strait will contribute to an extreme wave energy concentration across the Singapore Strait and Pekanbau. This tsunami will assist in swallowing up small islands located in the Malacca Straits due to extreme tsunami heights of 20 meters and fast speed of 100 km/h. This can occur within an arrival time of 180 minutes (Figure 16.10b).

A breaking tsunami is a wave whose amplitude turns up a perilous stage at which some procedure can unexpectedly begin to occur. In this view, it can create massive wave run-up with strength to be transformed into turbulent kinetic power (Figure 16.11). The maximum wave energy that can be input in coastal water of the Malacca Straits is 9E+014 J. This can be predicted easily along Thailand west coast, Penang, Aceh and Singapore. At this factor, easy physical fashions that describe wave dynamics often end up invalid, mainly those that expect linear behavior. Possibly the existences of sandbars inside the Malacca straits can cause a plunging wave. In this regard, the crest of the wave will become much steeper than a spilling wave, becomes vertical, then curls over and drops into the trough of the wave, liberating most of its strength straight away in a notably violent effect. A plunging wave, consequently, breaks with massive power than a substantially larger spilling wave. The wave can trap and compress the air underneath the lip, which generates the "crashing" sound associated with waves. With huge waves, this crash may be felt through the coastal zone.

When a plunging wave is not parallel to the shoreline (or the ocean bathymetry), the part of the wave which approaches the shallow water will vanish first, and the breaking crest (or curl) will move laterally across the face of the wave as the wave conserves. Then, as such, the surf will generate and form a barrel. The sequences of barrels will generate along the coastline of Johor, Singapore, Batam Island, and Pulau Bulan which lead to massive damages. This will sweep out the reclaimed lands and many of coastal zone infrastructures.

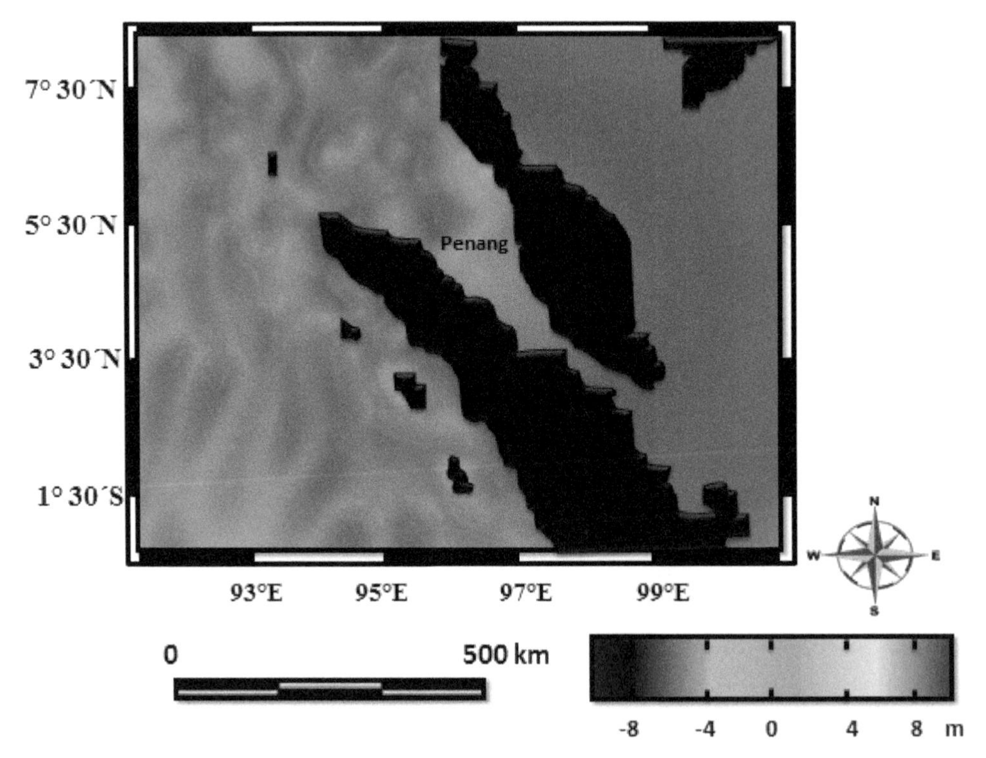

Figure 16.7 Tsunami wave propagation with 60 minutes.

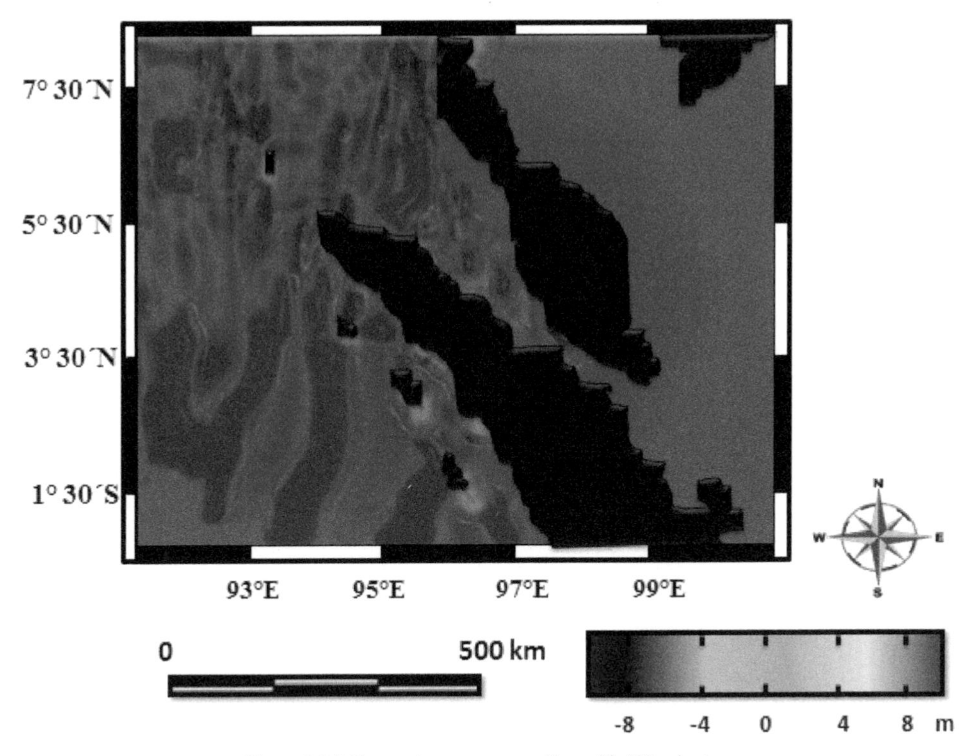

Figure 16.8 Tsunami wave propagation with 120 minutes.

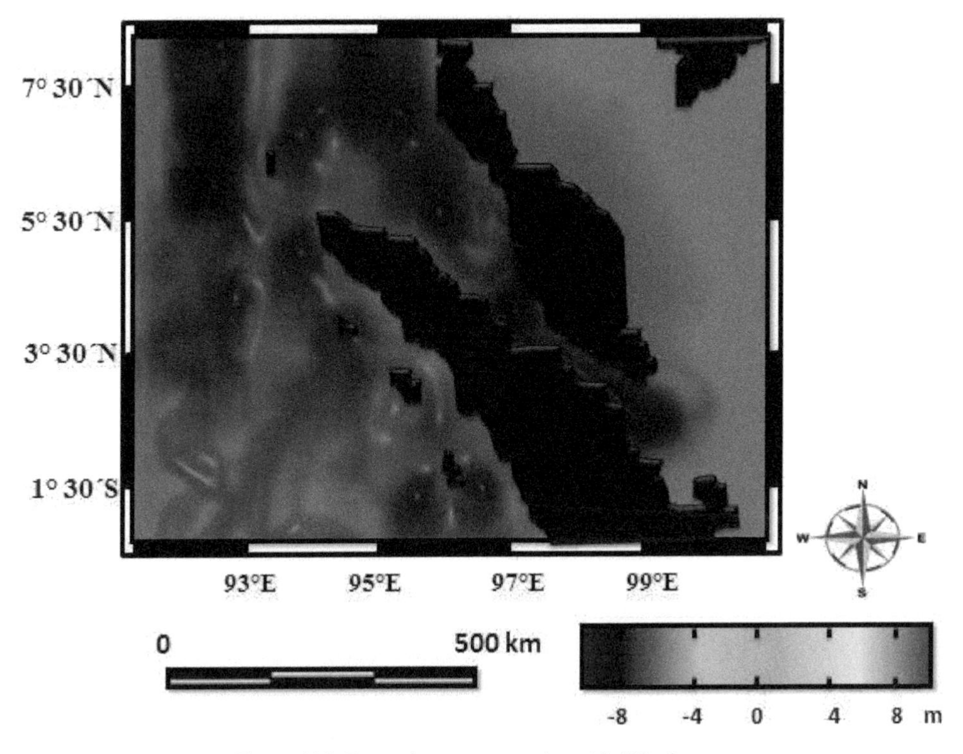

Figure 16.9 Tsunami wave propagation with 180 minutes.

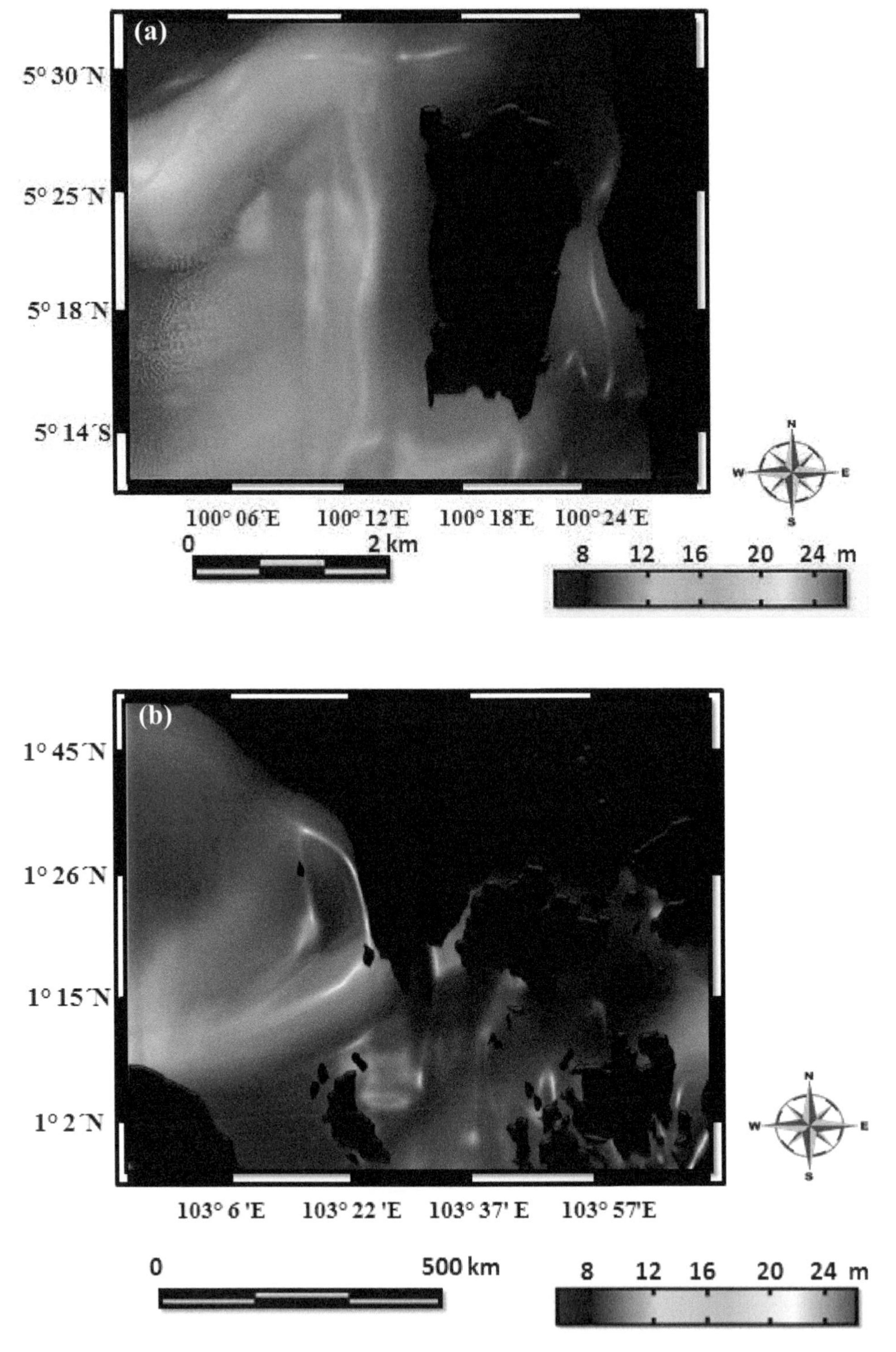

Figure 16.10 Tsunami wave breaking along (a) Penang and (b) Singapore Strait and Pekanbau.

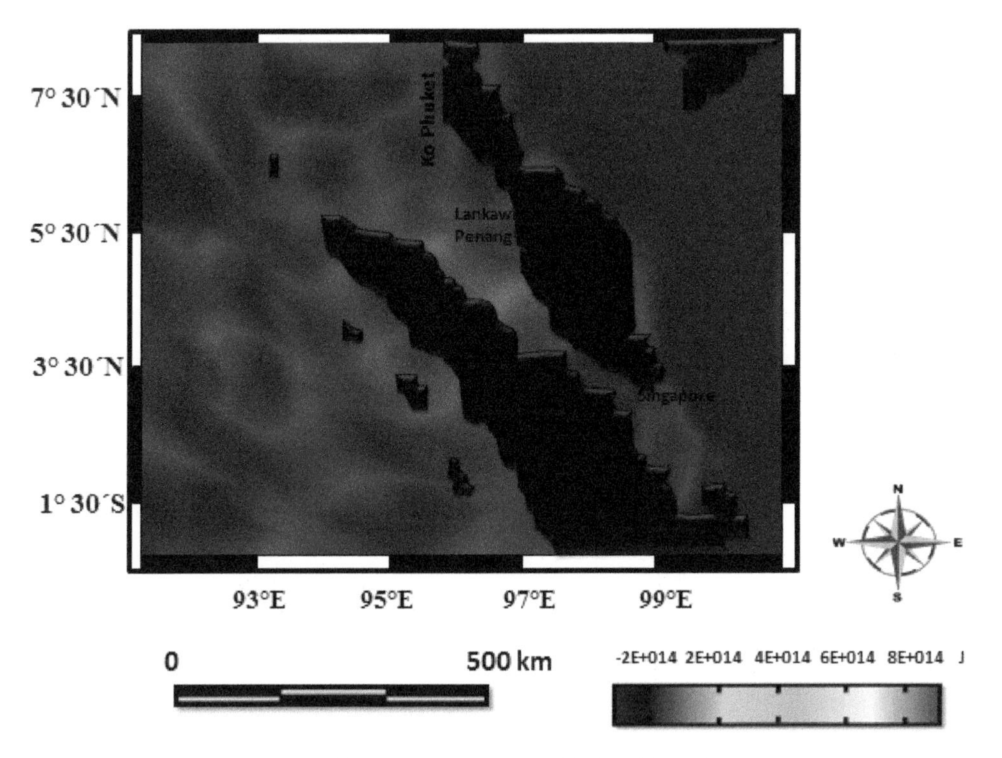

Figure 16.11 Energy input by tsunami in the Malacca Straits.

Earlier, Singapore has been quaked by Bengkulu earthquake in 2007 with magnitude of about 8.5. Even though, the epicentre was approximately 700 km away, various constructions in Singapore sensed the shock and some were even evacuated. Singapore, consequently, is dominates by soft soil and reclaimed land which can be amplified by a distant seismic wave. Under this circumstance, Singapore, Johor, and Batam Island can be hit by any future tsunami and even the main coastal infrastructure, without being too high, could get damaged.

16.4.2 Red Sea and Indian Ocean Tsunami by Grand Ethiopian Renaissance Dam (GERD)

It is well recognized that GERD can cause a massive destruction for Sudan, and Egypt. Consequently, can GERD cause a tsunami through the Red Sea and Indian Ocean? This section tries to answer this serious question. In fact, GERD is located in an active plate tectonic zone which is known as the African Horn (Figure 16.12). This zone is dominated by metamorphic and sedimentary rocks, respectively. The dangerous issue is that shear stress zones and approximately four faults cross-cut these rocks. In this view, the dam site is included in the Nubian Block which is constrained by the African Rift. In this regard, the faults indicate instability zone where there are huge tensions in the plate. In fact, this tension is the main factor for inducing faults. Therefore, there are plenty of earthquakes that have occurred in this area with magnitude larger than 6 Richter. Volcanic activity is also found in this zone and numerous of the conical hills in the mountains are volcanic.

On 31 August 2012, a quake of 5.3 Richter struck the Gulf of Aden because of the plate displacement of the west Shoba ridge (Figure 16.13). It has a depth of 10.2 km. Previously, this zone was dominated by the earthquake of 5.0 to 6.3 Richter on 31 October 2011, and 13 July 2015, respectively. The Ethiopian,

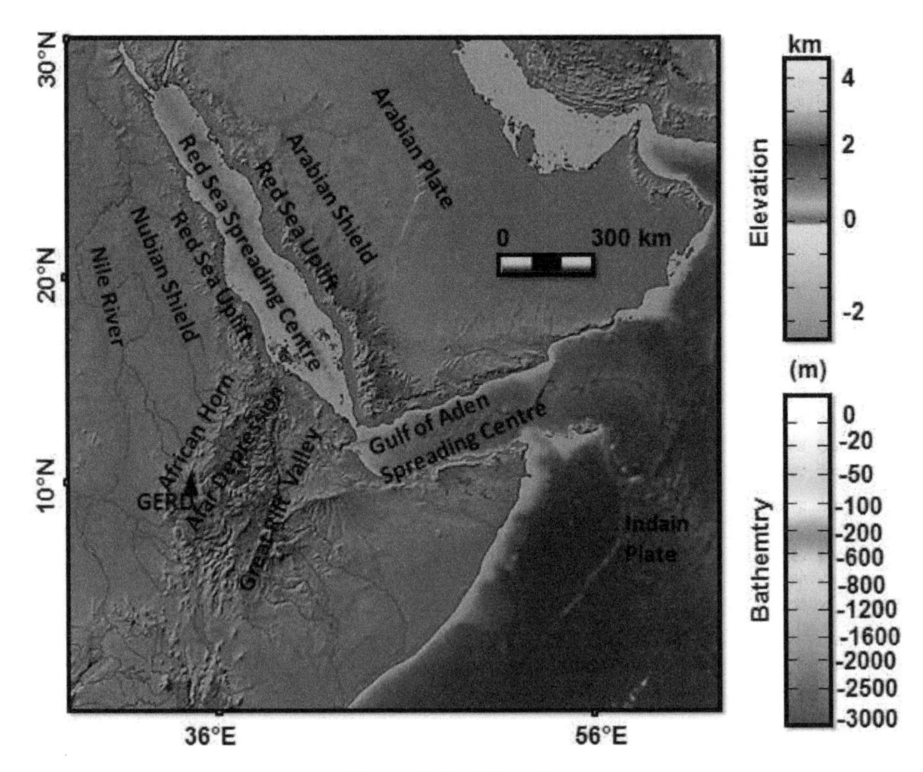

Figure 16.12 African Horn.

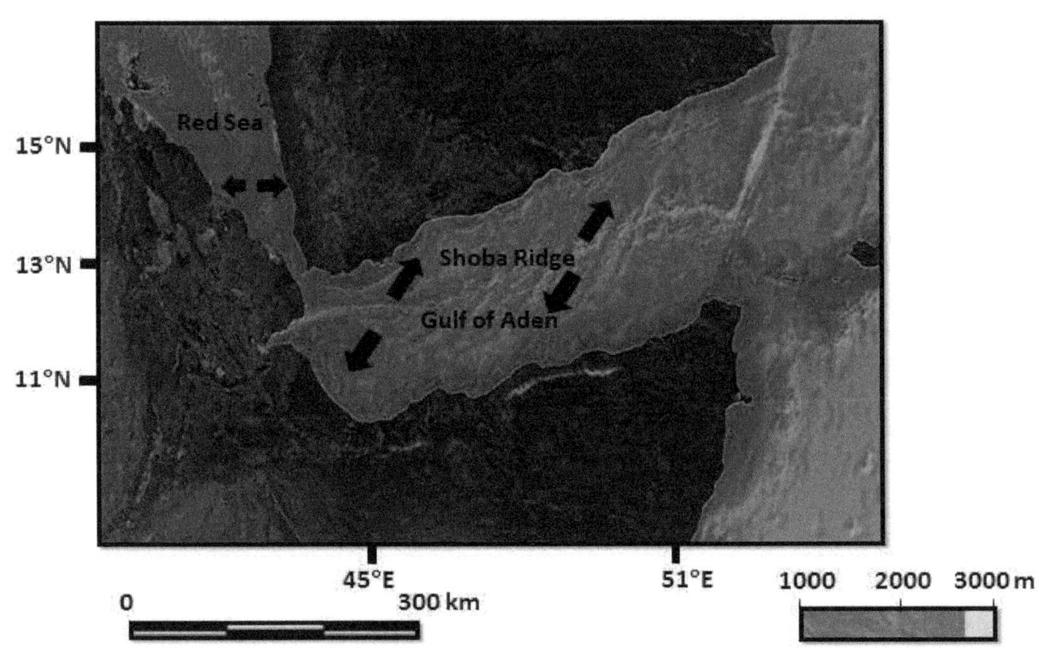

Figure 16.13 Shoba ridge along Gulf of Aden.

Somalian and Yemen plateaux, consequently, express the trilateral form of the Afar Depression (AD), and involve numerous foremost tectonic units, specifically the East African Rift, the Gulf of Aden and the Red Sea. Further, AD is clockwise rotation for the southern Afar owing to 'bookshelf' left-lateral faulting between left-stepping overlapping rifts [279]. The seismic activity is uppermost along the axes of the Red Sea, which is intense at the northern and the southern part. Furthermore, it spreads to the Gulf of Aden and the Afar triple joint through the foremost Ethiopian rift.

By the year's end, one of the world's largest dams will begin filling up, inducing a greater shear stress along the African Horn. Indeed, the heavy volume of 74 billion cubic meters will store at the dam water, which can allow tectonic plates to move apart without much thinning of the crust, as this is 300% oversized. In other words, the surface area of the dam is 1,680 square kilometres and its height is 145 m high with a length of 1.8 km which would accelerate the collapsing of the dam. Further, the dam will induce shear stress on the four faults cross-cut [279].

In this understanding, the earthquake can be generated with an approximately 9 Richter. The main question is how can GERD induce tsunami in the Red Sea and the Gulf of Aden, despite the GERD being located 1000 km away from the Shoba ridge?

Recently, scientists demonstrated that the 2014 Chile quake shook other faults and made them extremely likely to slip. The Chile quake of 8.3 magnitude triggered aftershocks as far as 1000 kilometres away. In fact, seismic waves, either strong or weak, can shake up particles of rock compressed inside remote faults. The earthquakes occur when two tectonic plates that have been pressed together suddenly slip. In contrast, 1992's Landers earthquake in California sent out seismic waves that triggered copycat quakes 1000 kilometres away, although the waves become lower as they move [280].

Two scenarios will lead to a massive earthquake along the African rift which are stressed due to the massive volume of water inside the dam and its collapsing. Under these circumstances, the seismic waves with a magnitude of 9 Richter would actually push in the direction against a slip, and induce an earthquake. This earthquake will vibrate each faults which exist along the African rift and the one located in the Red Sea causing a tsunami. When a signal approaches at acoustic resonance frequency (AF), it will trigger an earthquake and tsunami in the Red Sea and the Gulf of Aden (Figure 16.14). The initial wave height would be around 4 m (Figure 16.14) because of the narrow width of the Gulf of Aden, i.e., 250 km width. The tsunami would travel from the Gulf of Aden towards the Red Sea, Arabian Gulf, India and Sri Lanka. The time of arrival in India and Sri Lanka is approximately two hours. In less than two hours, the wave height inside the Red sea and Arabian Gulf will range from –8 m to 8 m. These sequences of the wave will reflect and diffract along the coastal waters of the Red Sea (Figure 16.15). It is worth mentioning that the Indian coastal waters will be struck by tsunami wave height of 8 m in two hours (Figure 16.15).

According to the AF mechanism, seismic fracture produces elastic waves that diffuse and scatter inside the fault, generate a normal stress, contrasting with the confining one, and thus promote seismic failure. The same mechanism could be also activated by transient seismic waves generated by other earthquakes. AF, indeed, has been also proposed to explain why seismic activity is observed to increase within minutes after big earthquakes, in areas at a distance of thousand kilometres from the mainshock epicentre [279]. Further, the depth of the Indian Ocean then allowed the wave to travel great distances without losing much energy.

The Digital Elevation Model from GERD is extremely steep as it builds top DEM of 5000 m with very steep reach of 170 to 300 kilometers downstream of the dam. In this view, the collapsing of GERD or heavy weight after an initial filling can cause a megatsunami in which 74 million cubic metres of water will overtop the dam in a wave about 850 metres high, similar to Vajont catastrophe, Northern Italy [281]. This can drive catastrophic collapse in Sudan and Egypt. The steep DEM from GERD to Sudan passing to Egypt will accelerate the damage and vanish Sudan and Egypt in 1 and 3 hours, respectively. The train of wave in more than 100 m height, and wavelength in more than 1000 m with vortices will run across the Egypt and will accelerate sinking of the most Egyptian territory under massive wave (Figure 16.16). The overtoping water wavelength is about initial water overtoping a dam times the constant value of 7.042254 [268]. In addition, the High Dam will collapse and its wave superposes the massive wave propagation post collapsing of GERD. In the area of High Dam, upper Egypt, the great vortex will be generated. It is clear

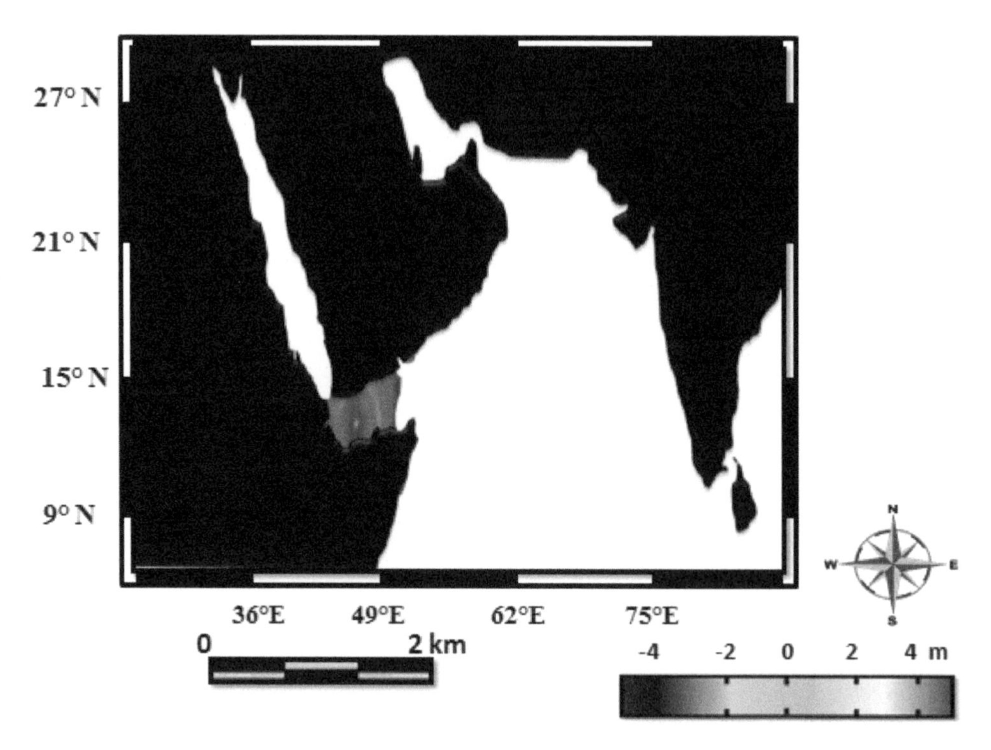

Figure 16.14 Initial tsunami height in Gulf of Aden.

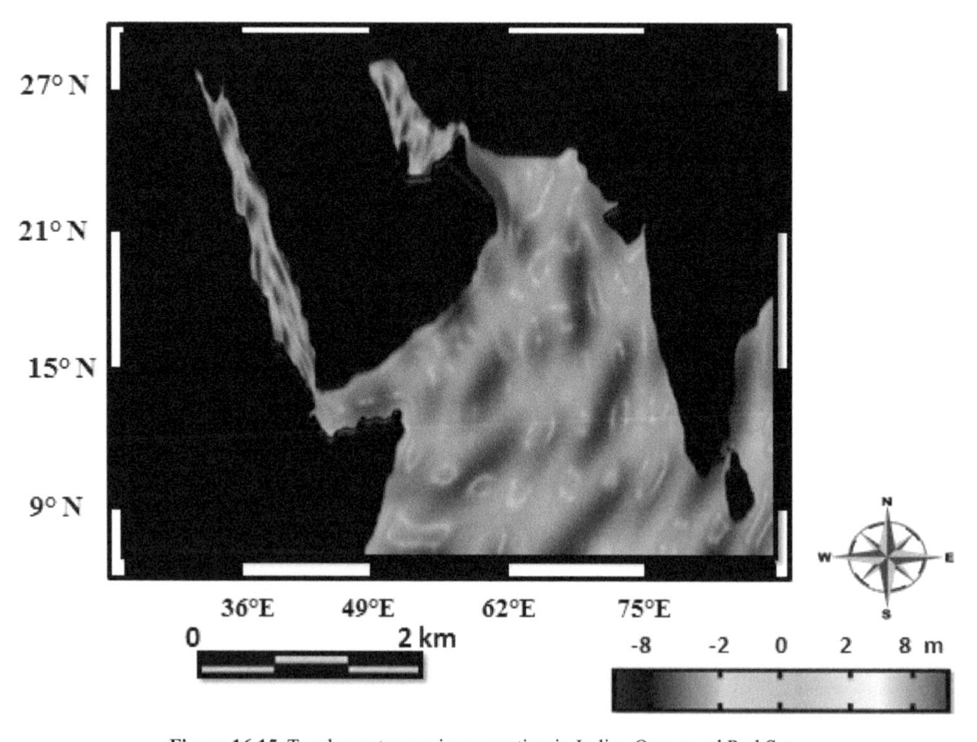

Figure 16.15 Two hours tsunami propagation in Indian Ocean and Red Sea.

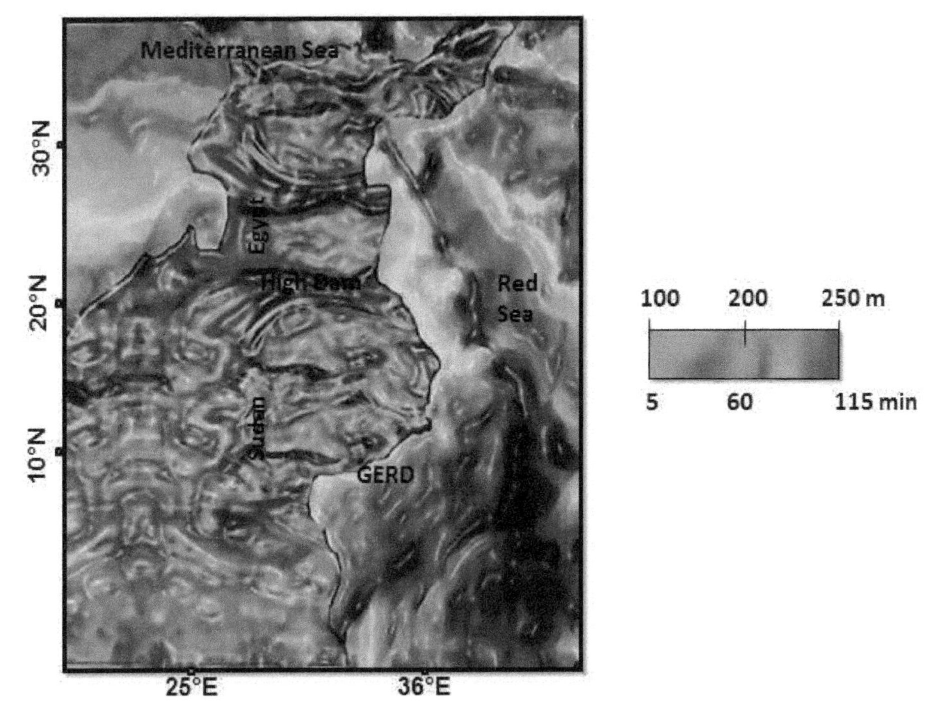

Figure 16.16 Massive tsunami time propagations due to GERD collapsing.

that the Nile Delta will submerge under the water height. Finally, the water will run into Medditerrian Sea forming a great eddy which will be moved up to the Cyprus and Syrian coastal waters. In contrast, the Red Sea mountain zone will be prevented from the disaster of GERD collapsing as the height of these series mountains is higher than 1000 m above the sea level.

The foremost cause of failure as cited in the catalogues is overlapping. More water flows into the reservoir than the reservoir can hold or pass through its spillway. The excess water has to go somewhere, and the most likely place is over the top of the dam. This does serious damage to the dam, especially to an embankment dam, which is likely to fail.

The risk of dam failure is also not uniform over the life of the dam. Like most engineered products, the chance that a dam will fail is the highest during the first use, which for a dam is first-filling, the first time that the reservoir is filled to capacity. If something was overlooked, or if some adverse geological detail was not found during exploration, then this is usually the time that it will first become apparent. As a result, about half of all dam failures occur during first filling. The other half occurs more or less uniformly in time during the remaining life of the dam. So, if the rate of failure averaged over the whole life of a dam is about 1/10,000 per dam-year, the rate during the first, say, five years reaches almost 1/1,000 per dam-year, or ten times higher. This is exactly what the historical record shows.

That about half of all dam failures occur during the first filling is a troubling observation, for the following reason. In the arid areas, which use dams primarily for irrigation and only secondarily for flood control, reservoirs are often kept full. If a heavy storm is forecast, the reservoir is lowered to make room for the larger inflows coming from upstream. But in temperate regions, where dams primarily serve flood control needs and irrigation is not an important benefit, reservoirs are typically kept low. If a flood comes, either its entire flow is caught behind the dam, or if it is a very large storm, at least its peak flow is caught. But since most flood control reservoirs are designed for floods of a size that essentially never comes—the probable maximum flood (PMF)—many dams in temperate regions, such as the eastern US, have never experienced design pool levels, they have never seen first filling, and thus have never been proof tested.

The probability of failure of these dams, should an extreme flood come, could be ten times greater than that of a load-tested dam. Of course, the chance of PMF is purposely remote.

16.5 Tsunami from Point View of Quantum Mechanics

Schrödinger equation is implemented in this chapter to derive the sea surface modulation due to the deformation of submarine fault. This makes the possible description of tsunami wave propagation by the quantum mechanics theory. Even the dam derives catastrophic collapse for the Sudanese and Egyptian territories also can be by explicated by the quantum theory.

The most important property in the quantum mechanics theory is wave function. Consequently, the wave function pronounces the spatial distribution of tsunami across the space. In other words, the magnitude of tsunami propagation at each Cartesian position is proportional to the probability of locating a tsunami wave around any point in the propagation direction. Like a particle moving in the space-dimension, the tsunami quantity $|\psi(x, y, z, t)|^2 \, \partial M_{obs}$ is the probability that the tsunami wave can be located at the time t within height changes ∂M_{obs} around the Cartesian coordinate (x, y, z). Therefore, the tsunami is most likely to be located in regions where $|\psi|^2$ is large which is required for $|\psi|^2$ to be normalized. In other words, the integral of $|\psi|^2 \, \partial M_{obs}$ overall space must match precisely 1. Under this circumstance, the probability of tsunami propagation at each point must accurately correspond to 100%.

$|\psi(x, y, z, t)|^2 \, \partial M_{obs}$ is also considered as the probability distribution function (pdf) of the tsunami wave propagation. In other words, it identifies the probability of locating the tsunami wave at different locations and times, which varies over the tsunami travelling space. This makes sense: as modulated sea surface travels from the main epicentre towards the shallow waters in certain time, the height of tsunami wave grows larger. Under this circumstance, the Schrödinger equation must be dependent on time, otherwise, the tsunami turns into a state of a definite energy. This occurs as long as the tsunami does not move into shallow waters. In other words, a valid definite energy of tsunami must satisfy the Schrödinger equation with constant potential energy or equals zero. Thus, the transition between offshore to shallow waters is required for the time-dependent Schrödinger equation.

Now suppose that the tsunami is moving on the ocean's surface. Whereas for tsunami wave, the crest and trough of the wave (such as a resonating violin string) would exchange, the nodal array would continue unchanged. In a travelling tsunami wave, nevertheless, the entire wave, nodes and all, are travelling in time. The mathematical description of the tsunami wave time dependent is

$$\psi = M_{obs} \, e^{2\pi i v t} \tag{16.25}$$

Equation 16.25 demonstrates that a wave function Ψ is possessing both space and time t variables where v is velocity of tsunami propagation across the sea surface. In view of the Hamiltonian operator H, the tsunami energy is considered as eigenvalue E. The development of the Schrödinger equation can lead to extension of H. In this view, H includes both kinetic and potential energy of tsunami propagation. In this regard, the various *quantum states* of the tsunami propagation through its wave function and energy pairs of $\psi_1, E_1; \psi_2, E_2; \psi_3, E_3; \psi_4, E_4; \ldots$ can be expressed as:

$$H\Psi(x, y, z) = E\Psi(x, y, z) \tag{16.26}$$

The time evolution of the tsunami motion can be identified by splitting of the energy level when approaching shallow waters and hits the shoreline. This can introduce a new approach for run-up energy. Prior to run-up, there is water receding which has a lower energy function $E_0\Psi_0(x, y, z)_0$ than second wave propagation $E_i\Psi_i(x, y, z)_i$. However, the run-up wave function can be considered as the superimposed of the second and third wave function $E_{i+1}\Psi_{i+1}(x, y, z)_{i+1}$. This leads to awkward run-up wave due to superimposed large wavelengths with different wave number and heights which is only localized along different shoreline positions (Figure 16.17). Under these circumstances, wave's momentum no longer has a definite energy.

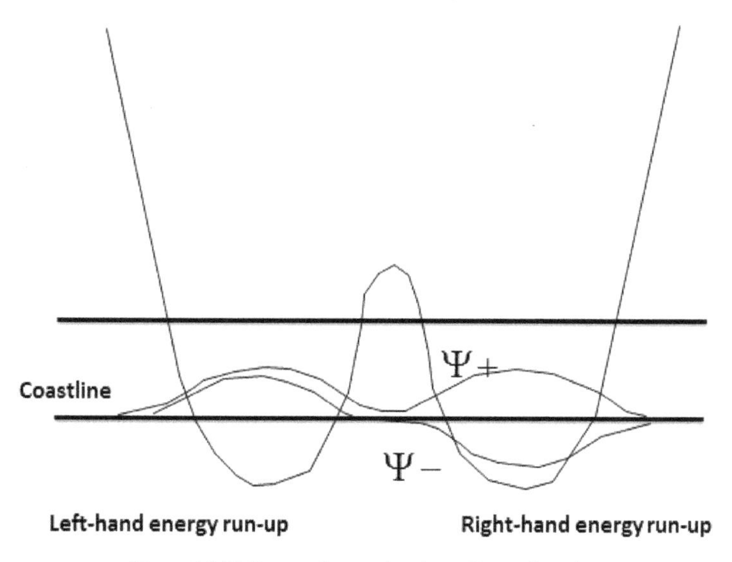

Figure 16.17 Run-up from point view of wave function.

In this view, we can represent a run-up as both particle and a wave. The run-up is a particle in the sense that it is well localized along shoreline. However, run-up also has a periodic structure which is characteristic of waves. Consequently, a narrow function of run-up amplitude $\Psi(A)$ will lead to a sharp peak of run-up and give a wave function $\Psi(x)$ with a broad spatial inundation extent. On the contrary, a wave function with a narrow spatial inundation extent is delivered by a wide function $\Psi(A)$.

Since tsunami height is related to water depth, the deeper the water, the larger the potential size of a tsunami. It is believed that the very large size of the Indonesian tsunami was due not only to the magnitude of the earthquake and seafloor rupture, but also because the seafloor displacement took place near a deep sea trench, in very deep water. The depth of the Indian Ocean then allowed the wave to travel great distances without losing much energy.

16.6 Quantum Viewpoints of GERD Impacts

Consistent with plate tectonic theory, the surface of the Earth is broken into a chain of comparatively thin, but huge plates. Consequently, the mass and location of these plates revolutionize over time. For instance, Pacific Plates are the rapidest plates, which exchange at over 10 cm/year. On the contrary, the African plate changes approximately 25 mm/year. Conversely, the Australian plate changes more rapidly than the African plate with 60 mm/year. The boundaries of these tectonic plates move away from each other. In this regard, this mechanism is responsible for inducing intense geologic activities of earthquakes, volcanoes, and mountain building.

The Great Rift Valley is the one of the longest fault systems in the world. It splits Africa into two new separate plates of the Nubian Plate and the Somali Plate. In addition, Afar Depression is a geological depression in Ethiopia caused by the Afar Triple Junction which is a part of the Great Rift Valley in East Africa. In this view, the depression overlaps the borders of Eritrea, Djibouti and the Afar area of Ethiopia which consists of the lowest point in Africa, Lake Asal, Djibouti, at level of 155 m beneath the sea level. In 2006, it was rocked by a series of earthquakes in which the depression widened by about three meters and sank by a further 100 m. GRED geometry was designed to store water volume of 74 billion cubic which would cause horizontal pressure along the edge of a rug. Under this circumstance, folds are performed upright to the path of shear stress. While horizontal shear stress pertains to rocks, ridges and valleys develop steeply to the route of the shear stress.

Consistent with quantum theory, every mass m is proportional directly with momentum E, then the wavelength is proportional inversely with E. In other words, $L = \frac{h}{E}$ where h is a plank constant. As long we are dealing with wave function $|\psi|^2$, h is ignored. In this regard, $L \sim E^{-1}$ can describe the shear stress created by GRED on the basis of wave-particle interpretation of quantum mechanics. Let the total energy of a particle be

$$E = mgy_0 \tag{16.27}$$

Equation 16.27 demonstrates that the particle energy is a function of the gravitational potential g.

Equation 16.27 can explain the impact of GRED on the plate tectonic along the Horn of Africa.

At GRED prior full water volume storage, the horizontal distance along the fault $x = 0$, which leads to the wave function $|\psi|^2$ is zero owing to the vast potential. Under this circumstance, the particle velocity is high and wavelength is small with small amplitudes. When the GRED is completely full by water volume of 75 billion cubic which is considered as over-sizes its geometry designing, $x = L$. In other words, the particle's wave gets huge and overtops the GERD of 155 m height. This would increase the size of x and wavelength L. In this context, the wavelength would be the seismic wave propagation along the fault of African Horn and tidal wave generated due to collapsing and overtopping the water in the dam. Based on the Schrödinger equation, the maximum wave height would be 350 m, as distance increases to 400 km (Figure 16.18), which is shown in Figure 16.17. In other words, the allied wave of fault and flooded wave will have the elongated wavelength and the largest amplitude.

The fault boundaries would be dominated by the total energy of $E_T = \sum_0^\infty E^2 2m$ which is approaching the infinite. Under this circumstance, the east of African Horn can be separated from Africa and moved away to the Indian Ocean (Figure 16.19). In this regard, $|\psi|^2$ is considered as a probability of realizing the geologic fragmentation at any point along the African Horn due to the shear stress inputs by GERD. The higher the shear stress input along the African Horn, the maximum $|\psi|^2$, namely, the probability of the destructive tsunami across Sudan, Egypt, Red Sea, Indian Ocean and splitting of the east side of the African Horn of Africa and ending of the Nile River life time (Figure 16.19).

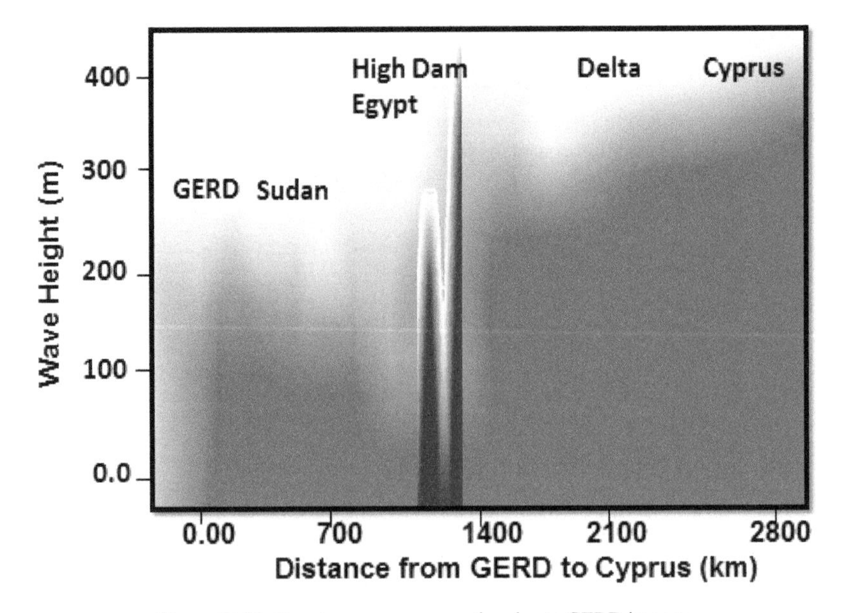

Figure 16.18 Quantum wave propagation due to GERD impact.

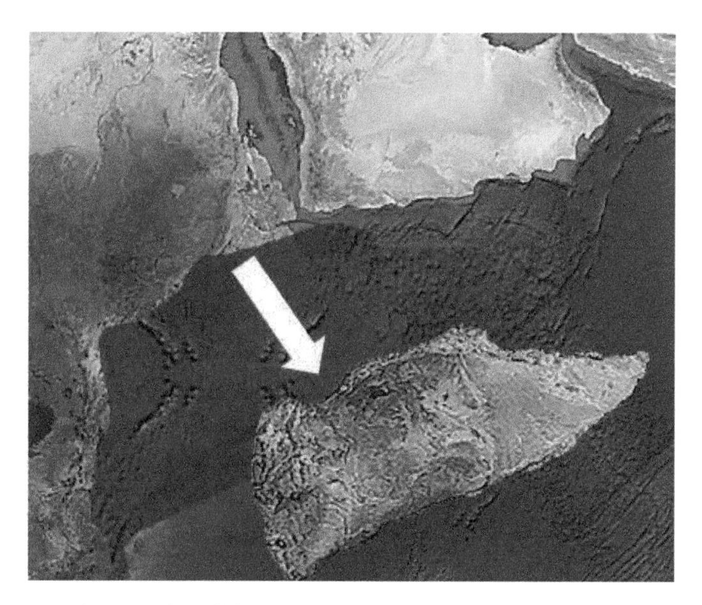

Figure 16.19 Splitting mechanism of the east side of African Horn.

Finally, GERD faces a serious and unprecedented risk of catastrophic failure with no warning. The flood wave would resemble an inland tidal wave between Sudan and Lower Egypt, and would sweep downstream anything in its path. Flooding upper Egypt, specially with collapsing of the High Dam, would resemble that of Hurricane Katrina, with standing water that pervades all Egyptian territory for few months.

The resulting "tsunami-like wave" would crush nearly all Sudanese territory, flattening major cities. A breach in the country's biggest dam would unleash an "inland tidal wave" which could reach tens of feet in height and knock out Egypt's entire power grid. Sudan and Egypt could be under 45 feet of water within hours of a dam breach. The flood water would even reach Syrian and Cyprus coastal waters within a few hours and bring chaos to the Sudan and Egypt capital with "increased health hazards, limited to no mobility, and losses of homes, buildings, and services".

References

1 Dawson A. and Stewart I. 2007. Tsunami geoscience. Progress in Physical Geography 31(6): 575–590.
2 University of Washington. 2013. Earth & Space science: The Physics of Tsunamis. http://earthweb.ess.washington.edu/tsunami/general/physics/physics.html (accessed 11 November 2013).
3 Taru T. 2013. Japan Experts Warn of Future Risk of Giant Tsunami. http://www.cosmostv.org/2012/04/japan-experts-warn-of-future-risk-of.html (accessed 11 November 2013).
4 Halabrin N. and Lamb R. 2013. How Tsunamis Work. http://science.howstuffworks.com/nature/natural-disasters/tsunami.html (accessed 11 November 2013).
5 Zahibo N., Pelinovsky E., Talipova T., Kozelkov A. and Kurkin A. 2006. Analytical and numerical study of nonlinear effects at tsunami modelling. Applied Mathematics and Computation 174(2): 795–809.
6 Zaitsev A., Kurkin A., Levin B., Pelinovsky E., Yalciner A.C., Troitskaya Y. and Ermakov S. 2005. Modeling of propagation of the catastrophic tsunami (December 26, 2004) in the Indian Ocean. Doklady Earth Sciences 3: 403.
7 Titov V.V. and Synolakis C.E. 1996. Numerical modeling of 3-D long wave runup using VTCS-3. pp. 242–248. *In*: Liu P., Yeh H. and Synolakis C. (eds.). Long Wave Runup Models. World Scientific Publishing Co. Pte. Ltd., Singapore.
8 Haugen K.B., Løvholt F. and Harbitz C.B. 2005. Fundamental mechanisms for tsunami generation by submarine mass flows in idealised geometries. Marine and Petroleum Geology 22: 209–217.
9 Liu P.L.F., Wu T.R., Raichlen F., Synolakis C. and Borrero J. 2005. Runup and rundown generated by three-dimensional sliding masses. Journal of Fluid Mechanics 536: 107–144.
10 Kelmelis J.A., Schwartz L., Christian C., Crawford M. and King D. 2006. Use of geographic information in response to the Sumatra-Andaman earthquake and Indian Ocean tsunami of December 26, 2004. Photogrammetric Engineering & Remote Sensing 72(8): 862–876.
11 Abbott P.L. 2008. Natural Disasters, 6th edition. McGraw Hill Higher Education, New York.
12 Catherine J.K., Gahalaut V.K., Ambikapathy A., Kundu B., Subrahmanyam C., Jade S., Bansal Amit, Chadha R.K., Narsaiah M., Premkishore L. and Gupta D.C. 2008. Little Andaman aftershock: Genetic linkages with the subducting 90° E ridge and 2004 Sumatra-Andaman earthquake. Tectono Physics 479: 271–276.
13 Shulgin A., Kopp H., Klaeschen D., Papenberg C., Tilmann F., Flueh E.R., Franke D., Barckhausen U., Krabbenhoeft A. and Djajadihardja Y. 2013. Subduction system variability across the segment boundary of the 2004/2005 Sumatra megathrust earthquakes. Earth and Planetary Science Letters 365: 108–119.
14 Titov V.V., Gonzalez F.I., Bernard E.N. et al. 2005. Real-time tsunami forecasting: challenges and solutions. Natural Hazards US National Tsunami Hazard Mitigation Program 35(1): 41–58.
15 Ahn Y.H., Shanmugam P., Moon J.E. and Ryu J.H. 2008. Satellite remote sensing of a low-salinity water plume in the East China Sea. Annals of Geophysics 26: 2019–2035.
16 Marghany M. 2009. Linear algorithm for salinity distribution modelling from MODIS data. Geoscience and Remote Sensing Symposium, 2009 IEEE International, IGARSS 2009, 12–17 July 2009, Cape Town, South Africa 3: III-365–III-368.
17 Marghany M. 2010. Examining the least square method to retrieve sea surface salinity from MODIS satellite data. European Journal of Research Science 40: 377–386.
18 Marghany M., Hashim M. and Cracknell A.P. 2010. Modelling sea surface salinity from MODIS satellite data. Computational Science and Its Applications—ICCSA 2010. Lecture Notes in Computer Science 6016: 545–556.
19 Marghany M. and Hashim M. 2011. A numerical method for retrieving sea surface salinity from MODIS satellite data. International Journal of the Physical Sciences 6(13): 3116–3125.
20 Hu C., Chen T., Clayton P., Swarnzenski J., Brock I. and Muller-Karger F. 2004. Assessment of estuarine water-quality indicators using MODIS medium-resolution bands: Initial results from Tampa Bay, fL. Remote Sensing of Environment 93: 423–441.
21 Wong M.S., Kwan L., Young J.K., Nichol J., Zhangging L. and Emerson N. 2007. Modelling of suspended solids and sea surface salinity in Hong Kong using Aqua/MODIS satellite images. Korian Journal of Remote Sensing 23: 161–169.
22 Font E., Nascimento C., Omira R., Baptista M.A. and Silva P.A. 2010. Identification of tsunami-induced deposits using numerical modelling and rock magnetism techniques: A study case of the 1755 Lisbon tsunami in Algarve, Portugal. Physics of the Earth and Planetary Interiors 182: 187–198.

23 Moore A., Nishimura Y., Gelfenbaum G., Kamataki T. and Triyono R. 2006. Sedimentary deposits of the 26 December 2004 tsunami on the northwest coast of Aceh, Indonesia. Earth Planets Space 58: 253–258.
24 Qing S., Zhang J., Cui T. and Bao Y. 2013. Retrieval of sea surface salinity with MERIS and MODIS data in the Bohai Sea. Remote Sensing of Environment 136: 117–125.
25 Anthony E.J. 2009. Developments in marine geology: Shore Processes and their palaeo environmental applications. Series Editor Chamley H 4: 415–420.
26 Saraf A., Choundhury K. and Dasgupta S. 2005. Satellite observation of the great megathrust Sumatra earthquake activities. International Journal of Geoinformatics 14: 67–74.
27 Oh H.J., Lee C.H., Kim H.J. and Jang M.G. 2005. June. Descriptive question answering in encyclopedia. pp. 21–24. *In*: Proceedings of the ACL 2005 on Interactive Poster and Demonstration Sessions Association for Computational Linguistics.
28 NOAA. 2016. Tsunami Terminology. http://nthmp history.pmel.noaa.gov/terms.html [access on July 7, 2016].
29 Goff J., Terry J.P., Chagué-Goff C. and Goto K. 2014. What is a mega-tsunami? Marine Geology 358: 12–17.
30 Shuto N. 1993. Tsunami intensity and disasters. pp. 197–216. *In*: Tsunamis in the World Springer, Netherlands.
31 Marghany M. 2014. Simulation of tsunami impact on sea surface salinity along Banda Aceh Coastal Waters, Indonesia. pp. 229–251. *In*: Marghany Maged (ed.). Advanced Geoscience Remote Sensing. INTECH Publisher, Croatia.
32 Appuhamy J.M.R.S. 2007. Numerical simulation of Tsunami in Indian Ocean. Master of Science Theses. Università degli Studi di Pavia.
33 NOAA. 2016. Tsunami Vocabulary and Terminology. http://www.tsunami.noaa.gov/terminology.html [Access date August 20, 2016].
34 Glossary T. and Intergovernmental Oceanographic Commission. 2013. UNESCO. IOC Technical Series, Paris, France.
35 USGS. 2005. The 26 December 2004 and 28 March 2005 Indian Ocean Tsunamis. USGS Scientists in Sumatra Study Impacts of 2 Tsunamis March 30–April 26, 2005. http://walrus.wr.usgs.gov/news/sumatra.html [Access date August 20, 2016].
36 Maine Geological Survey. 2016. Tsunamis in the Atlantic Ocean. http://www.maine.gov/dacf/mgs/hazards/tsunamis/ [Access date August 20, 2016].
37 Alday L., Gaiotto D. and Tachikawa Y. 2010. Liouville correlation functions from four-dimensional gauge theories. Letters in Mathematical Physics 91(2): 167–197.
38 Anne D.W. 2013. Moral Authority, Men of Science, and the Victorian Novel. Cambridge University Press.
39 De la Torre-Ibarra M.H., Santoyo F.M. and Moreno I. 2010. Digital holographic interferometer using simultaneously three lasers and a single monochrome sensor for 3D displacement measurements. Optics Express 18(19): 19867–19875.
40 DaneshPanah M., Zwick S., Schaal F., Warber M., Javidi B. and Osten W. 2010. 3D holographic imaging and trapping for non-invasive cell identification and tracking. Journal of Display Technology 6(10): 490–499.
41 Kohli D., Richard S., Norman K. and Stein A. 2012. An ontology of slums for image-based classification. Computers. Environment and Urban Systems 36(2): 154–163.
42 Hussein S.A., Gdeist M., Burton D. and Lalor M. 2005. Fast three-dimensional phase unwrapping algorithm based on sorting by reliability following a non-continuous path. Proc. SPIE 5856: 40.
43 Haupt R.L. and Haupt S.E. 2004. Practical Genetic Algorithms. John-Wiley & Sons, New York, United States.
44 Karout S. 2007. Two-Dimensional Phase Unwrapping. Ph.D Theses. Liverpool John Moores University.
45 Marghany M. 2012. Simulation of 3-D coastal spit geomorphology using differential synthetic aperture interferometry (DInSAR). pp. 83–94. *In*: Padron I. (ed.). Recent Interferometry Applications in Topography and Astronomy. Croatia: InTech—Open Access Publisher.
46 Marghany M. 2013. DinSAR technique for three-dimensional coastal spit simulation from radarsat-1 fine mode data. Acta Geophysica 61(2): 478–493.
47 Marghany M. 2014. Simulation of three-dimensional of coastal erosion using differential interferometric synthetic aperture radar. Global NEST Journal 16(1): 80–86.
48 Marghany M. 2014. Hybrid genetic algorithm of interferometric synthetic aperture radar for three-dimensional coastal deformation. Frontiers in Artificial Intelligence and Applications: New Trends in Software Methodologies, Tools and Technique 265: 116–31.
49 Marghany M. 2011. Modelling shoreline rate of changes using holographic interferometry. International Journal of Physical Sciences 6: 7694–7698.
50 Marghany M. 2003. Polarised AIRSAR along track interferometry for shoreline change modeling. In Geoscience and Remote Sensing Symposium, 2003. IGARSS'03. Proceedings. 2003 IEEE International 2: 945–947.
51 Marghany M. 2014. Hologram interferometric SAR and optical data for fourth-dimensional urban slum reconstruction. CD of 35th Asian Conference on Remote Sensing (ACRS 2014), Nay Pyi Taw, Myanmar 27–31, October 2014, http:// www. a-a-r-s.org/acrs/administrator/components/com.../OS-303%20.pdf [Access on August 2 2016].
52 Marghany M. 2015. Fourth dimensional optical hologram interferometry of RapidEye for Japan's tsunami effects CD of 36th Asian Conference on Remote Sensing (ACRS 2015), Manila, Philippines, 24–28 October 2015, http://www.a-a-r-s.org/acrs/index.php/acrs/acrs-overview/proceedings-1?view=publication&task=show&id=1691. Access on August 2 2016.

53 Pepe A. 2012. Advanced multitemporal phase unwrapping techniques for DInSAR analyses. pp. 57–82. *In*: Padron I. (ed.). Recent Interferometry Applications in Topography and Astronomy. InTech—Open Access Publisher, University Campus STeP Ri, Croatia.
54 Smith J. 2010. Optimizing remote sensing for disaster monitoring, retrieved May 13, 2005, from http://www.remotesening.net/index.htm.
55 Saravana S.S., Ponnanbalam S.G. and Rajendran C.A. 2003. Multiobjective genetic algorithm for scheduling a flexible manufacturing system. International Journal of Advanced Manufacturing Technology 22: 229–236.
56 Ray I. 1992. p. 319. Introducing Einstein's Relativity. Geometry of 3-spaces of Constant Curvature, Clarendon Press. ISBN 0-19-859653-7.
57 Schwarz O. 2004. Hybrid Phase Unwrapping in Laser Speckle Interferometry with Overlapping Windows. Shaker Verlag, German.
58 Takeda M., Ina H. and Kobayashi S. 1982. Fourier-transform method of fringe-pattern analysis for computer-based topography and interferometry. JosA 72(1): 156–160.
59 Saxby G. 1987. Parctical Holography. Prentice Hall, NJ, New Jersy.
60 Falvey D.A. 1974. The development of continental margins in plate tectonic theory. APEA J. 14(1): 95–106.
61 Dickinson W.R. 1972. Evidence for plate-tectonic regimes in the rock record. American Journal of Science 272(7): 551–576.
62 Nitecki M.H., Lemke J.L., Pullman H.W. and Johnson M.E. 1978. Acceptance of plate tectonic theory by geologists. Geology 6(11): 661–664.
63 Turcotte D.L. and Schubert G. 2002. Plate Tectonics. Geodynamics (2 ed.). Cambridge University Press, London, pp. 1–21.
64 Mitri Giuseppe, Bland Michael T., Showman Adam P., Radebaugh Jani, Stiles Bryan, Lopes Rosaly M.C., Lunine Jonathan I. and Pappalardo Robert T. 2010. Mountains on Titan: Modeling and observations. Journal of Geophysical Research 115(E10).
65 Wikipedia. 2016. List of tectonic plate interactions. https://en.wikipedia.org/wiki/List_of_tectonic_plate_interactions [Access September 25 2016].
66 Frankel H.R. 2012. The Continental Drift Controversy: Volume 2, Paleomagnetism and Confirmation of Drift. Cambridge University Press.
67 Uyeno G. and St. Writer. 2016. Buried Tectonic Plate Reveals Hidden Dinosaur-Era Sea. http://www.livescience.com/55855-newfound-tectonic-plate-east-asian-sea.html [Access September 25 2016].
68 Moore G.F., Bangs N.L., Taira A., Kuramoto S., Pangborn E. and Tobin H.J. 2007. Three-dimensional splay fault geometry and implications for tsunami generation. Science 318(5853): 1128–1131.
69 WARD S.N. 1980. Relationships of tsunami generation and an earthquake source. Journal of Physics of the Earth 28(5): 441–474.
70 Horbitz C.B., Lovholt F., Pedersen G. and Masson D.G. 2006. Mechanisms of tsunami generation by submarine landslides: A short review. Norsk Geologisk Tidsskrift 86(3): 255.
71 Erickson J. and Gates A.E. 2014. Quakes, Eruptions, and other Geologic Cataclysms: Revealing the Earth's Hazards. Infobase Publishing, New York.
72 Levin B.W. and Nosov M.A. 2016. The physics of tsunami formation by sources of nonseismic origin. pp. 263–309. *In*: Physics of Tsunamis. Springer International Publishing.
73 Wünnemann K., Collins G.S. and Weiss R. 2010. Impact of a cosmic body into Earth's ocean and the generation of large tsunami waves: insight from numerical modeling. Reviews of Geophysics 48(4).
74 Wu J., Suppe J., Lu R. and Kanda R. 2016. Philippine Sea and East Asian plate tectonics since 52 Ma constrained by new subducted slab reconstruction methods. Journal of Geophysical Research: Solid Earth 121(6): 4670–4741.
75 Wikipedia. 2016. List of historical tsunamis. https://en.wikipedia.org/wiki/List_of_historical_tsunamis [Access September 25 2016].
76 Maps of World. 2016. Tsunami dates and history. http://www.mapsofworld.com/tsunami/ [Access September 25 2016].
77 Earth Observatory of Singapore. 2016. Subduction zone beneath Sumatra, Indonesia. http://www.earthobservatory.sg/resources/images/subduction-zone-under-sumatra-indonesia [Access September 25 2016].
78 Socquet A., Vigny C., Chamot-Rooke N., Simons W., Rangin C. and Ambrosius B. 2006. India and Sunda plates motion and deformation along their boundary in Myanmar determined by GPS. Journal of Geophysical Research: Solid Earth 111(B5).
79 Ioualalen M., Asavanant J., Kaewbanjak N., Grilli S.T., Kirby J.T. and Watts P. 2007. Modeling the 26 December 2004 Indian Ocean tsunami: Case study of impact in Thailand. Journal of Geophysical Research: Oceans 112(C7).
80 Simoes M., Avouac J.P., Cattin R. and Henry P. 2004. The Sumatra subduction zone: A case for a locked fault zone extending into the mantle. Journal of Geophysical Research: Solid Earth 109(B10).
81 Sibuet J-C., Rangin C., Le Pichon X., Singh S., Cattaneo A., Graindorge D., Klingelhoefer F., Lin J-Y., Malod J., Maury T., Schneider J.-L., Sultan N., Umber M., Yamaguchi H. and the Sumatra aftershocks team. 2004. Great Sumatra–Andaman earthquake: Co-seismic and post-seismic motions in northern Sumatra. Earth and Planetary Science Letters 263(1-2): 88–103.

82 Vallée M. 2007. Rupture Properties of the Giant Sumatra Earthquake Imaged by Empirical Green's Function Analysis (PDF). Bulletin of the Seismological Society of America. Seismological Society of America 97(1A): S103–S114.
83 Bilham R. 2005. A flying start, then a slow slip. Science 308(5725): 1126–1127.
84 Gaia V. 2005. Tsunami seabed shows massive disruption. Daily News. https://www.newscientist.com/article/dn7465-tsunami-seabed-shows-massive-disruption/ [Access September 25 2016].
85 Brudzinski M.R., Thurber C.H., Hacker B.R. and Engdahl E.R. 2007. Global prevalence of double Benioff zones. Science 316(5830): 1472–1474.
86 Engdahl E.R., Villasenor A., DeShon H.R. and Thurber C.H. 2007. Teleseismic relocation and assessment of seismicity (1918–2005) in the region of the 2004 Mw 9.0 Sumatra–Andaman and 2005 Mw 8.6 Nias Island great earthquakes. Bulletin of the Seismological Society of America 97(1A): S43–S61.
87 Knight K. 2005. Asian tsunami seabed pictured with sonar. New Scientist 10.
88 Titov V., Rabinovich A.B., Mofjeld H.O., Thomson R.E. and González F.I. 2005. The global reach of the 26 December 2004 Sumatra tsunami. Science 309(5743): 2045–2048.
89 National Geography News. 2005. The Deadliest Tsunami in History? http://news.nationalgeographic.com/news/2004/12/1227_041226_tsunami.html [Access September 25 2016].
90 Michael Schirber. 2005. A new spin on Earth's rotation. http://www.livescience.com/178-spin-earth-rotation.html [Access September 25 2016].
91 CAIN F. 2015. How long is day on Earth? http://www.universetoday.com/123218/how-long-is-a-day-on-earth-2 [Access September 25 2016].
92 Wikipedia. 2016. Day. https://en.wikipedia.org/wiki/Day [Access September 25 2016].
93 Roland A. Madden and Paul R.J. 1994. Observations of the 40–50-day tropical oscillation—A review. Monthly Weather Review 122: 814–837.
94 Mueller I.I. 1969. Spherical and Practical Astronomy as Applied to Geodesy. Frederick Ungar Publishing, NY, pp. 80.
95 Gross R.S. 2000. The Excitation of the Chandler Wobble. Geophysical Research Letters 27(15): 2329–2332.
96 Hopkin M. 2004. Sumatran quake sped up Earth's rotation. Nature 30: 041229–6.
97 Dutch S. 2005. The 2004 Indonesian and 2010 Chilean Earthquakes and Earth's Rotation. https://www.uwgb.edu/dutchs/PLATETEC/RotationQk2004.HTM [Access September 25 2016].
98 Bakshi U.A. and Godse A.P. 2009. Basic Electronics Engineering. Technical Publications. pp. 8–10.
99 Lillesand T.M., Kiefer Ralph W. and Chipman J. 2007. Remote Sensing and Image Interpretation (6th ed.). New York: John Wiley and Sons, New York, 750 p.
100 Campbell James B. and Wynne Randolph H. 2011. Introduction to Remote Sensing (5th ed.). The Guilford Press, 667 p.
101 Hecht E. 2001. Optics, 4th Edition, Addison Wesley.
102 Chelton D.B., Ries J.C., Haines B.J., Fu L.L. and Callahan P.S. 2001. Satellite altimetry. *In*: Fu L.L. and Cazenave A. (eds.). Satellite Altimetry and Earth Sciences, Academic Press.
103 Brown G.S. 1977. The average impulse response of a rough surface and its applications. IEEE Transactions on Antennas and Propagation 25(1): 67–74.
104 Hayne G.S. 1980. Radar altimeter mean return waveforms from near-normal-incidence ocean surface scattering. IEEE Transactions on Antennas and Propagation 28(5): 687–692.
105 Robinson I.S. 2004. Measuring the Oceans from Space: The Principles and Methods of Satellite Oceanography. Springer Praxis Books, 669 pp.
106 Godin O.A., Irisov V.G., Leben R.R., Hamlington B.D. and Wick G.A. 2009. Variations in sea surface roughness induced by the 2004 Sumatra-Andaman tsunami. Natural Hazards and Earth System Sciences 9(4): 1135–1147.
107 Wikipedia. 2016. Jason-1. https://en.wikipedia.org/wiki/Jason-1.
108 Kulikov E. 2006. Dispersion of the Sumatra Tsunami waves in the Indian Ocean detected by satellite altimetry. Russian Journal of Earth Sciences 8(4): 1–5.
109 Arcas D. and Titov V. 2006. Sumatra tsunami: lessons from modeling. Surveys in Geophysics 27(6): 679–705.
110 Rosmorduc V., Benveniste J., Bronner E., Dinardo S., Lauret O., Maheu C., Milagro M., Picot N., Ambrozio A., Escolà R., Garcia-Mondejar A., Restano M., Schrama E. and Terra-Homem M. 2016. Radar altimetry tutorial. *In*: Benveniste J. and Picot N. (eds.). http://www.altimetry.info.
111 Hayashi Y. 2008. Extracting the 2004 Indian Ocean tsunami signals from sea surface height data observed by satellite altimetry. Journal of Geophysical Research: Oceans 113(C1).
112 Ray R.D. 1999. A global ocean tide model from TOPEX/POSEIDON altimetry: GOT99 2.
113 Schrama E.J.O. and Ray R.D. 1994. A preliminary tidal analysis of TOPEX/POSEIDON altimetry. Journal of Geophysical Research: Oceans 99(C12): 24799–24808.
114 Matsumoto K., Takanezawa T. and Ooe M. 2000. Ocean tide models developed by assimilating TOPEX/POSEIDON altimeter data into hydrodynamical model: a global model and a regional model around Japan. Journal of Oceanography 56(5): 567–581.

115 Larnicol G., Le Traon P.Y., Ayoub N. and De Mey P. 1995. Mean sea level and surface circulation variability of the Mediterranean Sea from 2 years of TOPEX/POSEIDON altimetry. Journal of Geophysical Research: Oceans 100(C12): 25163–25177.
116 USGS. 2005. Tsunami Generation from the 2004 M=9.1 Sumatra-Andaman Earthquake. https://walrus.wr.usgs.gov/tsunami/sumatraEQ/jason.html.
117 Smith W.H., Scharroo R., Titov V.V., Arcas D. and Arbic B.K. 2005. Satellite altimeters measure tsunami. Oceanography 18(2): 11–13.
118 Gower J. 2007. The 26 December 2004 tsunami measured by satellite altimetry. International Journal of Remote Sensing 28: 13–14.
119 Chen J. 2005. Finite fault model. Available online at: http://neic.usgs.gov/neis/eq_depot/2004/ eq_041226/neic_slav_ff.html (accessed March 2005).
120 Song Y.T., Ji C., Fu L.-L., Zlotnicki V., Shum C.K., Yi Y. and Hjorleifsdottir V. 2005. The 26 December 2004 tsunami source estimated from satellite radar altimetry and seismic waves. Geophysical Research Letters 32(20).
121 Gibbons H. and Gelfenbaum G. 2005. Astonishing wave heights among the findings of an international tsunami survey team in Sumatra. Sound Waves, March.
122 Di K., Wang J., Ma R. and Li R. 2003, May. Automatic shoreline extraction from high-resolution IKONOS satellite imagery. In Proceeding of ASPRS 2003 Annual Conference (Vol. 3).
123 Canny J. 1986. Computational approach to edge detection. IEEE Transactions on Pattern Analysis and Machine Intelligence. PAMI 8(6): 679–698.
124 Deriche R. 1987. Using Canny's criteria to derive a recursively implemented optimal edge detector. International Journal of Computer Vision 1(2): 167–187.
125 Gonzalez R. and Woods R. 1992. Digital Image Processing, 3rd edition, Addison-Wesley Publishing Company, pp. 200–229.
126 Deriche R. 1987. Using Canny's criteria to derive a recursively implemented optimal edge detector. International Journal of Computer Vision 1(2): 167–187.
127 Green B. 2010. Canny edge detection tutorial, 2002. DOI = http://www. pages. drexel. edu/~weg22/can_tut. html, referred on, 9.
128 Moore A. 2016. Coastline changes to Aceh from the great 2004 Sumatra-Andaman earthquake. http://serc.carleton.edu/vignettes/collection/25462.html.
129 Bayas J.C.L., Marohn C., Dercon G., Dewi S., Piepho H.P., Joshi L., van Noordwijk M. and Cadisch G. 2011. Influence of coastal vegetation on the 2004 tsunami wave impact in west Aceh. Proceedings of the National Academy of Sciences 108(46): 18612–18617.
130 Yan Z. and Tang D. 2009. Changes in suspended sediments associated with 2004 Indian Ocean tsunami. Advances in Space Research 43(1): 89–95.
131 ICMAM (Integrated Coastal and Marine Area Management). 2005. Preliminary Assessment of Impact of Tsunami in Selected Coastal Areas of India. Chennai, India: Department of Ocean Development Post Tsunami Field Report.
132 Sarangi R.K. 2011. Remote sensing of chlorophyll and sea surface temperature in Indian water with impact of 2004 Sumatra Tsunami. Marine Geodesy 34(2): 152–166.
133 Anil N., Sarma Y.V.B., Babu K.N., Madhukar M. and Pandey P.C. 2006. Post-tsunami oceanographic conditions in southern Arabian Sea and Bay of Bengal. Current Science 90(3): 421–427.
134 Singh R.P., Cervone G., Kafatos M., Prasad A.K., Sahoo A.K., Sun D., Tang D.L. and Yang R. 2007. Multi-sensor studies of the Sumatra earthquake and tsunami of 26 December 2004. International Journal of Remote Sensing 28(13-14): 2885–2896.
135 Singh R.P., Bhoi S. and Sahoo A.K. 2001. Significant changes in ocean parameters after the Gujarat earthquake. Current Science 80(11): 1376–1377.
136 Kundu S.N., Sahoo A.K., Mohapatra S. and Singh R.P. 2001. Change analysis using IRS-P4 OCM data after the Orissa super cyclone. International Journal of Remote Sensing 22(7): 1383–1389.
137 Gautam R., Singh R.P. and Kafatos M. 2005. Changes in ocean properties associated with Hurricane Isabel. International Journal of Remote Sensing 26(3): 643–649.
138 Pennish E. 2005. Powerful tsunami's impacts on coral reefs was hit and miss. Science 307–657.
139 Santek D.A. and Winguth A. 2007. A satellite view of internal waves induced by the Indian Ocean tsunami. International Journal of Remote Sensing 28(13-14): 2927–2936.
140 Ruddick K., Ovidio F. and Rijkeboer M. 2000. Atmospheric correction of SeaWiFS imagery for turbid coastal and inland waters. Applied Optics 39(6): 897–912.
141 Zheng G.M. and Tang D.L. 2007. Offshore and near shore chlorophyll increases induced by typhoon and typhoon rain. Marine Ecology Progress Series 333: 61–74.
142 Saraf A.K., Choudhury S. and Dasgupta S. 2005. Satellite observations of the great mega thrust Sumatra earthquake activities. International Journal of Geoinformatics 1(4): 67–74.
143 Ilayaraja K. and Krishnamurthy R.R. 2010. Sediment characterisation of the 26 December 2004 Indian ocean tsunami in Andaman group of islands, Bay of Bengal, India. Journal of Coastal Conservation 14(3): 215–230.

144 Cannizzaro J.P. and Carder K.L. 2006. Estimating chlorophyll a concentrations from remote-sensing reflectance in optically shallow waters. Remote Sensing of Environment 101(1): 13–24.
145 Miller R.L. and McKee B.A. 2004. Using MODIS Terra 250 m imagery to map concentrations of total suspended matter in coastal waters. Remote Sensing of Environment 93(1): 259–266.
146 Datt B. 1998. Remote sensing of chlorophyll a, chlorophyll b, chlorophyll a+b, and total carotenoid content in eucalyptus leaves. Remote Sensing of Environment 66(2): 111–121.
147 O'Reilly J.E., Maritorena S., Siegel D.A., O'Brien M.C., Toole D., Mitchell B.G., Kahru M., Chavez F.P., Strutton P., Cota G.F. and Hooker S.B. 2000. Ocean color chlorophyll a algorithms for SeaWiFS, OC2, and OC4: Version 4. pp. 9–23. *In*: Hooker S.B. and Firestone E.R. (eds.). SeaWiFS Postlaunch Calibration and Validation Analyses: Part 3. NASA Tech. Memo. 2000-206892, Vol. 11. Greenbelt' NASA Goddard Space Flight Center.
148 Feng H., Campbell J.W., Dowell M.D. and Moore T.S. 2005. Modeling spectral reflectance of optically complex waters using bio-optical measurements from Tokyo Bay. Remote Sensing of Environment 99(3): 232–243.
149 Aiken J., Moore G.F., Trees C.C., Hooker S.B. and Clark D.K. 1996. The SeaWiFS CZCS-type pigment algorithm. Oceanographic Literature Review 3(43): 315–316.
150 Marghany M. and Hashim M. 2010. MODIS satellite data for modeling chlorophyll-a concentrations in Malaysian coastal waters. International Journal of Physical Sciences 5(10): 1489–1495.
151 Montres-Hugo M.A., Vernet M., Smith R. and Carders K. 2008. Phytoplankton size-structure on the western shelf of the antarctic peninsula: A remote sensing approach. International Journal of Remote Sensing 29(3-4): 801–829.
152 Bissett W.P., Patch J.S., Carder K.L. and Lee Z.P. 1997. Pigment packaging and Chl a-specific absorption in high-light oceanic waters. Limnology and Oceanography 42(5): 961–968.
153 Tan C.K., Ishizaka J., Manda A., Siswanto E. and Tripathy S.C. 2007. Assessing post-tsunami effects on ocean colour at eastern Indian Ocean using MODIS Aqua satellite. International Journal of Remote Sensing 28(13-14): 3055–3069.
154 Tang D., Zhao H., Satyanarayana B., Zheng G., Singh R.P., Lv J. and Yan Z. 2009. Variations of chlorophyll-a in the northeastern Indian Ocean after the 2004 South Asian tsunami. International Journal of Remote Sensing 30(17): 4553–4565.
155 Maurer J. 2002. Infrared and microwave remote sensing of sea surface temperature. Remote Sensing Seminar of University of Colorado at Boulder, 9 October 2002.
156 Eugenio F., Marcello J., Rovaris E. and Hernández A. 2002. Accurate retrieval of sea surface temperature in the Canary Islands-Azores-Gibraltar area using AVHRR/3 and MODIS data. pp. 2129–2131. *In*: Geoscience and Remote Sensing Symposium, 2002. IGARSS'02. 2002 IEEE International (Vol. 4). IEEE.
157 Robert W. 2004. Moderate Resolution Imaging Spectroradiometer (MODIS) calibration, geolocation, production and direct broadcast software. National Aeronautics and Space Administration, 24 June 2004.
158 Sibson R.H. 1984. Roughness at the base of the seismogenic zone: contributing factors. Journal of Geophysical Research: Solid Earth 89(B7): 5791–5799.
159 Pedlosky J. 2013. Geophysical Fluid Dynamics. Springer Science & Business Media.
160 Pelinovsky E. 2006. Hydrodynamics of tsunami waves. pp. 1–48. *In*: Waves in Geophysical Fluids. Springer Vienna.
161 Plant W.J. 1990. Bragg scattering of electromagnetic waves from the air/sea interface. pp. 41–168. *In*: Geemaet G.L. and Plant W.J. (eds.). Surface Waves and Fluxes, Vol. II. Remote Sensing. Kluwer Acad., Norwell, Mass.
162 Zheng Q., Clemente-Colón P., Yan X.H., Liu W.T. and Huang N.E. 2004. Satellite synthetic aperture radar detection of Delaware Bay plumes: Jet-like feature analysis. Journal of Geophysical Research: Oceans 109: C03031.
163 Pan J., Jay D.A. and Orton P.M. 2007. Analysis of dynamic characteristics of internal solitary waves generated by the Columbia River plume with SAR images. Journal of Geophysical Research 112(C07014.3): 1–11.
164 Da Silva J.C.B., New A.L. and Magalhaes J.M. 2011. On the structure and propagation of internal solitary waves generated at the Mascarene Plateau in the Indian Ocean. Deep Sea Research Part I: Oceanographic Research Papers 58(3): 229–240.
165 Ermakov S.A., da Silva J.C.B. and Robinson I.S. 1998. Role of surface films in ERS SAR signatures of internal waves on the shelf. 2. Internal Tidal Waves. Journal of Geophysical Research 103: 8032–8043.
166 da Silva J.C.B. and Magalhaes J.M. 2016, October. Internal solitons in the Andaman Sea: a new look at an old problem. pp. 999907–999907. *In*: SPIE Remote Sensing. International Society for Optics and Photonics.
167 Stéphane Mallat. 1998. A Wavelet Tour of Signal Processing, 2nd Edition, Academic Press.
168 Robi. The engineering ultimate guide to wavelet analysis: The Wavelet Tutorial. Rowan University, College of Engineering.
169 Ghobadi M. Wavelet-based coding and its application in JPEG2000. webhome.cs.uvic.ca/~pan/csc461s06/.../csc561-monia-wavelet-v2.doc [Acess on February, 16, 2017].
170 Wu S.Y. and Liu A.K. 2003. Towards an automated ocean feature detection, extraction and classification scheme for SAR imagery. International Journal of Remote Sensing 24(5): 935–951.
171 Paris R., Lavigne F., Wassimer P. and Sartohadi J. 2007. Coastal sedimentation associated with the December 26, 2004 tsunami in Lhok Nga, west Banda Aceh (Sumatra, Indonesia). Marine Geology pp. 93–106.
172 Nalbant S., Steacy S., Sieh K., Natawidjaja D. and McCloskey J. 2005. Seismology: Earthquake risk on the Sunda trench. Nature 435(7043): 756–757.

173 Marghany M. and Sufian M. 2005. Simulation of the successive tsunami waves along kalatura coastline from Quickbird-1 satellite. Asian Journal of Geoinformatics 5(2): 73–77.
174 Lovholt F., Bungum H., Harbitz C.B., Glimsal S., Lindholm C.D. and Pedersen G. 2006. Earthquake related tsunami hazard along the western coast of Thailand. Nature Hazard and Earth System Science 6(6): 979–997.
175 Sibuet J.C., Rangin C., Le Pichon X., Singh S., Cattaneo A., Graindorge D., Klingelhoefer F., Lin J.Y., Malod J., Maury T., Schneider J.L., Sultan N., Umber M., Yamaguchi H. and the Sumatra aftershocks team. 2007. December 26, 2004 great Sumatra–Andaman earthquake: Co-seismic and post-seismic motions in northern Sumatra. Earth and Planetary Science Letters 263(1-2): 88–103.
176 Walter H.F., Smith R.S., Vasily V., Titov D.A. and Brian K.A. 2005. Satellite altimeters measure tsunami. Oceanography 18(2): 11–13.
177 Nakano M., Hori T., Araki E., Takahashi N. and Kodaira S. 2016. Ocean floor networks capture low-frequency earthquake event. Eos 97 https://doi.org/10.1029/2016EO052877.
178 Oo K.S., Mehdiyev M. and Samarakoon L. 2005. Potential of satellite data in assessing coastal damage caused by South-Asia tsunami in December 2005—A field survey report. Asian Journal of Geoinformatics 5(2): 16–37.
179 Salinas S.V., Low J.K.K. and Liew S.C. 2005. Quick analysis of wave patterns generated by tsunami waves and captured by SPOT imagery. Proc. IEEE International Geosciences and Remote Sensing Symposium 2005, 25–29 July 2005, Seoul, Korea 5: 3634–3636.
180 Salinas S., Cortijo V., Chen P. and Liew S.C. 2006. Tsunami effects on shallow waters: From wave scattering to land inundation. Proc. IEEE International Geosciences and Remote Sensing Symposium 2006, 31 July–4 August, Denver, Colorado, USA, pp. 3357–3360.
181 Kouchi K. and Yamazaki F. 2007. Characteristics of tsunami-affected areas in moderate-resolution satellite Images. IEEE Transactions on Geoscience and Remote Sensing 45(6): 1650–1657.
182 Ibrahim S.M., Nor Z.W., Mohammed S.A.R. and Kok F.L. 2009. Tsunami Hazard Evaluation using spatial tools for Kuala Muda, Kedah, Malaysia. Disaster Advances 2(3): 5–14.
183 Populus J., Aristaghes C., Jonsson L., Augustin J.M. and Pouliquen E. 1991. The use of SPOT data for wave analysis. Remote Sensing of Environment 36: 55–65.
184 Marghany M.M. 2001. Opertional of Canny algorithm on SAR data for modeling shoreline change. Photogrammetrie, Fenerkundung, Geoinformatik 2: 93–102.
185 Marghany M. 2004. Velocity bunching model for modelling wave spectra along east coast of Malaysia. Journal of the Indian Society of Remote Sensing 32(2): 185–198.
186 Gedik N., Irtem E. and Kabdasli S. 2005. Laboratory investigation on tsunami run-up. Ocean Engineering 32: 513–528.
187 Synolakis C.E. 1987. The runup of solitary waves. Journal of Fluid Mechanics 185: 523–545.
188 Mehdiyev M., Oo K.S. and Rajapaksha J. 2005. Tsunami disaster damage detection and assessment using high resolution satellite data, GIS and GPS–Case study in Sri Lanka. In 26th Asian Conference on Remote Sensing and 2nd Asian Space Conference (ACRS2005).
189 Goff J., Liu P.L., Higman B., Morton R., Jaffe B.E., Fernando H., Lynett P., Fritz H., Synolakis C. and Fernando S. 2006. Sri Lanka field survey after the December 2004 Indian Ocean tsunami. Earthquake Spectra 22(S3): 155–172.
190 USGS. 2004. http://Earthquake.usgs.gov/eqinthenews/2004/usslav/ [Accessed 30 December 2004].
191 Goto C. and Ogawa Y. 1992. Numerical Method of Tsunami Simulation with the Leap-frog Scheme. Dept. of Civil Engineering, Tohoku University. Translated for the TIME Project by N. Shuto, pp. 1–15.
192 Marghany M., Hashim M. and Moradi F. 2011. Object recognitions in RADARSAT-1 SAR data using fuzzy classification. International Journal of Physical Sciences 6(16): 3933–3938.
193 Marghany M., Mazlan H. and Cracknell A.P. 2010. 3-D visualizations of coastal bathymetry by utilization of airborne TOPSAR polarized data. International Journal of Digital Earth 3(2): 187–206.
194 Marghany M. 2012. Three-dimensional lineament visualization using fuzzy B-spline algorithm from multispectral satellite data. Remote Sensing–Advanced Techniques and Platforms, p. 213.
195 Anile A.M. 1997. Report on the activity of the fuzzy soft computing group. Technical Report of the Dept. of Mathematics, University of Catania, March 1997: 10.
196 Anile A.M., Deodato S. and Privitera G. 1995. Implementing fuzzy arithmetic. Fuzzy Set and System 72: 123–156.
197 Anile A.M., Gallo G. and Perfilieva I. 1997. Determination of membership function for cluster of geographical data. Genova: Institute for Applied Mathematics, National Research Council, October 1997, Tech. Report No. 26/97: 20–25.
198 Hiroshi O., Toshiro N. and Genki Y. 1994. Basic study on element-free gelerkin method (2nd report, application to two-dimensional potential problems). JSME International Journal 95: 1825–1845.
199 Singh I.V. 2004. Meshless EFG method in three-dimensional heat transfer problems: A numerical Comparison, Cost and Errors Analysis. Numerical Heat Transfer, Part A 46: 199–220.
200 Cingoski V., Miyamoto N. and Yamashita H. 1988. Element-free galerkin method for electromagnetic field computations. IEEE Transactions on Magnetics 34: 3236–3239.
201 Belyschko T., Lu Y.Y. and Gu L. 1994. Element-free galerkin method. International Journal for Numerical Methods in Engineering 37: 229–256.

202 Lancaster P. and Salkauskas K. 1981. Surfaces generated by moving least squares methods. Mathematics of Computation 37: 141–158.
203 Stein S. and Okal E.A. 2005. Speed and size of the Sumatra earthquake. Nature 434: 581–582.
204 Joe F., Alexander B., Ranjit G., Janaka R. and Chan S.W. 2012. The Indian Ocean Tsunami of December 26, 2004: A Journey through the Disaster with an Emphasis on Sri Lanka schulich.ucalgary.ca/civil/files/civil/IndianOceanTsunami.doc. pp. 1–30. Accessed [23 January 2012].
205 Fuchs H.Z., Kedem M. and Uselton S. 1977. Optimal surface reconstruction from planar contours. *In*: Communications of the ACM 20(10): 693–702.
206 Keppel E. 1975. Approximation complex surfaces by triangulations of contour lines. IBM Journal of Research and Development 19: 2–11.
207 Rövid A., Várkonyi A.R. and Várlaki, P. 2004. 3D model estimation from multiple imasges. IEEE Int. Conf. on fuzzy Syst., FUZZ-IEEE'2004, July 25–29, 2004, Budapest, Hungary, 1661–1666.
208 Russo F. 1998. Recent advances in fuzzy techniques for image enhancement. IEEE Transactions on Instrumentation and Measurement 47: 1428–1434.
209 Brain K. 2014. What It Means to Live in a Holographic Universe. http://nautil.us/blog/what-it-means-to-live-in-a-holographic-universe [Access February, 18, 2017].
210 de Haro S., Skenderis K. and Solodukhin S.N. 2001. Holographic reconstruction of spacetime and renormalization in the AdS/CFT correspondence. Communications in Mathematical Physics 217(3): 595–622.
211 Caulfield H.J. 2012. Handbook of Optical Holography. Elsevier, Amsterdam, The Netherlands.
212 Collier R. 2013. Optical Holography. Elsevier, Amsterdam, The Netherlands.
213 Hariharan P. 1996. Optical Holography: Principles, Techniques and Applications. Cambridge University Press, London, England.
214 Hong J. and Kim M.K. 2013. Single-shot self-interference incoherent digital holography using off-axis configuration. Optics Letters 38(23): 5196–5199.
215 Working of Hologram. 2010. How Holograms Work.http://how-does-things-work.blogspot.my/2010/02/working-of-hologram.html [Access on February 19 2017].
216 Shaked N.T. and Rosen J. 2008, April. Holography of incoherently illuminated 3D scenes. pp. 69830Q–69830Q. *In*: SPIE Defense and Security Symposium. International Society for Optics and Photonics.
217 Ghiglia D.C. and Pritt M.D. 1998. Two-dimensional Phase Unwrapping: Theory, Algorithms, and Software. New York: Wiley.
218 Loecher M., Schrauben E., Johnson K.M. and Wieben O. 2015. Phase unwrapping in 4D MR flow with a 4D single-step laplacian algorithm. Journal of Magnetic Resonance Imaging 43(4): 833–842.
219 Murty T.S. 1984. Storm Surges—Meteorological Ocean Tides, Bull. 212, Fisheries Res. Board Canada, Ottawa.
220 Pelinovsky E.N. 1996. Tsunami Wave Hydrodynamics, Russian Academy of Sciences, Nizhni Novgorod, 275 pp.
221 Imamura F. 1996. Review of tsunami simulation with a finite difference method. pp. 25–42. *In*: Yeh H., Liu P. and Synolakis C. (eds.). Long-Wave Runup Models. World Scientific.
222 Titov V.V. and Synolakis C.E. 1998. Numerical modeling of tidal wave runup. J. Waterway, Port, Coastal and Ocean Eng. 124(4): 157–171.
223 Kowalik Z. and Murty T.S. 1993. Numerical simulation of two-dimensional tsunami runup. Marine Geodesy 16: 87–100.
224 Troshina E.N. 1996. Tsunami Waves Generated by Mt. St. Augustine Volcano, Alaska. MS thesis, University of Alaska Fairbanks.
225 Reid R.O. and Bodine B.R. 1968. Numerical model for storm surges in Galveston Bay. Journal of the Waterways and Harbors Division 94(WWI): 33–57.
226 Kowalik Z. and Murty T.S. 1993. Numerical Modeling of Ocean Dynamics, World Scientific, 481 pp.
227 Suleimani E.N., Hansen R.A. and Kowalik Z. 2003. Inundation modeling of the 1964 tsunami in Kodiak Island, Alaska. pp. 191–201. *In*: Submarine Landslides and Tsunamis. Springer Netherlands.
228 MacDonald James. Spring 2001. Course Notes for PHYS306: Computational Methods of Physics University of Delaware. ONLINE: http://www.physics.udel.edu/faculty/macdonald/Ordinary%20Differential%20Equations/Euler's%20Method.htm.
229 Shampine Lawrence F. and Gordon M.K. 1975. Computer Solution of Ordinary Differential Equations: The Initial Value Problem. W.H. Freemand and Company: San Francisco, CA.
230 Patel V.M., Patel H.S. and Singh A.P. 2010. Tsunami propagation in Arabian sea and its effect on Porbandar, Gujarat, India. Journal of Engineering Research and Studies E-ISSN 976, p. 7916.
231 Chubarov L.B. and Fedotova Z.I. 2003. A method for mathematical modelling of tsunami runup on a shore. pp. 203–216. *In*: Yalçıner A.C., Pelinovsky E., Okal E. and Synolakis C.E. (eds.). Submarine Landslides and Tsunamis. Springer Netherlands.
232 Le Traon P.Y., Klein P., Hua B.L. and Dibarboure G. 2008. Do altimeter wavenumber spectra agree with the interior or surface quasigeostrophic theory? Journal of Physical Oceanography 38(5): 1137–1142.

233 Wang X. and Liu P.L.F. 2007. Numerical simulations of the 2004 Indian Ocean tsunamis—coastal effects. Journal of Earthquake and Tsunami 1(03): 273–297.
234 Ahern F.J. 1995. Fundamental concepts of imaging radar: basic level; unpublished manual, Canada Centre for Remote Sensing, Ottawa, Ontario, 87 p.
235 Chan Y.K. and Koo V.C. 2008. An introduction to synthetic aperture radar (SAR). Progress in Electromagnetics Research B 2: 27–60.
236 Bamler R. 2000. Principles of synthetic aperture radar. Surveys in Geophysics 21(2): 147–157.
237 CCRS. 1993. Radar basics—introduction to synthetic aperture radar; unpublished manual. Canada Centre for Remote Sensing, Ottawa, Ontario, 75 p.
238 Moreira A. 1992. Real-time synthetic aperture radar (SAR) processing with a new subaperture approach. IEEE Transactions on Geoscience and Remote Sensing 30(4): 714–722.
239 Ramirez A.B., Rivera I.J. and Rodriguez D. 2005, August. Sar image processing algorithms based on the ambiguity function. pp. 1430–1433. *In*: Circuits and Systems, 2005. 48th Midwest Symposium on IEEE.
240 Zhang S., Long T., Zeng T. and Ding Z. 2008. Space-borne synthetic aperture radar received data simulation based on airborne SAR image data. Advances in Space Research 41(11): 1818–1821.
241 Hovanessian A. 1984. Radar System Design and Analysis. Artech, pp. 5, ISBN: 0-89006-147-5.
242 Kingsley S. and Quegan S. 1999. Understanding Radar Systems. Scitech Publishing, Inc, New York.
243 Lillesand T., Kiefer R.W. and Chipman J. 2014. Remote Sensing and Image Interpretation. Sixth Edition. John Wiley & Sons, New York.
244 Ahern F.J. 1995. Fundamental concepts of imaging radar: basic level; unpublished manual. Canada Centre for Remote Sensing, Ottawa, Ontario, 87 p.
245 Kornei K. 2017. Indonesian cave reveals nearly 5,000 years of tsunamis, Eos 98, https://doi.org/10.1029/2017EO079283. Published on 07 August 2017.
246 Rubin C.M., Benjamin P., Horton P.M., Sieh K., Pilarczyk J.E., Daly P., Ismail N. and Parnell A.C. 2017. Highly variable recurrence of tsunamis in the 7,400 years before the 2004 Indian Ocean tsunami. Nature Communications 8: 1–12.
247 Johnson H.P., Gomberg J.S., Hautala S.L. and Salmi M.S. 2017. Sediment Gravity Flows Triggered by Remotely-generated Earthquake Waves. Journal of Geophysical Research: Solid Earth.
248 Wendel J. 2016. Subtle seismic movements may help forecast large earthquakes, Eos 97, doi: 10.1029/2016EO045137. Published on 1 February 2016.
249 Uchida N., Iinuma T., Nadeau R.M., Bürgmann R. and Hino R. 2016. Periodic slow slip triggers megathrust zone earthquakes in northeastern Japan. Science 351(6272): 488–492.
250 Geoscience Australia Earthquakes. 2016. Strong and shallow M6.5 earthquake hits Sumatra, 103 dead [Http://www.Ga.Gov.Au/earthquakes/].
251 Newman A.V., Hayes G., Wei Y. and Convers J. 2011. The 25 October 2010 Mentawai tsunami earthquake, from real-time discriminants, finite-fault rupture, and tsunami excitation. Geophysical Research Letters 38(5).
252 Rivera P.C. 2006. Modeling the Asian tsunami evolution and propagation with a new generation mechanism and a non-linear dispersive wave model. Science of Tsunami Hazards 25(1): 18–33.
253 Burkov V. and Melker A.I. 2012. Molecular hydrodynamics of tsunami waves. Materials Physics and Mechanics 13: 157–161.
254 Gray R. 2016. Large earthquakes could be triggered by a full moon: Biggest shakes in history occurred around spring tides. Mail online. http://www.dailymail.co.uk/sciencetech/article-3781711/Large-earthquakes-triggered-MOON-Biggest-shakes-history-occurred-spring-tides.html.
255 Miura H., Yamazaki F. and Matsuoka M. 2007, April. Identification of damaged areas due to the 2006 Central Java, Indonesia earthquake using satellite optical images. pp. 1–5. *In*: Urban Remote Sensing Joint Event, 2007. IEEE.
256 Kouchi K.I. and Yamazaki F. 2007. Characteristics of tsunami-affected areas in moderate-resolution satellite images. IEEE Transactions on Geoscience and Remote Sensing 45(6): 1650–1657.
257 McAdoo B.G., Richardson N. and Borrero J. 2007. Inundation distances and run-up measurements from ASTER, QuickBird and SRTM data, Aceh coast, Indonesia. International Journal of Remote Sensing 28(13-14): 2961–2975.
258 Chien N.Q. 2007. Simulating shoreline retreat in a 2-D hydro-morphologic model. Msc. Theses, Institute of UNESCO-IHE for water education, Delft, The Netherlands.
259 Giannini M.B. and Parente C. 2015. An object based approach for coastline extraction from Quickbird multispectral images. International Journal Engineering Technology 6: 2698–2704.
260 Guariglia A., Buonamassa A., Losurdo A., Saladino R., Trivigno M.L., Zaccagnino A. and Colangelo A. 2006. A multisource approach for coastline mapping and identification of shoreline changes. Annals of Geophysics 49(1): 295–304.
261 Paris R., Lavigne F., Wassmer P. and Sartohadi J. 2007. Coastal sedimentation associated with the december 26, 2004 tsunami in lhok nga, west banda aceh (sumatra, indonesia). Marine Geology 238(1): 93–106.
262 Voosen P. 2017. How the Himalayas primed the Indonesian tsunami. Science 356(6340): 794.

263 Hüpers A., Torres M.E., Owari S., McNeill L.C., Dugan B., Henstock T.J., Milliken K.L., Petronotis K.E., Backman J., Bourlange S. and Chemale F. 2017. Release of mineral-bound water prior to subduction tied to shallow seismogenic slip off Sumatra. Science 356(6340): 841–844.
264 Whitley D. 1994. A genetic algorithm tutorial. Statistics and Computing 4(2): 65–85.
265 Yang J. and Honavar V. 1998. Feature subset selection using a genetic algorithm. IEEE Intelligent Systems and their Applications 13(2): 44–49.
266 Marghany M. and Mansor S. 2016. Genetic Algorithm for South China Sea Water Mass Variations using MODIS Satellite Data. CD of 37th Asian Conference on Remote Sensing (ACRS), 37th ACRS from 17th–21st October 2016, Galadari Hotel, Colombo, Sri Lanka, pp. 1–6.
267 Sivanandam S.N. and Deepa S.N. 2008. Introduction to Genetic Algorithms, by Springer Berlin Heidelberg New York.
268 Bowden K.F. 1983. Physical Oceanography of Coastal Waters. Ellis Horwood Ltd., England.
269 Alejandro C. and Saadon M.N. 1996. Dynamic behaviour of the upper layers of the South China Sea. Proceedings of the National Conference on Climate Change, 12–13 August 1996, Universiti Pertanian Malaysia, Serdang, pp. 135–140.
270 Alejandro C. and Demmler M.I. 1997. Wind-driven circulation of Peninsular Malaysia's eastern continental shelf. Scientia Marina 61(2): 203–211.
271 Asgedom E.G., Cecconello E., Orji O.C. and Söllner W. 2017, June. Rough sea surface reflection coefficient estimation and its implication on hydrophone-only pre-stack deghosting. In 79th EAGE Conference and Exhibition 2017.
272 Thorsos E.I. 1987. The validity of the Kirchhoff approximation for rough surface scattering using a Gaussian roughness spectrum. Journal of the Acoustical Society of America 83: 78–92.
273 Kennedy James and Russell C. Eberhart 1997. A discrete binary version of the particle swarm algorithm. Systems, Man, and Cybernetics, 1997. Computational Cybernetics and Simulation, 1997 IEEE International Conference on Vol. 5. IEEE, 1997.
274 Marghany. 2015. Copper mine automatic detection from TerraSAR-X using particle swarm optimization. CD of 36th Asian Conference on Remote Sensing (ACRS 2015), Manila, Philippines, 24–28 October 2015, a-a-r-s.org/acrs/administrator/components/com.../files/.../TH3-2-1.pdf.
275 Xie X., Zhang W. and Yang L. 2003. Particle swarm optimization. Control and Decision 18: 129–134.
276 Stevenson D. 2005. Tsunamis and earthquakes: What physics is interesting? Physics Today 58(6): 10–11.
277 Dorigo M., de Oca M.A.M. and Engelbrecht A. 2008. Particle swarm optimization. Scholarpedia 3(11): 1486.
278 Ibrahim S., Abdul Khalid N.E. and Manaf M. 2010. Computer aided system for brain abnormalities segmentation. Malaysian Journal of Computing (MJOC) 1(1): 22–39.
279 Naila M. and Babiker O. 2015. Seismic hazard assessment of the red sea region. American Journal of Earth Sciences 2(6): 230–235.
280 Giacco F., Saggese L., de Arcangelis L., Lippiello E. and Ciamarra M.P. 2015. Dynamic weakening by acoustic fluidization during stick-slip motion. Physical Review Letters 115(12): 128001.
281 Kilburn C.R. and Petley D.N. 2003. Forecasting giant, catastrophic slope collapse: lessons from Vajont, Northern Italy. Geomorphology 54(1): 21–32.

Index

T

U

V

W

X

Y

Z

Author's Biography

Assoc., Prof. Dr. Maged Marghany is a microwave remote sensing expert. He was awarded European Space Agency post-doctoral fellowship by the International Institute of Aerospace and Earth Observation (ITC) in Enschede, the Netherlands, funded by the European Space Agency (ESA) for one year from March 2000 to March 2001. He was awarded his Ph.D from Universiti Putra Malaysia in field of Environment Remote Sensing. His Ph.D theses was titled "Wave Spectra Studies and Shoreline Change by Remote Sensing". Since his Ph.D, his research work is in the field of microwave remote sensing. Dr. Marghany has more than 18 years working experience in the field of microwave remote sensing both in teaching and research work. He has been leading several projects related to the application of synthetic aperture radar (SAR) funded by Ministry of Science and Technology, Malaysia (MOSTE) and Ministry of Malaysian High Education (MOHE).

Dr. Marghany was the editor of other three books (i) Applied Studies of Coastal and Marine Environments; (ii) Environmental Applications of Remote Sensing; and (iii) Advanced Geoscience Remote Sensing. In addition, he has more than 250 of publications in refereed and indexed conferences and journals.

He has also been a referee for top indexed journals such as Remote Sensing, International Journal of Remote Sensing, Environment Remote Sensing, and IEEE transactions on geoscience and remote sensing.